T0135582

Augsburger Schriften zur Mathematik, Physik und Informatik
Band 21

herausgegeben von:
Professor Dr. F. Pukelsheim
Professor Dr. W. Reif
Professor Dr. D. Vollhardt

Bibliografische Information der Deutschen Nationalbibliothek

Die Deutsche Nationalbibliothek verzeichnet diese Publikation in der
Deutschen Nationalbibliografie; detaillierte bibliografische Daten sind
im Internet über http://dnb.d-nb.de abrufbar.

ISBN 978-3-8325-3397-7
ISSN 1611-4256


Logos Verlag Berlin GmbH
Comeniushof, Gubener Str. 47,
10243 Berlin
Tel.: +49 030 42 85 10 90
Fax: +49 030 42 85 10 92
INTERNET: http://www.logos-verlag.de

# Multiscale modeling and homogenization of reaction-diffusion systems involving biological surfaces

Dissertation zur Erlangung des Doktorgrades der
Mathematisch-Naturwissenschaftlichen Fakultät der
Universität Augsburg

vorgelegt von Isabella M. Graf
Januar 2013

Erstgutachter:        Prof. Dr. Malte A. Peter,
                      Universität Augsburg

Zweitgutachter:       Prof. Dr. Dirk Blömker,
                      Universität Augsburg

Mündliche Prüfung:    5. März 2013

# Acknowledgements

At this point I would like to express my thanks to everybody who contributed to the success of my thesis.

First and foremost I am especially grateful to my supervisor Malte Peter for his perpetual help and bountiful support. I am thankful that he gave me the freedom to choose the applications of my theoretical work according to my fields of interest. Thank you also for making the research stay at UBC in Vancouver, Canada, possible. Above all I thank you for always taking time whenever I had any question or needed advice.

Thanks to all who encouraged and helped me by answering questions or stimulating discussions or for proofreading my thesis, especially Dirk Blömker, Melanie Jahny, Christoph Kawan, Bernhard Konrad, Christian Möller, Michaela Möller, Johannes Neher, Peter Quast and Bernd Schmidt.

I am grateful to Yue-Xian Li and Daniel Coombs for the opportunity to study at UBC, for the support and fruitful cooperation.

Finally, I am thankful to my parents and Bernhard for great encouragements and comfort.

# CONTENTS

# ONE

# INTRODUCTION

In the human body many complex chemical processes are responsible for its functionality. Various examinations of these occurrences are essential for a deep understanding of the functioning of the body and to intervene in case of a breakdown. Many failures already arise in faulty metabolisms of molecules that take place in the organelles of human cells.

All applications in this thesis have in common that metabolisms happening on the surface of the endoplasmic reticulum are central. The endoplasmic reticulum is an organelle formed by a multilayer network of membranes that acts as a workshop of the cell and converts molecules into other molecules with help of enzymes that stick to the membranes of the endoplasmic reticulum.

It is our aim to deduce mathematical models of biological events describing molecules living on the endoplasmic reticulum, simplify them and do numerical simulations.We are going to use partial differential equations.

There is previous work done to model the very fine structure of the endoplasmic reticulum and to be able to do simulations of partial differential equations defined on the endoplasmic reticulum (see [29, 7, 26]).

In this work we use the mathematical tool of periodic homogenization with different techniques to handle compartments that live on the membrane of the endoplasmic reticulum. This allows us to consider diffusion of these compartments on the membrane. In the context of periodic homogenization we are going to use two-scale convergence (see [2, 52]) and the periodic unfolding method (see [11]).

Homogenization is a method to transform a partial differential equation that is defined on a spatially very finely structured domain to another partial differential equation that is defined on a spatially homogeneous domain where the essential properties remain. The technique of periodic homogenization deduces another partial differential equation as an approximation of the initial problem. In the approximation the spatial complexity and fine structure disappear and the equation is defined on a homogeneous domain. This technique is described in section 2.1. To derive the homogeneous model one uses two-scale convergence also described in section 2.1 or the periodic unfolding operator presented in section 3.2.

We are going to explore three different scenarios appearing in cells with the aid of partial differential equations. The first one in chapter 2 considers the invasion of the carcinogenic molecule *Benzo[a]pyrene*. We use a system of partial differential equations to examine the processes inside a human cell when *Benzo[a]pyrene* molecules enter the cytoplasm, perform chains of chemical reactions, called metabolisms, to the rather aggressive form *Benzo[a]pyrene-7,8-diol-9,10-epoxide*, reach the nucleus and possibly bind to the DNA and make the cell malign. The partial differential equation is linear and the model simple. We find more details about this scenario in section 2.2.

In chapter 3 we keep the biological setting of the problem and use more complicated mathematical tools. We improve the model by regarding and including cleaning mechanisms of the cell and a natural threshold of receptors and enzymes. The treatment of nonlinear terms used for this improvements needs other mathematical tools, that are introduced in the beginning of chapter 3. This improved model contains nonlinear terms. More details are found in section 3.3.

Last we consider another problem regarding the functionality of signaling of T-lymphocytes. A highly nonlinear system of equations describes the correlation between the calcium household of a cell and the binding of *Stim1* molecules to the plasma membrane that causes the opening of so-called *CRAC channels*. This event induces the emission of messengers to help the immune system to protect the body from alien substances. This scenario is described in section 4.3.

In every chapter we first introduce the mathematics needed to derive the homogeneous model of the respective partial differential equations. Then we derive and model the system of partial differential equations and prove that the conditions for homogenization are satisfied. We show that there exists at least one solution of the differential equations. Then we derive the homogeneous system of equations and show uniqueness of its solution.

**Remark 1.1 (notation)**

- In proofs of the following sections we use constants $c_1, \ldots, c_9$ for the estimations. These constants can have different values in every single step of the estimation. However, $c_0$ always stands for the estimation constant coming from the trace inequality.

- Let $\Omega \subset \mathbb{R}^n$ be a bounded domain and

$$L^2(\Omega) = \{u : \Omega \to \mathbb{R} \,|\, u \text{ measurable}, \int_\Omega u(x)^2 dx < \infty\}.$$

With define the norm

$$\|u\|_\Omega := \|u\|_{L^2(\Omega)} = \left( \int_\Omega u(x)^2 dx \right)^{1/2}.$$

We are going to use this abbreviation of the norm notation in the following chapters.

CHAPTER

# TWO

# LINEAR CARCINOGENESIS MODEL

## Introduction

In the first section 2.1 of this chapter we introduce the method of periodic homogenization in more detail and also explain the mathematical tool of two-scale convergence. With this particular notion of convergence that exploits the periodicity of the finely structured domain we can derive the homogenized model. Finally, the new system of differential equations is defined on two different domains: the homogeneous macroscopic one and the microscopic unit cell.

In section 2.2 the derivation of the linear carcinogenesis model is described. We explain the biological and chemical background, model a preliminary system of equations and transform it to a system of equations that depends on a variable $\varepsilon > 0$. This small variable which tends to zero by finding the homogenized approximation specifies the size of the periodicity of the finely structured domain, called $\Omega_\varepsilon$. The system of equations consists of four partial differential equations; two of which are defined on the membrane of the endoplasmic reticulum.

Further, in section 2.3, we show that the functions remain non-negative, if the initial values are nonnegative, and we prove that

5

the conditions to use two-scale convergence are fulfilled. This is the case, if the functions are bounded in the function space $L^2$, independent of $\varepsilon$.

Since the two differential equations that are defined on the membrane of the endoplasmic reticulum are solvable analytically given the remaining two functions, we are able to abbreviate the system to a partial differential equation consisting of two equations. We perform this abbreviation in section 2.4.

In section 2.5 we show that there exists at least one solution of the system of differential equations for every $\varepsilon > 0$. For the proof we use the abbreviated model derived in the previous section.

With this, in section 2.6 we are able to deduce the two-scale limit of the system of equations with help of two-scale convergence. We find that we need to solve the so-called *cell problem* that describes the dynamics on the microscopic domain. Afterwards we establish the macroscopic limit model with form and coefficients, such that the essential properties of the dynamics remain.

Further, we show in section 2.7 that the solutions of the cell problem and the macroscopic problem are unique.

At last in section 2.8 we solve the differential equations numerically and illustrate the results. We interpret the results in both mathematical and biological sense.

## 2.1 Two-scale convergence

The idea of periodic homogenization starts with the approximation of finely structured geometries by a periodic pattern. Let $\Omega \subset \mathbb{R}^n$ be an open domain.

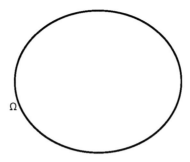

We define a domain $\Omega_\varepsilon \subset \Omega$, which is paved by a so-called unit cell $Y$ of size $\varepsilon > 0$, $\varepsilon$ small. This unit cell $Y$ should look like a characteristical small part of the structure to be approximated. Then the domain $\Omega_\varepsilon$ gets a periodic pattern. The smaller $\varepsilon$, the finer is the structure of the domain $\Omega_\varepsilon$. We define $\Omega_\varepsilon := \bigcup_{k \in \mathbb{Z}^n} \varepsilon(k + Y) \cap \Omega$.

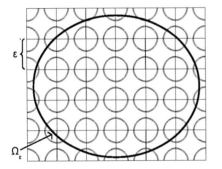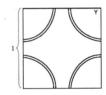

Usually it is impossible to find an analytical solution to partial differential equations defined on nontrivial domains. In default thereof one wants to solve the problem numerically. But also here one encounters difficulties because of the fine structure of the domain. To get reasonable results the numerical resolution in space ought to be unfeasible high. Furthermore, in most cases one is only interested in the global dynamics and not in the precise behavior in every part of the fine structure. Hence, such accurate numerical calculations are too costly and not even necessary. Since the periodic structure is an approximation to the real world anyway, it does not make sense to spend more efford

7

in the accuracy than the structure of the domain reaches.
The technique of periodic homogenization is a dexterous escape
from this predicament. The idea is to approximate the domain $\Omega_\varepsilon$
by the infinitely finely structured domain $\Omega_{\varepsilon \to 0}$. By doing so, the
domain $\Omega_\varepsilon$ loses its fine structure and becomes homogeneous,
where the crucial properties remain. This means for example,
we are going to find averaged values for the coefficients or find
an effective form and effective coefficients of the given partial
differential equation.

Functions that satisfy partial differential equations on such
domains keep essential characteristics. The idea is described in
detail in the books [6, 51, 52].

Now let $Y = [0,1]^n$ and $\Omega \subset \mathbb{R}^n$. With $\Omega_\varepsilon = \bigcup_{k \in \mathbb{Z}^n} \varepsilon(k + Y) \cap$
$\Omega$ we can build a domain with a special periodic fine structure
depending on $Y$.
For a given partial differential equation with operator $L : X \to$
$X^*$, Banach space $X$, its dual $X^*$, and a right-hand side functional
$f \in X^*$, $L_\varepsilon$ and $f_\varepsilon$ also depend on $\varepsilon$ with the domain $\Omega_\varepsilon$ depending
on $\varepsilon$. Hence, we would need to solve

$$L_\varepsilon u_\varepsilon = f_\varepsilon \quad \text{in } \Omega_\varepsilon$$

to find the solution $u_\varepsilon$ depending on $\varepsilon$ of the partial differential
equation. This gives us a family of differential operators $L_\varepsilon$. The
important question is, how does the limit operator

$$L_\varepsilon \xrightarrow{\varepsilon \to 0} L_0,$$

look like, such that $L_0 u_0 = f_0$ in $\Omega$? Which convergence $L_\varepsilon \xrightarrow{\varepsilon \to 0} L_0$
leads to $L_0 u_0 = f_0$? In general, such a kind of convergence is un-
known.
We use another trick to find the limit differential operator $L_0$
by using another weak convergence, called *two-scale convergence*,
adapted for oscillation sequences of functions $u_\varepsilon$.
We apply two-scale convergence to every term of the partial differential equation and find in the limit for $\varepsilon \to 0$ a limit PDE,
defining the limit differential operator $L_0$. This is, we calculate the

8

limits $u_\varepsilon$ in the various terms of the PDE and only then find the operator $L_0$.

Two-scale convergence is a kind of weak convergence in $L^2$ spaces using special test functions, called *admissible test functions for two-scale convergence*.

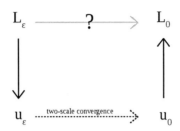

With the development of two-scale convergence by G. Nguetseng and G. Allaire, published in the articles [2, 44, 40], the limit of a sequence of functions depending on the period $\varepsilon$ can be calculated conveniently.

In the limit the function $u_0 = \lim_{\varepsilon \to 0} u_\varepsilon$ does not depend on one structured domain $\Omega_\varepsilon$ any more, but on two separated domains: the homogeneous domain $\Omega$ and the unit cell $Y$. The domain $\Omega$ contains the macroscopical variable, that states what is happening in the coarse view, and the unit cell $Y$ contains the microscopical variable, that states what transpires in the infinitesimally small cell $Y$.

Two-scale convergence is stronger than weak convergence. It holds that

$$\text{strong convergence} \Rightarrow \text{two-scale convergence} \Rightarrow \text{weak convergence.}$$

We recall the most important definitions and theorems of two-scale convergence. We are going to apply them in the following sections. Henceforth, let $\Omega \subset \mathbb{R}^n$ and $Y = [0,1]^n$.

At first we cite the definition of an admissible test function for the two-scale convergence from [2, 43].

9

**Definition 2.1** *Let $\varphi \in L^2(\Omega \times Y)$ be a function that can be approximated by a sequence of functions $\varphi_n \in C^\infty(\Omega, C_\#^\infty(Y))$ in the following sense:*

- *$\|\varphi_n - \varphi\|_{L^2(\Omega \times Y)} \to 0$,*

- *$\sup_{\varepsilon > 0} \|(\varphi_n - \varphi)\left(x, \frac{x}{\varepsilon}\right)\|_{L^2(\Omega)} \to 0$,*

*for $n \to \infty$. Then we call $\varphi$ an* admissible test function *for two-scale convergence.*

*The index # means the $Y$-periodicity of the function in its second argument living in the function space. This means, that $\varphi_n(x,y) = \varphi_n(x, y + k)$ for all $x \in \Omega$, $y \in Y$ and $k \in \mathbb{Z}^n$.*

More information about test functions are found in [56].

Whenever we use two-scale convergence, we use test functions of this kind. The next five statements are cited from [2]. Initially we define the concept of two-scale convergence.

**Definition 2.2** *A sequence of functions $u_\varepsilon$ in $L^2(\Omega)$ is said to two-scale converge to a limit $u_0$ belonging to $L^2(\Omega \times Y)$, if we have for any admissible test function $\psi$ in $C^\infty(\Omega, C_\#^\infty(Y))$*

$$\lim_{\varepsilon \to 0} \int_\Omega u_\varepsilon(x)\psi\left(x, \frac{x}{\varepsilon}\right) dx = \int_\Omega \int_Y u_0(x,y)\psi(x,y)dxdy.$$

The main theorem of two-scale convergence ensures the existence of a two-scale limit of a sequence bounded in the $L^2$-norm.

**Theorem 2.3** *From each bounded sequence $u_\varepsilon$ in $L^2(\Omega)$ we can extract a subsequence and there exists a limit $u_0 \in L^2(\Omega \times Y)$ such that this subsequence two-scale converges to $u_0$.*

If we consider a sequence of function products of two two-scale convergent sequences of functions, the next theorem tells us conditions for weak convergence.

**Theorem 2.4** *Assume that $u_\varepsilon$ and $v_\varepsilon$ are two bounded sequences of functions in $L^2(\Omega)$ which two-scale converge to limits $u_0$ and $v_0$ in $L^2(\Omega \times Y)$, respectively. Assume further that*

$$\lim_{\varepsilon \to 0} \|u_\varepsilon\|_\Omega = \|u_0\|_{\Omega \times Y}.$$

10

*Then we have*

$$u_\varepsilon v_\varepsilon \rightharpoonup \int_Y u_0(x,y)v_0(x,y)dy$$

*weakly in* $C_0^\infty(\Omega)'$.
*Further, if* $u_0$ *belongs to* $L^2(\Omega, C_\#(Y))$, *we have*

$$\lim_{\varepsilon \to 0} \|u_\varepsilon(x) - u_0\left(x, \frac{x}{\varepsilon}\right)\|_{L^2(\Omega)} = 0.$$

Working with partial differential equations often the gradient of a functions occurs. Here we find a correlation between the limit of a sequence of functions and the limit of the sequence of the gradients of the functions.

**Proposition 2.5** (i) *Let* $u_\varepsilon$ *be a bounded sequence in* $H^1(\Omega)$ *that converges weakly to a limit* $u_0$ *in* $H^1(\Omega)$. *Then* $u_\varepsilon$ *two-scale converges to* $u_0$ *and there exists a function* $u_1$ *in* $L^2(\Omega, H_\#^1(Y))$ *such that up to a subsequence* $\nabla u_\varepsilon$ *two-scale converges to* $\nabla_x u_0 + \nabla_y u_1$.

(ii) *Let* $u_\varepsilon$ *and* $\varepsilon \nabla u_\varepsilon$ *be two bounded sequences in* $L^2(\Omega)$. *Then there exists a function* $u_0$ *in* $L^2(\Omega, H_\#^1(Y))$ *such that, up to a subsequence,* $u_\varepsilon$ *and* $\varepsilon \nabla u_\varepsilon$ *two-scale converge to* $u_0(x,y)$ *and to* $\nabla_y u_0(x,y)$, *respectively.*

In the following theorem we find conditions for strong convergence in $H^1$.

**Theorem 2.6** *If* $u_1$, $\nabla_x u_1$ *and* $\nabla_y u_1$ *are* $Y$-*periodic in the second argument* $y$ *and satisfy the equation*

$$\lim_{\varepsilon \to 0} \int_\Omega \varphi\left(x, \frac{x}{\varepsilon}\right)^2 dx = \int_\Omega \int_Y \varphi(x,y)^2 dy dx,$$

*for* $\varphi = u_1$, $\nabla_x u_1$ *and* $\nabla_y u_1$, *then we have*

$$\left[u_\varepsilon(x) - u_0(x) - \varepsilon u_1\left(x, \frac{x}{\varepsilon}\right)\right] \to 0 \qquad \text{strongly in } H^1(\Omega).$$

Let $\Gamma \subset Y$ be a smooth, periodic $n-1$ dimensional surface, such that

$$\Gamma_\varepsilon := \bigcup_{k \in \mathbb{Z}^n} \varepsilon(k + \Gamma) \cap \Omega \qquad (2.1)$$

also is a smooth, $n - 1$ dimensional surface.
For structures like this manifold $\Gamma_\varepsilon$, Maria Neuss-Radu proved in [43] the following statements. Initially we define the concept of two-scale convergence on manifolds.

**Definition 2.7** *A sequence of functions $u_\varepsilon \in L^2(\Gamma_\varepsilon)$ is said to two-scale converge to a limit $u_0 \in L^2(\Omega \times \Gamma)$ iff for any $\varphi \in C^\infty(\Omega, C_\#^\infty(\Gamma))$ we have*

$$\lim_{\varepsilon \to 0} \varepsilon \int_{\Gamma_\varepsilon} u_\varepsilon(x) \varphi\left(x, \frac{x}{\varepsilon}\right) d\sigma_x = \int_\Omega \int_\Gamma u_0(x,y)\varphi(x,y)d\sigma_y dx,$$

*where $dx$ has to be understood as the Hausdorff measure on $\Omega$ and $d\sigma_x$ and $d\sigma_y$ are the measures build by the Riemannian tensors on the manifold $\Gamma_\varepsilon$ and $\Gamma$, respectively.*

**Remark 2.8** The $\varepsilon$ in front of the integral stems from the scalar product on $\Gamma_\varepsilon$ and is necessary because the area of a periodic surface grows with decreasing $\varepsilon$ and constant size of $\Omega_\varepsilon$ proportional to $1/\varepsilon$. Hence, we need to compensate the growing factor with $\varepsilon$ to keep the integral bounded. For more details see [43].

It follows a general limit result about functions periodic in its second argument.

**Lemma 2.9** *For any function $f$ in $C(\Omega, C_\#(\Gamma))$ it holds*

$$\lim_{\varepsilon \to 0} \varepsilon \int_{\Gamma_\varepsilon} f\left(x, \frac{x}{\varepsilon}\right) d\sigma_x = \int_\Omega \int_\Gamma f(x,y)d\sigma_y dx.$$

The following theorem is similar to theorem 2.3. It ensures the existence of a two-scale limit of a sequence, bounded in the $L^2$-norm.

**Theorem 2.10** *Let $u_\varepsilon$ be a bounded sequence in $L^2(\Gamma_\varepsilon)$ such that $\varepsilon \|u_\varepsilon\|_{L^2(\Gamma_\varepsilon)}^2 < C$ for $C > 0$ independent of $\varepsilon$. Then we can extract a subsequence which two-scale converges to $u_0 \in L^2(\Omega \times \Gamma)$. Further, we have*

$$\lim_{\varepsilon \to 0} \varepsilon \|u_\varepsilon\|_{L^2(\Gamma_\varepsilon)} \geq \|u_0\|_{L^2(\Omega \times \Gamma)}.$$

12

We also cite the generalized trace inequalities useful when dealing with periodic microstructures.

**Lemma 2.11** *For $\partial\Omega$, $\Gamma_\varepsilon$ being Lipschitz boundaries, the following trace estimations are valid for a constant $c_0 > 0$.*

1. *If $u \in H^1(\Omega)$, then $\|u\|_{\partial\Omega}^2 \leq c_0 \left( \|u\|_\Omega^2 + \|\nabla u\|_\Omega^2 \right)$,*

2. *If $u_\varepsilon \in H^1(\Omega_\varepsilon)$, then $\varepsilon\|u_\varepsilon\|_{\Gamma_\varepsilon}^2 \leq c_0 \left( \|u_\varepsilon\|_{\Omega_\varepsilon}^2 + \varepsilon^2 \|\nabla u_\varepsilon\|_{\Omega_\varepsilon}^2 \right).$*

More information about theory of two-scale convergence on surfaces can be found in [3].

At last, we cite the lemma of Gronwall in [21], because we will use it frequently in the following sections.

**Lemma 2.12** *Let $u$ be a nonnegative, summable function on $[0, T]$ which satisfies for almost every $t \in [0, T]$ the integral inequality*

$$u(t) \leq C_1 \int_0^t u(s)ds + C_2$$

*for constants $C_1, C_2 \geq 0$. Then*

$$u(t) \leq C_2 \left( 1 + C_1 t e^{C_1 t} \right) \leq C_2 \left( 1 + C_1 T e^{C_1 T} \right)$$

*for almost every $0 \leq t \leq T$.*

With help of these theorems we are able to homogenize the following linear problem.

## 2.2 Linear carcinogenesis problem

One of the longest known and best understood reasons for carcinogenesis is the molecule *Benzo[a]pyrene*, abbreviated called **BP**. It is found, for example, in coal tar, automobile exhaust fumes, cigarette smoke and charbroiled food.

One of the main reason for lung cancer (caused by inhaling cigarette smoke), testicular cancer and skin cancer is the contact with the molecule *Benzo[a]pyrene*. Often chimney sweepers are infested because of the frequent touch with coal (see [35, 27]). *Benzo[a]pyrene* is a polycyclic aromatic hydrocarbon (PAH) with five benzene rings, its chemical formula is $C_{20}H_{12}$.

**BP**

At every edge of the hexagons a carbon atom is located and hydrogen atoms are attached such that every carbon atom has four binding partners. The molecule itself is not dangerous. But chemical reactions in the human cell can transform it to the molecule *Benzo[a]pyrene-7,8-diol-9,10-epoxide*, abbreviated called **DE**, which can bind to and damage the human DNA (see [27]). The metabolism works as follows:

1. **BP**

   Oxidization at carbon 7 and 8:

2. **BP-7,8-oxide**

   Adding $H_2O$ to carbon 7 and 8:

3. **trans-BP-7,8-dihydrodiol**

14

Oxidization at carbon 9 and 10:

4.  **BP-7,8-diol-9,10-epoxide**

Binding to DNA:

5.

The chemical reactions mostly take place on the surface of the endoplasmic reticulum induced by the enzyme system called *MFO (microsomal mixed-function oxidases).* One of the more important and active enzymes of MFO is *P-450.* It mostly is bound to the membrane of the endoplasmic reticulum. A few of the P-450 molecules also live in the cytosol, but we neglect this minor impact. Normally the MFOs serve a detoxification role, but unfortunately not in this case (see [24, 50, 54, 45]). For convenience we abbreviate *endoplasmic reticulum* by ER.

Hence, the process of toxification is simplified by the following scenario: BP molecules pass the plasma membrane from the intercellular space to the cytosol inside of a human cell, where they diffuse freely. But they cannot enter the nucleus. They can bind to the surface of the endoplasmic reticulum. There a series of chemical reactions takes place caused by the enzyme system MFO as we see in the figure above; from the mathematical point of view we summarize the reactions to just one metabolism from BP to DE. Newly created DE molecules unbind from the surface of the endoplasmic reticulum and diffuse again in the cytosol of the cell. There they may enter the nucleus. Hence, BP cannot pass the nuclear membrane, DE cannot pass the plasma membrane, which describes a worst case scenario.

15

Here, when DE molecules enter the nucleus, we stop our consideration. A similar carcinogenesis model is used in "Mathematical modelling and numerical solution of chemical reactions and diffusion of carcinogenic compounds in cells" by Donald O Besong. [7]. Besong's model does not consider the metabolism from BP to DE molecules, but already starts with DE molecules and regards linear cleaning reactions and chemical bonds between DE molecules and DNA.

Besong does not use periodic homogenization, but also incorporates the fine structure of the endoplasmic reticulum using spatial averaging and deduces an averaged value for the diffusion coefficient in cytoplasm.

Now we want to formulate a mathematical model of this process. Let $\Omega \subset \mathbb{R}^n$ be a human cell with a Lipschitz boundary $\partial\Omega$. The process shall happen in the time interval $[0, T]$ for fixed $0 < T < \infty$. Let $\Gamma^{ER} \subset \Omega$ be a smooth manifold, the surface of the endoplasmic reticulum. Further, the concentration of BP molecules in cytosol is denoted by $u : [0, T] \times \Omega \to \mathbb{R}$ and the concentration of DE molecules in cytosol is $v : [0, T] \times \Omega \to \mathbb{R}$. The concentration of BP molecules bound to the surface of the endoplasmic reticulum is denoted by $s : [0, T] \times \Gamma^{ER} \to \mathbb{R}$ and the concentration of DE molecules bound to the surface of the endoplasmic reticulum is denoted by $w : [0, T] \times \Gamma^{ER} \to \mathbb{R}$.

In the following figure we see a sketch of a human cell with annotations.

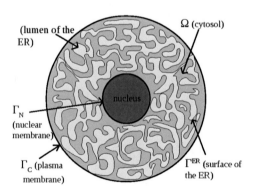

In the cytosol the BP molecules $u$ move by free diffusion without impact of an outer force,

$$\partial_t u - D_u \Delta u = 0 \qquad \text{in } \Omega,$$

with diffusion coefficient $D_u > 0$. We assume a constant concentration $u_{\text{Boundary}} \geq 0$ of BP molecules in the intercellular space and hence apply a constant Dirichlet boundary condition at the plasma membrane $\Gamma_C$ of the cell,

$$u = u_{\text{Boundary}} \qquad \text{on } \Gamma_C.$$

Since BP molecules are not able to enter the nucleus there is a no-flux condition at the nuclear membrane $\Gamma_N$,

$$-D_u \nabla u \cdot n = 0 \qquad \text{on } \Gamma_N.$$

Here, $n$ denotes the outer normal. At the surface of the endoplasmic reticulum BP molecules can bind to the membrane with the rate $l_s > 0$ and so be transformed to BP molecules bound to the ER membrane

$$-D_u \nabla u \cdot n = l_s(u - s) \qquad \text{on } \Gamma^{\text{ER}}.$$

On the surface of the ER the metabolism of the BP molecules to DE molecules takes place. We assume a linear transformation rate $f > 0$, with $f$ depending on the enzyme system MFO. BP molecules $u$ binding to the ER act as the source term $l_s u$. BP molecules bound to the ER $s$ can unbind with rate $l_s$ and diffuse freely on the surface of the ER. Therefore we need the Laplace-Beltrami operator with diffusion tensor $D_s > 0$,

$$\partial_t s - D_s \Delta_\Gamma s = -fs + l_s(u - s) \qquad \text{on } \Gamma^{\text{ER}}.$$

The molecules, which we loose by the term $-fs$, appear in the equation for the DE molecules $w$. As above there is a diffusion term with the Laplace-Beltrami operator and diffusion coefficient $D_w > 0$. DE molecules can bind and unbind with rate $l_w > 0$.

$$\partial_t w - D_w \Delta_\Gamma w = fs + l_w(v - w) \qquad \text{on } \Gamma^{\text{ER}}.$$

In the cytosol DE molecules diffuse freely with diffusion coefficient $D_v > 0$,

$$\partial_t v - D_v \Delta v = 0 \qquad \text{in } \Omega.$$

They bind and unbind to the membrane of the ER with rate $l_w$,

$$-D_v \nabla v \cdot n = l_w(v - w) \qquad \text{on } \Gamma^{\text{ER}}.$$

We assume that DE molecules cannot leave the cell. Hence, we apply a no-flux condition at the plasma membrane,

$$-D_v \nabla v \cdot n = 0 \qquad \text{on } \Gamma_C.$$

In the worst case all DE molecules touching the nuclear membrane enter the nucleus which implies the Dirichlet boundary condition

$$v = 0 \qquad \text{on } \Gamma_N.$$

We set the initial values to be

$$
\begin{aligned}
u(x,t) &= u_I(x), &\text{for } t = 0,\ x \in \Omega, \\
v(x,t) &= v_I(x), &\text{for } t = 0,\ x \in \Omega, \\
s(x,t) &= s_I(x), &\text{for } t = 0,\ x \in \Gamma^{\text{ER}}, \\
w(x,t) &= w_I(x), &\text{for } t = 0,\ x \in \Gamma^{\text{ER}}.
\end{aligned}
\qquad (2.2)
$$

We assume that the initial values are smooth, bounded and non-negative.

To illustrate the situation we show a microscopic view of a cell. In the following figure we see a section of a human cell. The lighter region in the bottom right is the nucleus, the dark nuclear membrane separates it from the cytoplasm. The oval circles are mitochondria and the thin lines are part of the surface of the endoplasmic reticulum, [36].

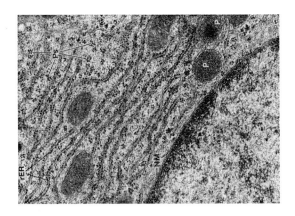

We see that the membrane of the ER gives the cell a finely structured pattern.
Because of this very fine structure of the ER our system of equations has a high level of complexity. In [20] we find the data

$L \approx 20\mu m$ (diameter of a cell) - magnitude of the coarse view

$l \approx 0.6\mu m$ (ER thickness) - magnitude of the small view

We suppose that the endoplasmic reticulum has a periodic structure. For this purpose we define a model district $Y = [0,1]^n$ with
smooth manifold $\Gamma \subset Y$, where $\Gamma$ does not touch the boundary of
$Y$. Let $Y^* \subset Y$ be the inside of $\Gamma$, this is the lumen of the ER. We
pave the cell with the model districts with size $\varepsilon > 0$, $\varepsilon \ll 1$, for
example as illustrated in the figures below. In this 2-dimensional
figure we form the ER as small bubbles and the cytosol as the connected gray area. In real life, the 3-dimensional ER also is connected. But in two dimensions it is impossible to build a periodic
pattern with two connected areas.

19

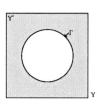
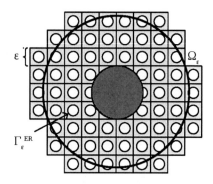

We keep denoting the boundary of the cell with $\Gamma_C$ and the boundary of the nucleus with $\Gamma_N$. Furthermore, $\Omega_\varepsilon$ and $\Gamma_\varepsilon^{ER}$ are defined as

$$\Omega_\varepsilon := \bigcup_{k \in \mathbb{Z}^n} \varepsilon(Y^* + k) \cap \Omega \quad \text{and} \quad \Gamma_\varepsilon^{ER} := \bigcup_{k \in \mathbb{Z}^n} \varepsilon(\Gamma + k) \cap \Omega.$$

Another way to homogenize a cell with the endoplasmic reticulum as a finely structured membrane can be found in "Homogenization of the cell cytoplasm: the calcium bidomain equations" by P. Goel, J. Sneyd and A. Friedman, [26]. There, they consider a mathematical model about calcium dynamics and use homogenization to remove the complexity of the ER for various shapes of the ER. They do not regard diffusion on the surface of the ER.

Next we prepare our model to make it suitable for the homogenization technique. Since the domains $\Omega_\varepsilon$ and $\Gamma_\varepsilon^{ER}$ are dependent on $\varepsilon$, also the functions $u$, $v$, $s$, $w$ become $\varepsilon$ dependent. We indicate that with an $\varepsilon$ in the index of the function.

Further, at the membrane of the ER, $\Gamma_\varepsilon^{ER}$, we multiply the binding-unbinding term with $\varepsilon^l$, for $l \geq 0$. The higher the exponent $l$, the slower is the exchange between bound and unbound molecules. An exchange in "normal" speed would be $l = 1$, because the binding process happens on a surface, see remark 2.8, but also other values would make sense for special binding-unbinding processes. For $l < 1$, the binding process is faster; for $l > 1$ it is slower.

20

Similarly, on the membrane of the ER, $\Gamma_\varepsilon^{ER}$, the diffusion term can be multiplied with $\varepsilon$ to the power of a number $k \geq 0$. The higher the exponent $k$, the slower is the movement of the molecule bound to the membrane compared to the molecules living in cytosol. One finds the coherences between the value of the exponents and their physical meanings by nondimensionalization, see for example [48].

We choose $k$ to be 2. We denote the exponent of $\varepsilon$ in the term that corresponds to the binding term for DE molecules by $m \geq 0$.

Another example of a system of equations with different choices of scaling one finds in [47]. More examples for homogenization of systems of equations describing chemical reactions are done in [48, 46].

With these modifications we arrive at the following model:

$$
\begin{aligned}
\partial_t u_\varepsilon - D_u \Delta u_\varepsilon &= 0 && \text{in } \Omega_\varepsilon \\
u_\varepsilon &= u_{\text{Boundary}} && \text{on } \Gamma_C \\
-D_u \nabla u_\varepsilon \cdot n &= 0 && \text{on } \Gamma_N \\
-D_u \nabla u_\varepsilon \cdot n &= \varepsilon^l l_s (u_\varepsilon - s_\varepsilon) && \text{on } \Gamma_\varepsilon^{ER} \\
u_\varepsilon(x,0) &= u_I(x) && \text{for } x \in \Omega_\varepsilon
\end{aligned}
$$

(2.3)

$$
\begin{aligned}
\partial_t v_\varepsilon - D_v \Delta v_\varepsilon &= 0 && \text{in } \Omega_\varepsilon \\
-D_v \nabla v_\varepsilon \cdot n &= 0 && \text{on } \Gamma_C \\
v_\varepsilon &= 0 && \text{on } \Gamma_N \\
-D_v \nabla v_\varepsilon \cdot n &= \varepsilon^m l_w (v_\varepsilon - w_\varepsilon) && \text{on } \Gamma_\varepsilon^{ER} \\
v_\varepsilon(x,0) &= v_I(x) && \text{for } x \in \Omega_\varepsilon
\end{aligned}
$$

$$
\begin{aligned}
\varepsilon \partial_t s_\varepsilon - D_s \varepsilon^3 \Delta_\Gamma s_\varepsilon &= -\varepsilon f s_\varepsilon + \varepsilon^l l_s (u_\varepsilon - s_\varepsilon) && \text{on } \Gamma_\varepsilon^{ER} \\
s_\varepsilon(x,0) &= s_I(x) && \text{for } x \in \Gamma_\varepsilon^{ER}
\end{aligned}
$$

$$
\begin{aligned}
\varepsilon \partial_t w_\varepsilon - D_w \varepsilon^3 \Delta_\Gamma w_\varepsilon &= \varepsilon f s_\varepsilon + \varepsilon^m l_w (v_\varepsilon - w_\varepsilon) && \text{on } \Gamma_\varepsilon^{ER} \\
w_\varepsilon(x,0) &= w_I(x) && \text{for } x \in \Gamma_\varepsilon^{ER}
\end{aligned}
$$

for $t \in [0, T]$.

## Weak formulation

We use the following function spaces

$$L^2(\Omega_\varepsilon) = \{u : \Omega_\varepsilon \to \mathbb{R} \mid u \text{ measurable}, \|u\|_{\Omega_\varepsilon}^2 < \infty\}$$

and

$$L^2([0,T], L^2(\Omega_\varepsilon)) = \{u : [0,T] \to L^2(\Omega_\varepsilon) \mid \|u\|_{\Omega_\varepsilon,T}^2 < \infty\}.$$

The scalar product on $L^2(\Omega_\varepsilon)$ is given by

$$(u,v)_{\Omega_\varepsilon} = \int_{\Omega_\varepsilon} u(x)v(x)\mathrm{d}x.$$

The scalar product on $L^2([0,T], L^2(\Omega_\varepsilon))$ is given by

$$(u,v)_{\Omega_\varepsilon,T} = \int_0^T (u(t), v(t))_{\Omega_\varepsilon}\mathrm{d}t.$$

Furthermore, we use the Sobolev space

$$H^1(\Omega_\varepsilon) = \{u : \Omega_\varepsilon \to \mathbb{R} \mid (u,u)_{\Omega_\varepsilon} + (\nabla u, \nabla u)_{\Omega_\varepsilon} < \infty\}$$

where $\nabla u$ is the weak derivative; the scalar product on $H^1(\Omega_\varepsilon)$ is given by

$$(u,v)_{H^1(\Omega_\varepsilon)} = (u,v)_{\Omega_\varepsilon} + (\nabla u, \nabla v)_{\Omega_\varepsilon}.$$

Furthermore, we need the function spaces

$$
\begin{aligned}
\mathcal{V}(\Omega_\varepsilon) &= L^2([0,T], H^1(\Omega_\varepsilon)), \\
\mathcal{V}_\mathrm{N}(\Omega_\varepsilon) &= \{u \in \mathcal{V}(\Omega_\varepsilon) \mid u = 0 \text{ on } \Gamma_\mathrm{N}\}, \\
\mathcal{V}_\mathrm{C0}(\Omega_\varepsilon) &= \{u \in \mathcal{V}(\Omega_\varepsilon) \mid u = 0 \text{ on } \Gamma_\mathrm{C}\}, \\
\mathcal{V}_\mathrm{C}(\Omega_\varepsilon) &= \{u \in \mathcal{V}(\Omega_\varepsilon) \mid u = u_\mathrm{Boundary} \text{ on } \Gamma_\mathrm{C}\},
\end{aligned}
\tag{2.4}
$$

where $0$ means the equivalence relation in $L^2$ of the zero function.

Analogously we define

$$\mathcal{V}(\Gamma_\varepsilon^\mathrm{ER}) = L^2([0,T], H^1(\Gamma_\varepsilon^\mathrm{ER})).$$

After homogenization we need the function spaces

$$\mathcal{V}(\Omega, Y) = L^2([0,T] \times \Omega, H^1_\#(Y))),$$

$$\mathcal{V}(\Omega, \Gamma) = L^2([0,T] \times \Omega, H^1_\#(\Gamma))).$$

For the test functions we define the function spaces

$$
\begin{aligned}
V_{C0}(\Omega_\varepsilon) &= \{u \in H^1(\Omega_\varepsilon)| \ u = 0 \text{ on } \Gamma_C\} \\
V_N(\Omega_\varepsilon) &= \{u \in H^1(\Omega_\varepsilon)| \ u = 0 \text{ on } \Gamma_N\} \\
V(\Gamma_\varepsilon^{ER}) &= H^1(\Gamma_\varepsilon^{ER}) \\
V(\Omega, Y) &= L^2(\Omega, H^1_\#(Y)) \\
V(\Omega, \Gamma) &= L^2(\Omega, H^1_\#(\Gamma)).
\end{aligned}
\tag{2.5}
$$

If we want to restrict a function $u \in H^1(\Omega_\varepsilon)$ to the boundary $\Gamma_\varepsilon^{ER}$, we use the trace operator

$$\gamma: H^1(\Omega_\varepsilon) \to L^2(\Gamma_\varepsilon^{ER})$$

$$u \mapsto u|_{\Gamma_\varepsilon^{ER}}.$$

For $u, v \in L^2(\Gamma_\varepsilon^{ER})$ there is the scalar product

$$\langle u, v \rangle_{\Gamma_\varepsilon^{ER}} := \int_{\Gamma_\varepsilon^{ER}} g_\varepsilon uv \mathrm{d}\sigma_x,$$

where $g_\varepsilon$ is the metric tensor on $\Gamma_\varepsilon^{ER}$, see [19].

**Remark 2.13** We only use the notation $\gamma u$ for the trace, if we want to draw attention explicitly of the restriction to the boundary $\Gamma_\varepsilon^{ER}$. Commonly we use the notation $\langle u, v \rangle_{\Gamma_\varepsilon^{ER}}$, when we mean $\langle \gamma u, \gamma v \rangle_{\Gamma_\varepsilon^{ER}}$ for functions $u, v \in \mathcal{V}(\Omega_\varepsilon)$.

**Remark 2.14** In spaces satisfying the Gelfand triple $N \subset H' = H \subset N'$, the dual pairing $\langle \cdot, \cdot \rangle_{N,N'}$ can be identified with the scalar product on $H$ such that $\langle x, y \rangle_{N,N'} = (x,y)_{H,H}$, where $x \in N \subset H$ and $y \in N'$.
We use this fact for the dual pairing $\langle \cdot, \cdot \rangle_{H^1(\Omega_\varepsilon)' \times H^1(\Omega_\varepsilon)} = (\cdot, \cdot)_{L^2(\Omega_\varepsilon) \times L^2(\Omega_\varepsilon)} = (\cdot, \cdot)_{\Omega_\varepsilon}$, for example for $\partial_t u_\varepsilon(t), \partial_t v_\varepsilon(t) \in H^1(\Omega_\varepsilon)'$ for almost every $t \in [0,T]$ and the test function $\varphi \in$

$H^1(\Omega_\varepsilon)$ in the weak formulation.

Analogously for $H^1(\Gamma_\varepsilon^{ER}) \subset L^2(\Gamma_\varepsilon^{ER}) \subset H^1(\Gamma_\varepsilon^{ER})'$ and $\partial_t s_\varepsilon$ and $\partial_t w_\varepsilon$.

In proposition 2.15 it is shown that the time-derivative of functions satisfying equation (2.6) are elements of the function space $H^1(\Omega_\varepsilon)'$ and $H^1(\Gamma_\varepsilon^{ER})'$, respectively. Hence, the dual pairing is an appropriate representation of the time-derivative in the weak formulation.

So we have the tools at hand to state the model equations in their weak form:

Find $(u_\varepsilon, v_\varepsilon, s_\varepsilon, w_\varepsilon) \in \mathcal{V}_C(\Omega_\varepsilon) \times \mathcal{V}_N(\Omega_\varepsilon) \times \mathcal{V}(\Gamma_\varepsilon^{ER}) \times \mathcal{V}(\Gamma_\varepsilon^{ER})$ with $(u_\varepsilon(x,0), v_\varepsilon(x,0), s_\varepsilon(x,0), w_\varepsilon(x,0)) = (u_I(x), v_I(x), s_I(x), w_I(x))$ such that

$$(\partial_t u_\varepsilon, \varphi_1)_{\Omega_\varepsilon} + D_u(\nabla u_\varepsilon, \nabla \varphi_1)_{\Omega_\varepsilon} + \varepsilon^l l_s \langle u_\varepsilon - s_\varepsilon, \varphi_1 \rangle_{\Gamma_\varepsilon^{ER}} = 0$$

$$(\partial_t v_\varepsilon, \varphi_2)_{\Omega_\varepsilon} + D_v(\nabla v_\varepsilon, \nabla \varphi_2)_{\Omega_\varepsilon} + \varepsilon^m l_w \langle v_\varepsilon - w_\varepsilon, \varphi_2 \rangle_{\Gamma_\varepsilon^{ER}} = 0$$

$$\varepsilon \langle \partial_t s_\varepsilon, \psi_3 \rangle_{\Gamma_\varepsilon^{ER}} + \varepsilon D_s \langle \varepsilon \nabla_\Gamma s_\varepsilon, \varepsilon \nabla_\Gamma \psi_3 \rangle_{\Gamma_\varepsilon^{ER}} + \varepsilon f \langle s_\varepsilon, \psi_3 \rangle_{\Gamma_\varepsilon^{ER}}$$
$$-\varepsilon^l l_s \langle u_\varepsilon - s_\varepsilon, \psi_3 \rangle_{\Gamma_\varepsilon^{ER}} = 0$$

$$\varepsilon \langle \partial_t w_\varepsilon, \psi_4 \rangle_{\Gamma_\varepsilon^{ER}} + \varepsilon D_w \langle \varepsilon \nabla_\Gamma w_\varepsilon, \varepsilon \nabla_\Gamma \psi_4 \rangle_{\Gamma_\varepsilon^{ER}} - \varepsilon f \langle s_\varepsilon, \psi_4 \rangle_{\Gamma_\varepsilon^{ER}}$$
$$-\varepsilon^m l_w \langle v_\varepsilon - w_\varepsilon, \psi_4 \rangle_{\Gamma_\varepsilon^{ER}} = 0$$

$$(2.6)$$

for any $(\varphi_1, \varphi_2, \psi_3, \psi_4) \in V_{C0}(\Omega_\varepsilon) \times V_N(\Omega_\varepsilon) \times V(\Gamma_\varepsilon^{ER}) \times V(\Gamma_\varepsilon^{ER})$.

**Proposition 2.15** Let $\varepsilon > 0$. For functions $(u_\varepsilon, v_\varepsilon, s_\varepsilon, w_\varepsilon) \in \mathcal{V}_C(\Omega_\varepsilon) \times \mathcal{V}_N(\Omega_\varepsilon) \times \mathcal{V}(\Gamma_\varepsilon^{ER}) \times \mathcal{V}(\Gamma_\varepsilon^{ER})$ satisfying equation (2.6) it holds that $\partial_t u_\varepsilon, \partial_t v_\varepsilon \in L^2([0,T], H^1(\Omega_\varepsilon)')$ and $\partial_t s_\varepsilon, \partial_t w_\varepsilon \in L^2([0,T], H^1(\Gamma_\varepsilon^{ER})')$.

**Proof** We consider equation (2.6) and see by using the Cauchy-Schwarz, Young's and trace inequality that

$$\|\partial_t u_\varepsilon\|_{H^1(\Omega_\varepsilon)'} = \sup_{\varphi \in H^1(\Omega_\varepsilon),\, \|\varphi\|_{H^1(\Omega_\varepsilon)}=1} \langle \partial_t u_\varepsilon, \varphi \rangle_{H^1(\Omega_\varepsilon)' \times H^1(\Omega_\varepsilon)}$$

$$= \sup_{\varphi \in H^1(\Omega_\varepsilon),\, \|\varphi\|_{H^1(\Omega_\varepsilon)}=1} \left( -D_u(\nabla u_\varepsilon, \nabla \varphi)_{\Omega_\varepsilon} - \varepsilon^l l_s \langle u_\varepsilon - s_\varepsilon, \varphi \rangle_{\Gamma_\varepsilon^{ER}} \right)$$

$$\leq \sup_{\varphi \in H^1(\Omega_\varepsilon),\, \|\varphi\|_{H^1(\Omega_\varepsilon)}=1} \Big( D_u \|\nabla u_\varepsilon\|_{\Omega_\varepsilon} \|\nabla \varphi\|_{\Omega_\varepsilon}$$

$$+ \varepsilon^l l_s \|u_\varepsilon\|_{\Gamma_\varepsilon^{ER}} \|\varphi\|_{\Gamma_\varepsilon^{ER}} + \varepsilon^l l_s \|s_\varepsilon\|_{\Gamma_\varepsilon^{ER}} \|\varphi\|_{\Gamma_\varepsilon^{ER}} \Big)$$

$$\leq \sup_{\varphi \in H^1(\Omega_\varepsilon),\, \|\varphi\|_{H^1(\Omega_\varepsilon)}=1} \Big( D_u \|\nabla u_\varepsilon\|_{\Omega_\varepsilon} \|\nabla \varphi\|_{\Omega_\varepsilon} + c_0 \varepsilon^{-1} l_s \|u_\varepsilon\|_{\Omega_\varepsilon}^2$$

$$+ c_0 \varepsilon^{l+1} l_s \|\nabla u_\varepsilon\|_{\Omega_\varepsilon}^2 + c_0 \varepsilon^{l-1} l_s \|\varphi\|_{\Omega_\varepsilon}^2 + c_0 \varepsilon^{l+1} l_s \|\nabla \varphi\|_{\Omega_\varepsilon}^2$$

$$+ \varepsilon^l l_s \|s_\varepsilon\|_{\Gamma_\varepsilon^{ER}}^2 + c_0 \varepsilon^{l-1} l_s \|\varphi\|_{\Omega_\varepsilon} + c_0 \varepsilon^{l+1} l_s \|\nabla \varphi\|_{\Omega_\varepsilon} \Big).$$

Since $u_\varepsilon \in \mathcal{V}_C(\Omega_\varepsilon)$ and $\|\varphi\|_{\Omega_\varepsilon} + \|\nabla \varphi\|_{\Omega_\varepsilon} = 1$, the right-hand side is bounded for almost every $t \in [0, T]$. Analogously we find that $\|\partial_t v_\varepsilon\|_{H^1(\Omega_\varepsilon)'}$ is bounded. For $\partial_t s_\varepsilon$ we estimate

$$\|\partial_t s_\varepsilon\|_{H^1(\Gamma_\varepsilon^{ER})'} = \sup_{\psi \in H^1(\Gamma_\varepsilon^{ER}),\, \|\psi\|_{H^1(\Gamma_\varepsilon^{ER})}=1} \varepsilon \langle \partial_t s_\varepsilon, \psi \rangle_{H^1(\Gamma_\varepsilon^{ER})' \times H^1(\Gamma_\varepsilon^{ER})}$$

$$= \sup_{\psi \in H^1(\Gamma_\varepsilon^{ER}),\, \|\psi\|_{H^1(\Gamma_\varepsilon^{ER})}=1} \Big( -\varepsilon D_s \langle \varepsilon \nabla_\Gamma s_\varepsilon, \varepsilon \nabla_\Gamma \psi \rangle_{\Gamma_\varepsilon^{ER}}$$

$$- \varepsilon f \langle s_\varepsilon, \psi \rangle_{\Gamma_\varepsilon^{ER}} + \varepsilon^l l_s \langle u_\varepsilon - s_\varepsilon, \psi \rangle_{\Gamma_\varepsilon^{ER}} \Big)$$

$$\leq \sup_{\psi \in H^1(\Gamma_\varepsilon^{ER}),\, \|\psi\|_{H^1(\Gamma_\varepsilon^{ER})}=1} \Big( \varepsilon^3 \|\nabla_\Gamma s_\varepsilon\|_{\Gamma_\varepsilon^{ER}} \|\nabla \psi\|_{\Gamma_\varepsilon^{ER}} + \varepsilon f \|s_\varepsilon\|_{\Gamma_\varepsilon^{ER}} \|\psi\|_{\Gamma_\varepsilon^{ER}}$$

$$+ c_0 \varepsilon^{l-1} l_s \|u_\varepsilon\|_{\Omega_\varepsilon}^2 + c_0 \varepsilon^{l+1} l_s \|\nabla u_\varepsilon\|_{\Omega_\varepsilon}^2 + \varepsilon^l l_s \|\psi\|_{\Gamma_\varepsilon^{ER}}^2$$

$$+ \varepsilon^l l_s \|s_\varepsilon\|_{\Gamma_\varepsilon^{ER}} \|\psi\|_{\Gamma_\varepsilon^{ER}} \Big).$$

Since $u_\varepsilon \in \mathcal{V}_C(\Omega_\varepsilon)$, $s_\varepsilon \in \mathcal{V}(\Gamma_\varepsilon^{ER})$ and $\|\psi\|_{\Gamma_\varepsilon^{ER}} + \|\nabla_\Gamma \psi\|_{\Gamma_\varepsilon^{ER}} = 1$, the right-hand side is bounded for almost every $t \in [0, T]$. Analogously we find that $\|\partial_t w_\varepsilon\|_{H^1(\Omega_\varepsilon)'}$ is bounded. $\qquad\square$

In this section we modeled a family of differential equations suitable to use periodic homogenization and describing the carcinogenesis in a human cell. In the next section we proof that the required conditions for using homogenization are satisfied.

## 2.3 Estimations for the linear carcinogenesis model

First we show that the concentration of molecules stay nonnegative for nonnegative initial values. This means that the functions $u_\varepsilon$ and $v_\varepsilon$ do not become negative if we have nonnegative initial values for almost every $x \in \Omega_\varepsilon$.

**Theorem 2.16 (Positivity)**
*The functions $u_\varepsilon$ and $v_\varepsilon$ are nonnegative for almost every $x \in \Omega_\varepsilon$ and $t \in [0, T]$. The functions $s_\varepsilon$ and $w_\varepsilon$ are nonnegative for almost every $x \in \Gamma_\varepsilon^{ER}$ and $t \in [0, T]$.*

**Proof** We define

$$u_{\varepsilon-} := \begin{cases} -u_\varepsilon, & \text{if } u_\varepsilon \leq 0 \\ 0, & \text{otherwise} \end{cases} \qquad u_{\varepsilon+} := \begin{cases} u_\varepsilon, & \text{if } u_\varepsilon \geq 0 \\ 0, & \text{otherwise,} \end{cases}$$

pointwise and analogously, $v_{\varepsilon-}$, $s_{\varepsilon-}$ and $w_{\varepsilon-}$.
We test the weak formulation for $u_\varepsilon$ and $s_\varepsilon$ with $-u_{\varepsilon-}$ and $-s_{\varepsilon-}$

$$(\partial_t u_{\varepsilon-}, u_{\varepsilon-})_{\Omega_\varepsilon} + (D_u \nabla u_{\varepsilon-}, \nabla u_{\varepsilon-})_{\Omega_\varepsilon} + \varepsilon^l l_s \langle u_{\varepsilon-} + s_\varepsilon, u_{\varepsilon-} \rangle_{\Gamma_\varepsilon^{ER}} = 0,$$

$$\varepsilon \langle \partial_t s_{\varepsilon-}, s_{\varepsilon-} \rangle_{\Gamma_\varepsilon^{ER}} + \varepsilon D_s \langle \varepsilon \nabla_\Gamma s_{\varepsilon-}, \varepsilon \nabla_\Gamma s_{\varepsilon-} \rangle_{\Gamma_\varepsilon^{ER}} + \varepsilon f \langle s_{\varepsilon-}, s_{\varepsilon-} \rangle_{\Gamma_\varepsilon^{ER}} + \varepsilon^l l_s \langle u_\varepsilon + s_{\varepsilon-}, s_{\varepsilon-} \rangle_{\Gamma_\varepsilon^{ER}} = 0.$$

We integrate from 0 to $t$ and add the equations. With the assumption that $u_\varepsilon(0) \geq 0$ and $s_\varepsilon(0) \geq 0$ we get

$$\frac{1}{2}\|u_{\varepsilon-}\|^2_{\Omega_\varepsilon} + D_u\|\nabla u_{\varepsilon-}\|^2_{\Omega_\varepsilon,t} + \frac{1}{2}\varepsilon\|s_{\varepsilon-}\|^2_{\Gamma_\varepsilon^{ER}}$$
$$+ D_s\varepsilon^3\|\nabla_\Gamma s_{\varepsilon-}\|^2_{\Gamma_\varepsilon^{ER},t} + \varepsilon f\|s_{\varepsilon-}\|^2_{\Gamma_\varepsilon^{ER},t} + \varepsilon^l l_s\|u_{\varepsilon-} - s_{\varepsilon-}\|^2_{\Gamma_\varepsilon^{ER},t}$$
$$+ \underbrace{\varepsilon^l l_s\langle s_{\varepsilon+}, u_{\varepsilon-}\rangle_{\Gamma_\varepsilon^{ER},t} + \varepsilon^l l_s\langle u_{\varepsilon+}, s_{\varepsilon-}\rangle_{\Gamma_\varepsilon^{ER},t}}_{\geq 0} = 0.$$

This yields

$$\frac{1}{2}\|u_{\varepsilon-}\|^2_{\Omega_\varepsilon} + D_u\|\nabla u_{\varepsilon-}\|^2_{\Omega_\varepsilon,t} + \frac{1}{2}\varepsilon\|s_{\varepsilon-}\|^2_{\Gamma_\varepsilon^{ER}} + D_s\varepsilon^3\|\nabla_\Gamma s_{\varepsilon-}\|^2_{\Gamma_\varepsilon^{ER},t}$$
$$+ \varepsilon f\|s_{\varepsilon-}\|^2_{\Gamma_\varepsilon^{ER},t} + \varepsilon^l l_s\|u_{\varepsilon-} - s_{\varepsilon-}\|^2_{\Gamma_\varepsilon^{ER},t} \leq 0.$$

We deduce that $u_{\varepsilon-}(x,t) = 0$ for almost every $x \in \Omega_\varepsilon$ and $t \in [0,T]$ and $s_{\varepsilon-}(x,t) = 0$ for almost every $x \in \Gamma_\varepsilon^{ER}$ and $t \in [0,T]$. This means that $u_\varepsilon(x,t) \geq 0$ for almost every $x \in \Omega_\varepsilon$ and $t \in [0,T]$ and $s_\varepsilon(x,t) \geq 0$ for almost every $x \in \Gamma_\varepsilon^{ER}$ and $t \in [0,T]$.

We continue with similar estimations for the equation for $v_\varepsilon$ and $w_\varepsilon$ and test the weak formulation with $-v_{\varepsilon-}$ and $-w_{\varepsilon-}$,

$$\frac{1}{2}\|v_{\varepsilon-}\|^2_{\Omega_\varepsilon} + D_v\|\nabla v_{\varepsilon-}\|^2_{\Omega_\varepsilon,t} + \frac{1}{2}\varepsilon\|w_{\varepsilon-}\|^2_{\Gamma_\varepsilon^{ER}} + D_w\varepsilon^3\|\nabla_\Gamma w_{\varepsilon-}\|^2_{\Gamma_\varepsilon^{ER},t}$$
$$+ \underbrace{\varepsilon f\langle s_\varepsilon, w_{\varepsilon-}\rangle_{\Gamma_\varepsilon^{ER},t} + \varepsilon^m l_w\langle w_{\varepsilon+}, v_{\varepsilon-}\rangle_{\Gamma_\varepsilon^{ER},t} + \varepsilon^m l_w\langle v_{\varepsilon+}, w_{\varepsilon-}\rangle_{\Gamma_\varepsilon^{ER},t}}_{\geq 0}$$
$$+ \varepsilon^m l_w\|v_{\varepsilon-} - w_{\varepsilon-}\|^2_{\Gamma_\varepsilon^{ER},t} = 0.$$

This leads to

$$\frac{1}{2}\|v_{\varepsilon-}\|^2_{\Omega_\varepsilon} + D_v\|\nabla v_{\varepsilon-}\|^2_{\Omega_\varepsilon,t} + \frac{1}{2}\varepsilon\|w_{\varepsilon-}\|^2_{\Gamma_\varepsilon^{ER}}$$
$$+ D_w\varepsilon^3\|\nabla_\Gamma w_{\varepsilon-}\|^2_{\Gamma_\varepsilon^{ER},t} + \varepsilon^m l_w\|v_{\varepsilon-} - w_{\varepsilon-}\|^2_{\Gamma_\varepsilon^{ER},t} \leq 0.$$

We deduce that $v_\varepsilon(x,t) \geq 0$ for almost every $x \in \Omega_\varepsilon$ and $w_\varepsilon(x,t) \geq 0$ for almost every $x \in \Gamma_\varepsilon^{ER}$ and almost every $t \in [0,T]$. Then the proof is complete. $\qquad\square$

We are going to show that the conditions of proposition 2.5 (i) are satisfied for $u_\varepsilon$ and $v_\varepsilon$ to use two-scale convergence on $\Omega_\varepsilon$. Further we want to use proposition 2.5 (ii) for $s_\varepsilon$ and $w_\varepsilon$ and we are going to show that the required conditions are fulfilled.
Hence, the next lemma.

**Lemma 2.17 (Boundedness in $L^2$)**
*There is a constant $C > 0$ independent of $\varepsilon$ such that*

$$\|u_\varepsilon\|_{\Omega_\varepsilon}^2 + D_u\|\nabla u_\varepsilon\|_{\Omega_\varepsilon,t}^2 + \|v_\varepsilon\|_{\Omega_\varepsilon}^2 + D_v\|\nabla v_\varepsilon\|_{\Omega_\varepsilon,t}^2$$
$$+ \varepsilon\|s_\varepsilon\|_{\Gamma_\varepsilon^{ER}}^2 + D_s\varepsilon^3\|\nabla_\Gamma s_\varepsilon\|_{\Gamma_\varepsilon^{ER},t}^2 + \varepsilon\|w_\varepsilon\|_{\Gamma_\varepsilon^{ER}}^2 + D_w\varepsilon^3\|\nabla_\Gamma w_\varepsilon\|_{\Gamma_\varepsilon^{ER},t}^2$$
$$+ \varepsilon^l l_s\|u_\varepsilon - s_\varepsilon\|_{\Gamma_\varepsilon^{ER},t}^2 + \varepsilon^m l_w\|v_\varepsilon - w_\varepsilon\|_{\Gamma_\varepsilon^{ER},t}^2 \leq C$$

*for almost every $t \in [0,T]$.*

**Proof** To prove the claim we start with the weak formulation of our problem. We test the equations with the functions $(u_\varepsilon, v_\varepsilon, s_\varepsilon, w_\varepsilon)$ and add them up:

$$(\partial_t u_\varepsilon, u_\varepsilon)_{\Omega_\varepsilon} + (\partial_t v_\varepsilon, v_\varepsilon)_{\Omega_\varepsilon} + D_u(\nabla u_\varepsilon, \nabla u_\varepsilon)_{\Omega_\varepsilon} + D_v(\nabla v_\varepsilon, \nabla v_\varepsilon)_{\Omega_\varepsilon}$$
$$+ \varepsilon\langle\partial_t s_\varepsilon, s_\varepsilon\rangle_{\Gamma_\varepsilon^{ER}} + \varepsilon\langle\partial_t w_\varepsilon, w_\varepsilon\rangle_{\Gamma_\varepsilon^{ER}} + D_s\varepsilon^3\langle\nabla_\Gamma s_\varepsilon, \nabla_\Gamma s_\varepsilon\rangle_{\Gamma_\varepsilon^{ER}}$$
$$+ D_w\varepsilon^3\langle\nabla_\Gamma w_\varepsilon, \nabla_\Gamma w_\varepsilon\rangle_{\Gamma_\varepsilon^{ER}} + \varepsilon f\langle s_\varepsilon, s_\varepsilon\rangle_{\Gamma_\varepsilon^{ER}} - \varepsilon f\langle s_\varepsilon, w_\varepsilon\rangle_{\Gamma_\varepsilon^{ER}}$$
$$+ \varepsilon^l l_s\langle u_\varepsilon - s_\varepsilon, u_\varepsilon - s_\varepsilon\rangle_{\Gamma_\varepsilon^{ER}} + \varepsilon^m l_w\langle v_\varepsilon - w_\varepsilon, v_\varepsilon - w_\varepsilon\rangle_{\Gamma_\varepsilon^{ER}} = 0.$$

With $\frac{1}{2}\frac{d}{dt}\|u_\varepsilon(t)\|_{\Omega_\varepsilon}^2 = (\partial_t u_\varepsilon, u_\varepsilon)_{\Omega_\varepsilon}$ and integration from 0 to $t$ it holds that

$$\frac{1}{2}\|u_\varepsilon\|_{\Omega_\varepsilon}^2 + \frac{1}{2}\|v_\varepsilon\|_{\Omega_\varepsilon}^2 + D_u\|\nabla u_\varepsilon\|_{\Omega_\varepsilon,t}^2 + D_v\|\nabla v_\varepsilon\|_{\Omega_\varepsilon,t}^2 + \varepsilon\frac{1}{2}\|s_\varepsilon\|_{\Gamma_\varepsilon^{ER}}^2$$
$$+ \varepsilon\frac{1}{2}\|w_\varepsilon\|_{\Gamma_\varepsilon^{ER}}^2 + D_s\varepsilon^3\|\nabla_\Gamma s_\varepsilon\|_{\Gamma_\varepsilon^{ER},t}^2 + D_w\varepsilon^3\|\nabla_\Gamma w_\varepsilon\|_{\Gamma_\varepsilon^{ER},t}^2 + \varepsilon f\|s_\varepsilon\|_{\Gamma_\varepsilon^{ER},t}^2$$
$$+ \varepsilon^l l_s\|u_\varepsilon - s_\varepsilon\|_{\Gamma_\varepsilon^{ER},t}^2 + \varepsilon^m l_w\|v_\varepsilon - w_\varepsilon\|_{\Gamma_\varepsilon^{ER},t}^2 = \varepsilon f\langle s_\varepsilon, w_\varepsilon\rangle_{\Gamma_\varepsilon^{ER},t}$$
$$+ \frac{1}{2}\|u_\varepsilon(0)\|_{\Omega_\varepsilon}^2 + \frac{1}{2}\|v_\varepsilon(0)\|_{\Omega_\varepsilon}^2 + \frac{1}{2}\varepsilon\|s_\varepsilon(0)\|_{\Gamma_\varepsilon^{ER}}^2 + \frac{1}{2}\varepsilon\|w_\varepsilon(0)\|_{\Gamma_\varepsilon^{ER}}^2.$$

With initial conditions lying in $L^2$ and the binomial theorem we deduce

$$\frac{1}{2}\|u_\varepsilon\|^2_{\hat{\Omega}_\varepsilon} + \frac{1}{2}\|v_\varepsilon\|^2_{\hat{\Omega}_\varepsilon} + D_u\|\nabla u_\varepsilon\|^2_{\Omega_\varepsilon,t} + D_v\|\nabla v_\varepsilon\|^2_{\Omega_\varepsilon,t} + \varepsilon\frac{1}{2}\|s_\varepsilon\|^2_{\Gamma^{ER}_\varepsilon}$$

$$+ \varepsilon\frac{1}{2}\|w_\varepsilon\|^2_{\Gamma^{ER}_\varepsilon} + D_s\varepsilon^3\|\nabla_\Gamma s_\varepsilon\|^2_{\Gamma^{ER}_\varepsilon,t} + D_w\varepsilon^3\|\nabla_\Gamma w_\varepsilon\|^2_{\Gamma^{ER}_\varepsilon,t} + \varepsilon^l l_s\|u_\varepsilon - s_\varepsilon\|^2_{\Gamma^{ER}_\varepsilon,t}$$

$$+ \varepsilon^m l_w\|v_\varepsilon - w_\varepsilon\|^2_{\Gamma^{ER}_\varepsilon,t} \leq \varepsilon f\|s_\varepsilon\|^2_{\Gamma^{ER}_\varepsilon,t} + \varepsilon f\|w_\varepsilon\|^2_{\Gamma^{ER}_\varepsilon,t} + c_1.$$

Now we deduce from Gronwall's lemma 2.12 that

$$\frac{1}{2}\|u_\varepsilon\|^2_{\hat{\Omega}_\varepsilon} + \frac{1}{2}\|v_\varepsilon\|^2_{\hat{\Omega}_\varepsilon} + D_u\|\nabla u_\varepsilon\|^2_{\Omega_\varepsilon,t} + D_v\|\nabla v_\varepsilon\|^2_{\Omega_\varepsilon,t}$$

$$+ \varepsilon\frac{1}{2}\|s_\varepsilon\|^2_{\Gamma^{ER}_\varepsilon} + \varepsilon\frac{1}{2}\|w_\varepsilon\|^2_{\Gamma^{ER}_\varepsilon} + D_s\varepsilon^3\|\nabla_\Gamma s_\varepsilon\|^2_{\Gamma^{ER}_\varepsilon,t} + D_w\varepsilon^3\|\nabla_\Gamma w_\varepsilon\|^2_{\Gamma^{ER}_\varepsilon,t}$$

$$+ \varepsilon^l l_s\|u_\varepsilon - s_\varepsilon\|^2_{\Gamma^{ER}_\varepsilon,t} + \varepsilon^m l_w\|v_\varepsilon - w_\varepsilon\|^2_{\Gamma^{ER}_\varepsilon,t} \leq C$$

and the proof is complete. □

Further, we need one more corollary.

**Corollary 2.18** *There is a constant $C > 0$ independent of $\varepsilon$ such that*

$$\varepsilon\|u_\varepsilon\|^2_{\Gamma^{ER}_\varepsilon,t} < C \qquad and \qquad \varepsilon\|v_\varepsilon\|^2_{\Gamma^{ER}_\varepsilon,t} < C.$$

**Proof** We assume $\varepsilon < 1$. With lemma 2.11 and lemma 2.17 we find

$$\varepsilon\|u_\varepsilon\|^2_{\Gamma^{ER}_\varepsilon,t} \leq c_0 \left( \|u_\varepsilon\|^2_{\Omega_\varepsilon,t} + \varepsilon^2\|\nabla u_\varepsilon\|^2_{\Omega_\varepsilon,t} \right)$$

$$\leq c_0 \left( \|u_\varepsilon\|^2_{\Omega_\varepsilon,t} + \|\nabla u_\varepsilon\|^2_{\Omega_\varepsilon,t} \right) < C.$$

The analogous inequalities hold for $v_\varepsilon$. □

Now we proved that we are allowed to use the assertion in proposition 2.5. In the next section we are going to build an abbreviation of the model that is helpful to show existence of a solution.

# 2.4 Abbreviation of the model

The partial differential equation for the functions $s_\varepsilon$ and $w_\varepsilon$ is a standard non homogeneous heat equation with an additional linear term on a domain without boundary conditions. For such equations analytical solutions exist and are unique, see [21]. The right-hand side of the PDE for $s_\varepsilon$ and $w_\varepsilon$ is depending on $u_\varepsilon$ and $v_\varepsilon$, respectively; so, the analytical solution also will depend on $u_\varepsilon$ or $v_\varepsilon$. We will need to deduce the solution on the Riemannian manifold $\Gamma_\varepsilon^{ER}$. We take a short excursion about Riemannian manifolds cited from "Riemannian Geometry" by Manfredo Perdigao do Carmo [19] Chapter 1.

## Excursion about Riemannian manifolds

Let $M \subset \mathbb{R}^n$ be a Riemannian manifold with dimension $\mathbb{R}^{n-1}$ and tangent bundle $TM = \bigcup_{p \in M}\{p\} \times T_pM$. Let $(g_{ij})_{i,j=1...n}$ be the Riemannian metric on $M$.

**Dual space $T_p^*M$**
Let $v$ be an element of $T_pM$ for a $p \in M$, then there exists an isomorphism $j_x$ with

$$j_x : T_pM \to T_pM^*$$
$$j_x : v \mapsto \langle v, \cdot \rangle$$

where $\langle \cdot, \cdot \rangle_p : T_pM \times T_pM \to \mathbb{R}$ is a scalar product on $T_pM$.

**Local basis on $T_pM$ and $T_p^*M$**
For every $p \in M$ a basis of $T_pM$ is denoted by

$$\left\{ \frac{d}{dx_1}, \frac{d}{dx_2}, \dots, \frac{d}{dx_n} \right\},$$

and the Riemannian metric is build by $g_{ij} = \left\langle \frac{d}{dx_i}, \frac{d}{dx_j} \right\rangle_p$.

A basis of $T_p^*M$ is denoted by

$$\{dx^1, dx^2, \dots, dx^n\}.$$

**Gradient as element of $T_pM$.**
For a function $f : M \to \mathbb{R}$ we define the derivative on $M$ as

$$df = \sum_{i=1}^{n} \partial_i f dx^i \in T_p^*M,$$

where $\partial_i$ is the derivative in direction $\frac{d}{dx_i}$. Now we define the gradient $\nabla_M f$ as the element of $T_p M$ such that

$$\nabla_M f \overset{jx}{\to} \langle \nabla_M f, \cdot \rangle_p = df(\cdot)$$

with

$$\nabla_M f = \sum_{i,j=1}^{n} g^{ij} \partial_j f \frac{d}{dx_i}.$$

The entries of the inverse of the matrix $(g_{ij})_{ij}$ are denoted by $g^{ij}$, that is $\sum_k g_{ik} g^{kj} = \delta_{ij}$.

**Integrals on $M$**
Let $(U, \alpha)$ be a chart on $M$ with $U \subset M$ and

$$\alpha : U \to V \subset \mathbb{R}^{n-1}.$$

Then it holds for an integral of $f : M \to \mathbb{R}$ on $\tilde{M} \subset U$ that

$$\int_{\tilde{M}} f = \int_{\alpha(\tilde{M})} f(x_1, \ldots, x_n) g^{1/2}(x) dx_1 \ldots dx_n$$

where $\alpha(\tilde{M}) \subset V$ and $g^{1/2} = \sqrt{\det(g_{ij})_{ij=1,\ldots,n}}$.

**Divergence on $M$**
Now let $\xi : M \to \mathbb{R}^n$ be a differentiable vector field and $f : M \to \mathbb{R}$ be a differentiable function. We consider the divergence $\nabla_M \cdot \xi$ on the manifold $M$. We know using integration by parts that

$$\int_M f \cdot (\nabla_M \cdot \xi) = - \int_M (\nabla_M f) \cdot \xi.$$

In the following equation, the scalar product $\langle \cdot, \cdot \rangle_x$ is respective to the manifold $M$. Hence, for any $\tilde{M} \subset M$ with appropriate chart $\alpha$

holds

$$\int_{\tilde{M}} f \cdot \underbrace{(\nabla_M \cdot \xi)}_{\star} = - \int_{\tilde{M}} (\nabla_M f) \cdot \xi$$

$$= - \int_{\alpha(\tilde{M})} \langle \xi, \nabla_M f \rangle_x g^{1/2}(x) \mathrm{d}x$$

$$= - \int_{\alpha(\tilde{M})} \sum_{i,k=1}^{n} g_{ik} (\nabla_M f)_i \xi_k g^{1/2}(x) \mathrm{d}x$$

$$= - \int_{\alpha(\tilde{M})} \sum_{i,k=1}^{n} g_{ik} \sum_{j=1}^{n} g^{ij} (\partial_j f) \xi_k g^{1/2}(x) \mathrm{d}x$$

$$= - \int_{\alpha(\tilde{M})} \sum_{i,k,j=1}^{n} \underbrace{g_{ik} g^{ij}}_{\delta_{kj}} (\partial_j f) \xi_k g^{1/2}(x) \mathrm{d}x$$

$$= - \int_{\alpha(\tilde{M})} \sum_{j=1}^{n} (\partial_j f) \xi_j g^{1/2}(x) \mathrm{d}x$$

$$= \int_{\tilde{M}} f \cdot \underbrace{\sum_{j=1}^{n} \partial_j (g^{1/2} \xi_j) g^{-1/2}}_{\star}.$$

This yields,

$$\nabla_M \cdot \xi = g^{-1/2} \sum_{j=1}^{n} \partial_j (g^{1/2} \xi_j).$$

**Laplace-Beltrami operator on $M$**
Now we use this equation to calculate the Laplace-Beltrami operator $\Delta_M f$ for $\Delta_M f = \nabla_M \cdot (\nabla_M f)$

$$\Delta_M f = \nabla_M \cdot (\nabla_M f) = g^{-1/2} \sum_{j=1}^{n} \partial_j \left( g^{1/2} \sum_{i=1}^{n} g^{ji} (\partial_i f) \right).$$

# Analytical Solutions of $s_\varepsilon$ and $w_\varepsilon$

With this differential geometric introduction, we are able to solve the PDE for $s_\varepsilon$ and $w_\varepsilon$ analytically on the Riemannian manifold

$\Gamma_\varepsilon^{ER}$. To solve

$$\partial_t s_\varepsilon - D_s \varepsilon^2 \Delta_\Gamma s_\varepsilon + (f + \varepsilon^{l-1} l_s) s_\varepsilon = \varepsilon^{l-1} l_s u_\varepsilon \qquad \text{on } \Gamma_\varepsilon^{ER}$$

from equation (2.3), we implicitly define the auxiliary function $\lambda$ by

$$s_\varepsilon(x,t) = \lambda(x,t) e^{-(f+\varepsilon^{l-1} l_s)t}.$$

This trick can be found in "Partial differential equations" by J. Wloka in the proof of Theorem 26.1 [57].

By plugging the auxiliary function into the PDE, it follows that

$$\partial_t \lambda(x,t) e^{-(f+\varepsilon^{l-1} l_s)t} - (f + \varepsilon^{l-1} l_s) \lambda(x,t) e^{-(f+\varepsilon^{l-1} l_s)t}$$
$$- D_s \varepsilon^2 \Delta_\Gamma \lambda(x,t) e^{-(f+\varepsilon^{l-1} l_s)t} + (f + \varepsilon^{l-1} l_s) \lambda(x,t) e^{-(f+\varepsilon^{l-1} l_s)t}$$
$$= \varepsilon^{l-1} l_s u_\varepsilon(x,t).$$

We simplify the equation to

$$\partial_t \lambda(x,t) - D_s \varepsilon^2 \Delta_\Gamma \lambda(x,t) = \varepsilon^{l-1} l_s u_\varepsilon(x,t) e^{(f+\varepsilon^{l-1} l_s)t} \qquad (2.7)$$

for all $x \in \Gamma_\varepsilon^{ER}$ and $t \in [0,T]$. Obviously, $\lambda$ solves a heat equation.

The solution of the heat equation is well known. Therefore, we cite the next theorem from [18, 21].

**Theorem 2.19** *Let $M$ be a complete Riemannian manifold. Then the function $\varphi : M \times [0,T] \to \mathbb{R}$ given by*

$$\varphi(x,t) = e^{\alpha \Delta_M t} \varphi_0(x) + \int_0^t e^{\alpha \Delta_M (t-s)} f(x,s) ds$$

*solves the heat equation*

$$\partial_t \varphi - \alpha \Delta_M \varphi = f(x,t)$$
$$\varphi(x,0) = \varphi_0(x),$$

*for $f : M \times [0,T] \to \mathbb{R}$ and $\alpha \in \mathbb{R}$.*

Using this theorem we find that

$$\lambda(x,t) = e^{D_s \varepsilon^2 \Delta_\Gamma t} \lambda_0(x) + \int_0^t e^{D_s \varepsilon^2 \Delta_\Gamma (t-s)} \varepsilon^{l-1} l_s u_\varepsilon(x,s) e^{(f+\varepsilon^{l-1} l_s)s} ds$$

solves equation (2.7). We conclude by substituting this in $s_\varepsilon(x,t)$,

$$F(u_\varepsilon) := s_\varepsilon(x,t) = e^{(D_s\varepsilon^2\Delta_\Gamma - f - \varepsilon^{l-1}l_s)t}s_I(x) +$$
$$e^{-(f+\varepsilon^{l-1}l_s)t}\varepsilon^{l-1}l_s \int_0^t e^{D_s\varepsilon^2\Delta_\Gamma(t-s)}u_\varepsilon(s,x)e^{(f+\varepsilon^{l-1}l_s)s}ds. \quad (2.8)$$

Analogously, the solution for

$$\partial_t w_\varepsilon - D_w\varepsilon^2\Delta_\Gamma w_\varepsilon + \varepsilon^{m-1}l_w w_\varepsilon = \varepsilon^{m-1}l_w v_\varepsilon + fs_\varepsilon$$

is given by

$$G(u_\varepsilon, v_\varepsilon) := w_\varepsilon(x,t) = e^{(D_w\varepsilon^2\Delta_\Gamma - \varepsilon^{m-1}l_w)t}w_I(x)$$
$$+ \int_0^t e^{-\varepsilon^{m-1}l_w(t-s) + D_w\varepsilon^2(t-s)\Delta_\Gamma}\left(\varepsilon^{m-1}l_w v_\varepsilon(s,x) + fF(u_\varepsilon)(s,x)\right)ds,$$
$$(2.9)$$

where we define

$$\tilde{G}(v_\varepsilon) = e^{(D_w\varepsilon^2\Delta_\Gamma - \varepsilon^{m-1}l_w)t}w_I(x)$$
$$+ e^{-\varepsilon^{m-1}l_w t}\varepsilon^{m-1}l_w \int_0^t e^{D_w\varepsilon^2(t-s)\Delta_\Gamma}v_\varepsilon(s,x)e^{\varepsilon^{m-1}l_w s}ds$$

and

$$\tilde{\tilde{G}}(F(u_\varepsilon)) = e^{-\varepsilon^{m-1}l_w t}f \int_0^t e^{D_w\varepsilon^2(t-s)\Delta_\Gamma}F(u_\varepsilon)(s,x)e^{\varepsilon^{m-1}l_w s}ds.$$

**Remark 2.20** The operators $F$, $\tilde{G}$, and $\tilde{\tilde{G}}$ are linear, because $e^{\Delta_\Gamma t}$ and integrals are linear operators. Consequently, also $G$ is linear.

To show existence of a solution in the next section, we need positivity and boundedness of the operators $F$ and $G$.

**Lemma 2.21** *The operators* $F \; : \; L^2([0,T] \times \Gamma_\varepsilon^{ER}) \; \rightarrow \; L^2([0,T], H^1(\Gamma_\varepsilon^{ER}))$ *and* $G \; : \; L^2([0,T] \times \Gamma_\varepsilon^{ER})^2 \; \rightarrow \; L^2([0,T], H^1(\Gamma_\varepsilon^{ER}))$ *defined in* (2.8) *and* (2.9) *are positive and bounded.*

**Proof** Davis proves in "Heat Kernels and Spectral Theory" [18] that for the Laplace operator $\Delta$, the function $e^{\Delta t}$ has a strictly positive $C^\infty$ kernel. This means there exists a smooth function $k_\varepsilon(t, x) > 0$ for each $t > 0$ (we denote $k_{\varepsilon,t} = k_\varepsilon(t, \cdot)$) such that

$$e^{D_s \varepsilon^2 \Delta t} f(x) = (f \star k_{\varepsilon,t})(x) = \int_{\Gamma_\varepsilon^{ER}} k_\varepsilon(t, x - y) f(y) d\sigma_y$$

for every $f \in L^2(\Gamma_\varepsilon^{ER})$. Hence, for every function $f > 0$ it yields

$$e^{\Delta t} f > 0.$$

The integral operator $f \mapsto \int f$ also is a positive operator. Hence, it holds

$$F(u_\varepsilon) \geq 0, \qquad \text{if } u_\varepsilon \geq 0,$$
$$G(u_\varepsilon, v_\varepsilon) \geq 0, \qquad \text{if } u_\varepsilon, v_\varepsilon \geq 0.$$

Since the integral operator and $e^{\Delta t}$ are linear and bounded [18], the inequalities

$$\|F(u_\varepsilon)\|_{\Gamma_\varepsilon^{ER}, t}^2 \leq \|F\| \|u_\varepsilon\|_{\Gamma_\varepsilon^{ER}, t}$$

and

$$\|G(u_\varepsilon, v_\varepsilon)\|_{\Gamma_\varepsilon^{ER}, t} = \|\tilde{G}(v_\varepsilon) + \tilde{\tilde{G}}(F(u_\varepsilon))\|_{\Gamma_\varepsilon^{ER}, t}$$
$$\leq \|\tilde{G}\| \|v_\varepsilon\|_{\Gamma_\varepsilon^{ER}, t} + \|\tilde{\tilde{G}}\| \|u_\varepsilon\|_{\Gamma_\varepsilon^{ER}, t}$$

hold true with $0 < \|F\|, \|G\|, \|\tilde{G}\|, \|\tilde{\tilde{G}}\| < \infty$. $\qquad \square$

**Strong formulation**
With the solutions $s_\varepsilon = F(u_\varepsilon)$ and $w_\varepsilon = G(u_\varepsilon, v_\varepsilon)$ the system of equations (2.3) can be abbreviated as the following partial differential equation.

$$\begin{aligned}
\partial_t u_\varepsilon - D_u \Delta u_\varepsilon &= 0 && \text{in } \Omega_\varepsilon \\
u_\varepsilon &= u_{\text{Boundary}} && \text{on } \Gamma_C \\
-D_u \nabla u_\varepsilon \cdot n &= 0 && \text{on } \Gamma_N \\
-D_u \nabla u_\varepsilon \cdot n &= \varepsilon^l l_s (u_\varepsilon - F(u_\varepsilon)) && \text{on } \Gamma_\varepsilon^{\text{ER}}
\end{aligned}$$

$$\begin{aligned}
\partial_t v_\varepsilon - D_v \Delta v_\varepsilon &= 0 && \text{in } \Omega_\varepsilon \\
v_\varepsilon &= 0 && \text{on } \Gamma_N \\
-D_v \nabla v_\varepsilon \cdot n &= 0 && \text{on } \Gamma_C \\
-D_v \nabla v_\varepsilon \cdot n &= \varepsilon^m l_w (v_\varepsilon - G(u_\varepsilon, v_\varepsilon)) && \text{on } \Gamma_\varepsilon^{\text{ER}}
\end{aligned}$$

$$(2.10)$$

and initial values $u_\varepsilon(x,t) = u_I(x)$ and $v_\varepsilon(x,t) = v_I(x)$ for $t = 0$, $x \in \Omega_\varepsilon$.

**Weak formulation**
The weak formulation of the abbreviated form is given by $(u_\varepsilon, v_\varepsilon) \in \mathcal{V}_C(\Omega_\varepsilon) \times \mathcal{V}_N(\Omega_\varepsilon)$ such that

$$(\partial_t u_\varepsilon, \varphi_1)_{\Omega_\varepsilon} + (\partial_t v_\varepsilon, \varphi_2)_{\Omega_\varepsilon} + D_u (\nabla u_\varepsilon, \nabla \varphi_1)_{\Omega_\varepsilon} + D_v (\nabla v_\varepsilon, \nabla \varphi_2)_{\Omega_\varepsilon}$$
$$+ \varepsilon^l l_s \langle u_\varepsilon - F(u_\varepsilon), \varphi_1 \rangle_{\Gamma_\varepsilon^{\text{ER}}} + \varepsilon^m l_w \langle v_\varepsilon - G(u_\varepsilon, v_\varepsilon), \varphi_2 \rangle_{\Gamma_\varepsilon^{\text{ER}}} = 0$$

$$(2.11)$$

for all $\varphi = (\varphi_1, \varphi_2) \in V_{C0}(\Omega_\varepsilon) \times V_N(\Omega_\varepsilon)$.

With the system of equations (2.11) we have an abbreviated version of the model (2.6). We are going to use this representation for showing existence of a solution.

## 2.5 Existence of a solution

To show existence of a solution of the system of equation (2.11) we use proposition 3.2 in "Monotone Operators in Banach Space and Nonlinear Partial Differential Equations" by R.E. Showalter [53]. The proposition is summarized in theorem 2.23. We briefly introduce the setting used in [53].

Let $V$ and $\mathcal{V} = L^2([0,T], V)$ be separable Hilbert spaces with duals $V'$ and $\mathcal{V}' = L^2([0,T], V')$. Let $W$ be a Hilbert space with

continuous injection $V \hookrightarrow W$ and $V$ dense in $W$.
We are given for every $t \in [0,T]$ an operator $\mathcal{A}(t) \in \mathcal{L}(V,V')$
such that $(\mathcal{A}(\cdot)u)v \in L^\infty([0,T])$ for each pair $u,v \in V$. Furthermore let $\mathcal{B}(t) \in \mathcal{L}(W,W')$ be another family of operators with
$(\mathcal{B}(\cdot)u)v \in L^\infty([0,T])$ for each pair $u,v \in W$. Finally, suppose
that $u_0 \in W$ and $f \in L^2([0,T],V')$ are given.

The problem is given by: Find a $u \in V$ such that

$$
\begin{aligned}
\tfrac{d}{dt}(\mathcal{B}(t)u(t)) + \mathcal{A}(t)u(t) &= f(t) \quad \text{in } V', \\
(\mathcal{B})(0) &= \mathcal{B}(0)u_0.
\end{aligned}
\tag{2.12}
$$

**Definition 2.22** *The family $\{\mathcal{B}(t)\mid t \in [0,T]\}$ of operators given as
above will be called regular if for each pair $u,v \in V$ the function
$(\mathcal{B}(\cdot)u)v$ is absolutely continuous on $[0,T]$ and there is a function
$K \in L^1([0,T])$ such that*

$$
\left| \frac{d}{dt}(\mathcal{B}(t)u)v \right| \le K(t)\|u\|\|v\|, \quad u,v \in V, \quad a.e.\, t \in [0,T].
$$

**Theorem 2.23** *Let the separable Hilbert spaces $V,W$ and linear opera-
tors $\mathcal{A}(t)$, $\mathcal{B}(t)$, $0 \le t \le T$, the data $u_0 \in W$ and $f \in L^2([0,T],V')$
be given as above, and assume further that $\mathcal{B}(t)$ is a regular family of
self-adjoint operators. Furthermore, let $\mathcal{B}(0)$ be monotone and there are
numbers $\kappa,\lambda > 0$ such that*

$$
2(\mathcal{A}(t)v)v + \lambda(\mathcal{B}(t)v)v + (\mathcal{B}'(t)v)v \ge \kappa\|v\|^2,
\tag{2.13}
$$

*for all $v \in V$ and $0 \le t \le T$. Then the problem (2.12) has at least one
solution which satisfies*

$$
\|u\|_V \le C(\lambda,\kappa) \left( \|f\|_{V'}^2 + (\mathcal{B}(0)u_0)u_0 \right)^{1/2}.
$$

**Transformation of the problem**
Before we can use the theorem 2.23, the problem to solve (2.11)
is going to be transformed to a suitable form. Therefore, the constant function $u_B$ is defined by extending the constant boundary
function $u_{\text{Boundary}}$ to the domain $\Omega_\varepsilon$ such that $\gamma(u_B) = u_{\text{Boundary}}$.
Further, we define $\tilde{u}_\varepsilon \in \mathcal{V}_{C0} = \{u \in L^2([0,T],H^1(\Omega_\varepsilon))\mid u =$

0 on $\Gamma_C$} by $\tilde{u}_\varepsilon = u_\varepsilon - u_B$. Then the function $\tilde{u}_\varepsilon$ satisfies

$$(\partial_t \tilde{u}_\varepsilon + \partial_t u_B, \varphi_1)_{\Omega_\varepsilon} + (\partial_t v_\varepsilon, \varphi_2)_{\Omega_\varepsilon} + D_u(\nabla \tilde{u}_\varepsilon + \nabla u_B, \nabla \varphi_1)_{\Omega_\varepsilon}$$
$$+ D_v(\nabla v_\varepsilon, \varphi_2)_{\Omega_\varepsilon} + \varepsilon^l l_s \langle \tilde{u}_\varepsilon + u_B - F(\tilde{u}_\varepsilon) - F(u_B), \varphi_1 \rangle_{\Gamma_\varepsilon^{ER}}$$
$$+ \varepsilon^m l_w \langle v_\varepsilon - G(\tilde{u}_\varepsilon, v_\varepsilon) - G(u_B, 0), \varphi_2 \rangle_{\Gamma_\varepsilon^{ER}} = 0.$$

Hence, the problem (2.11) is equivalent to finding $(\tilde{u}_\varepsilon, v_\varepsilon) \in \mathcal{V}_{C0} \times \mathcal{V}_N$ with

$$(\partial_t \tilde{u}_\varepsilon, \varphi_1)_{\Omega_\varepsilon} + (\partial_t v_\varepsilon, \varphi_2)_{\Omega_\varepsilon} + D_u(\nabla \tilde{u}_\varepsilon, \nabla \varphi_1)_{\Omega_\varepsilon} + D_v(\nabla v_\varepsilon, \nabla \varphi_2)_{\Omega_\varepsilon}$$
$$+ \varepsilon^l l_s \langle \tilde{u}_\varepsilon - F(\tilde{u}_\varepsilon), \varphi_1 \rangle_{\Gamma_\varepsilon^{ER}} + \varepsilon^m l_w \langle v_\varepsilon - G(\tilde{u}_\varepsilon, v_\varepsilon), \varphi_2 \rangle_{\Gamma_\varepsilon^{ER}}$$
$$= -\varepsilon^l l_s \langle u_B - F(u_B), \varphi_1 \rangle_{\Gamma_\varepsilon^{ER}} + \varepsilon^m l_w \langle G(u_B, 0), \varphi_2 \rangle_{\Gamma_\varepsilon^{ER}} \quad (2.14)$$

for all $(\varphi_1, \varphi_2) \in \mathcal{V}_{C0} \times \mathcal{V}_N$ with initial conditions $(\tilde{u}_\varepsilon(0), v_\varepsilon(0)) = (u_I - u_B, v_I)$ and then setting $u_\varepsilon = \tilde{u}_\varepsilon + u_B$.

### Identification of the setting
For every $\varepsilon > 0$ we now identify the spaces, operators and functions in our setting with the ones used in theorem 2.23.

We set $V = \{u \in H^1(\Omega_\varepsilon) | \ u = 0 \text{ on } \Gamma_C\} \times \{u \in H^1(\Omega_\varepsilon) | \ u = 0 \text{ on } \Gamma_N\}$ and $W = L^2(\Omega_\varepsilon) \times L^2(\Omega_\varepsilon)$. Then the conditions for the spaces are satisfied and the space $\mathcal{V}$ in equation 2.12 is equal to the space $\mathcal{V}_{C0} \times \mathcal{V}_N$.
Further we define $\mathcal{A}_\varepsilon(t) \in \mathcal{L}(V, V')$ by

$$(\mathcal{A}_\varepsilon(t)w)\varphi = \int_{\Omega_\varepsilon} D_u \nabla w_1 \nabla \varphi_1 dx + \int_{\Omega_\varepsilon} D_v \nabla w_2 \nabla \varphi_2 dx$$
$$+ \varepsilon^l l_s \int_{\Gamma_\varepsilon^{ER}} (w_1 - F(t)(w_1))\varphi_1 d\sigma_x$$
$$+ \varepsilon^m l_w \int_{\Gamma_\varepsilon^{ER}} (w_2 - G(t)(w_1, w_2))\varphi_2 d\sigma_x \quad (2.15)$$

for $w = (w_1, w_2) \in V$ and $\varphi = (\varphi_1, \varphi_2) \in V$. We are going to prove later that $(\mathcal{A}_\varepsilon(\cdot)u)v \in L^\infty([0, T])$. The operator $\mathcal{B}_\varepsilon(t)$ is independent of $t$ and given by

$$(\mathcal{B}_\varepsilon(t)w)\varphi = \int_{\Omega_\varepsilon} w_1 \varphi_1 dx + \int_{\Omega_\varepsilon} w_2 \varphi_2 dx \quad (2.16)$$

for $w, \varphi \in W$. Obviously, it holds that $(\mathcal{B}_\varepsilon(\cdot)u)v \in L^\infty([0,T])$. With $u_I$ and $v_I$ elements of $L^2(\Omega_\varepsilon)$, it follows that $u_0 = (u_I - u_B, v_I) \in W$. Finally, we identify the right-hand side $f \in L^2([0,T], V')$ by

$$f(t)\varphi = -\varepsilon^l l_s \langle u_B - F(t)(u_B), \varphi_1 \rangle_{\Gamma_\varepsilon^{ER}} + \varepsilon^m l_w \langle G(t)(u_B, 0), \varphi_2 \rangle_{\Gamma_\varepsilon^{ER}}$$

for $\varphi \in V$.

Then, problem (2.12) and problem (2.14) are equivalent. Now three preparing lemmas are stated before the existence of a solution is proven.

**Lemma 2.24** *For $w, \varphi \in V$ and $F(t)$ defined in (2.8), $G(t)$ defined in (2.9) it holds that*

$$\left| \int_{\Gamma_\varepsilon^{ER}} F(w_1)\varphi_1 d\sigma_x \right| \leq c(T) \|w_1\|_{L^2(\Gamma_\varepsilon^{ER})} \|\varphi_1\|_{L^2(\Gamma_\varepsilon^{ER})}$$

*and*

$$\left| \int_{\Gamma_\varepsilon^{ER}} G(w_1, w_2)\varphi_2 d\sigma_x \right|$$
$$\leq c(T) \left( \|w_1\|_{L^2(\Gamma_\varepsilon^{ER})} + \|w_2\|_{L^2(\Gamma_\varepsilon^{ER})} \right) \|\varphi_2\|_{L^2(\Gamma_\varepsilon^{ER})}$$

*for all $0 \leq t \leq T$.*

**Proof** We know from [18] that there exists a smooth function $k_\varepsilon(t, x) > 0$ for $t > 0$ with $k_{t,\varepsilon}(\cdot) = k_\varepsilon(t, \cdot) \in L^2(\Gamma_\varepsilon^{ER})$ such that

$$e^{D_s \varepsilon^2 \Delta t} f(x) = (f \star k_{t,\varepsilon})(x) = \int_{\Gamma_\varepsilon^{ER}} k_\varepsilon(t, x - y) f(y) d\sigma_y$$

for every $f \in L^2(\Gamma_\varepsilon^{ER})$. This yields for $F(t)(w_1)$ with $w, \varphi \in V$ and $t > 0$ that

$$\left| \int_{\Gamma_\varepsilon^{ER}} F(w_1)\varphi_1 d\sigma_x \right| = \left| \int_{\Gamma_\varepsilon^{ER}} \left( e^{-(f + \varepsilon^{l-1} l_s)t} (w_1 \star k_{t,\varepsilon}) \right. \right.$$
$$\left. \left. + e^{-(f + \varepsilon^{l-1} l_s)t} \varepsilon^{l-1} l_s \int_0^t (w_1 \star k_{t-s,\varepsilon}) e^{(f + \varepsilon^{l-1} l_s)s} ds \right) \varphi_1 d\sigma_x \right|$$

$$\leq \left| \int_{\Gamma_\varepsilon^{ER}} (w_1 \star k_{t,\varepsilon}) \varphi_1 d\sigma_x \right| + \left| \int_0^t (w_1 \star k_{t-s,\varepsilon}) e^{(f+\varepsilon^{l-1} l_s)s} \varepsilon^{l-1} l_s \varphi_1 ds d\sigma_x \right|$$

$$\leq \| w_1 \star k_{t,\varepsilon} \|_{L^2(\Gamma_\varepsilon^{ER})} \| \varphi_1 \|_{L^2(\Gamma_\varepsilon^{ER})}$$

$$+ \| w_1 \star k_{t-\cdot,\varepsilon} \|_{L^2([0,T] \times \Gamma_\varepsilon^{ER})} \| e^{(f+\varepsilon^{l-1} l_s) \cdot} \varepsilon^{l-1} l_s \varphi_1 \|_{L^2([0,T] \times \Gamma_\varepsilon^{ER})}$$

$$\leq \| k_{t,\varepsilon} \|_{L^1(\Gamma_\varepsilon^{ER})} \| w_1 \|_{L^2(\Gamma_\varepsilon^{ER})} \| \varphi_1 \|_{L^2(\Gamma_\varepsilon^{ER})}$$

$$+ c(T) \| k_{t,\varepsilon} \|_{L^1([0,t] \times \Gamma_\varepsilon^{ER})} \| w_1 \|_{L^2(\Gamma_\varepsilon^{ER})} \| \varphi_1 \|_{L^2(\Gamma_\varepsilon^{ER})},$$

where we first used the Hölder inequality and then the Young inequality. With $\| k_{t,\varepsilon} \|_{L^1(\Gamma_\varepsilon^{ER})}$ and $\| k_{t,\varepsilon} \|_{L^1([0,t] \times \Gamma_\varepsilon^{ER})}$ bounded for $0 < t \leq T$, it hold that

$$\left| \int_{\Gamma_\varepsilon^{ER}} F(w_1) \varphi_1 d\sigma_x \right| \leq c(T) \| w_1 \|_{L^2(\Gamma_\varepsilon^{ER})} \| \varphi_1 \|_{L^2(\Gamma_\varepsilon^{ER})}.$$

For $t = 0$ we have $e^{D_s \varepsilon^2 \Delta t} f(x) = f(x)$ and it follows that

$$\left| \int_{\Gamma_\varepsilon^{ER}} F(w_1) \varphi_1 d\sigma_x \right| = \left| \int_{\Gamma_\varepsilon^{ER}} w_1 \varphi_1 d\sigma_y \right| \leq \| w_1 \|_{L^2(\Gamma_\varepsilon^{ER})} \| \varphi_1 \|_{L^2(\Gamma_\varepsilon^{ER})}.$$

Analogously we find that

$$\left| \int_{\Gamma_\varepsilon^{ER}} G(w_1, w_2) \varphi_2 d\sigma_x \right|$$

$$\leq c(T) \left( \| w_1 \|_{L^2(\Gamma_\varepsilon^{ER})} + \| w_2 \|_{L^2(\Gamma_\varepsilon^{ER})} \right) \| \varphi_2 \|_{L^2(\Gamma_\varepsilon^{ER})}.$$

$\square$

**Lemma 2.25** *The family of operators $\mathcal{B}_\varepsilon(t)$ (defined in 2.16) form a regular family of self-adjoint operators W. Furthermore, $\mathcal{A}_\varepsilon(t)$ and $\mathcal{B}_\varepsilon(t)$ are linear and $\mathcal{B}_\varepsilon(0)$ and $\mathcal{B}_\varepsilon(t)$ are monotone and $(\mathcal{A}_\varepsilon(\cdot)u)\varphi \in L^\infty([0,T])$ for $u, \varphi \in V$.*

**Proof** With $\mathcal{B}_\varepsilon(t)$ being the identity $Id$ on $W$, $\mathcal{B}_\varepsilon(t)$ is monotone and self-adjoint for every $0 \leq t \leq T$. It also follows that $\frac{d}{dt}(\mathcal{B}_\varepsilon(t)u)\varphi = 0 \leq \|\varphi\| \|u\|$ for every $u, \varphi \in V$ and hence, $\mathcal{B}_\varepsilon(t)$ is regular.

Because the Laplace-operator $\Delta$, the operators $F(t)$, $G(t)$ and the integral operator are linear, it follows that $\mathcal{A}_\varepsilon(t)$ and $\mathcal{B}_\varepsilon(t)$ are linear.

To show $(\mathcal{A}_\varepsilon(\cdot)w)\varphi \in L^\infty([0,T])$ for $w, \varphi \in V$, we consider

$$
\begin{aligned}
|(\mathcal{A}_\varepsilon(t)\ w)\varphi| = \Bigg| &\int_{\Omega_\varepsilon} D_u \nabla w_1 \nabla \varphi_1 dx + \int_{\Omega_\varepsilon} D_v \nabla w_2 \nabla \varphi_2 dx \\
&+ \varepsilon^l l_s \int_{\Gamma_\varepsilon^{\text{ER}}} (w_1 - F(t)(w_1))\varphi_1 d\sigma_x \\
&+ \varepsilon^m l_w \int_{\Gamma_\varepsilon^{\text{ER}}} (w_2 - G(t)(w_1, w_2))\varphi_2 d\sigma_x \Bigg| \\
\leq\ & D_u \|\nabla w_1\|_{\Omega_\varepsilon} \|\nabla \varphi_1\|_{\Omega_\varepsilon} + D_v \|\nabla w_2\|_{\Omega_\varepsilon} \|\nabla \varphi_2\|_{\Omega_\varepsilon} \\
&+ \varepsilon^l l_s \|w_1\|_{\Gamma_\varepsilon^{\text{ER}}} \|\varphi_1\|_{\Gamma_\varepsilon^{\text{ER}}} + \varepsilon^l l_s c(T) \|w_1\|_{\Gamma_\varepsilon^{\text{ER}}} \|\varphi_1\|_{\Gamma_\varepsilon^{\text{ER}}} \\
&+ \varepsilon^m l_w \|w_2\|_{\Gamma_\varepsilon^{\text{ER}}} \|\varphi_2\|_{\Gamma_\varepsilon^{\text{ER}}} \\
&+ \varepsilon^m l_w c(T) \left( \|w_1\|_{\Gamma_\varepsilon^{\text{ER}}} + \|w_2\|_{\Gamma_\varepsilon^{\text{ER}}} \right) \|\varphi_2\|_{\Gamma_\varepsilon^{\text{ER}}},
\end{aligned}
$$

where we use lemma 2.24. We continue by using Young's and trace inequality

$$
\begin{aligned}
|(\mathcal{A}_\varepsilon(t)w)\varphi| \leq\ & \|w_1\|_{\Omega_\varepsilon}^2 \left( c_0 \varepsilon^{l-1} l_s + c_0 c(T) \varepsilon^{l-1} l_s + c_0 \varepsilon^{m-1} l_w \right) \\
&+ \|\nabla w_1\|_{\Omega_\varepsilon}^2 \left( D_u + c_0 \varepsilon^{l+1} l_s + c_0 c(T) \varepsilon^{l+1} l_s + c_0 c(T) \varepsilon^{m+1} l_w \right) \\
&+ \|w_2\|_{\Omega_\varepsilon}^2 \left( c_0 \varepsilon^{m-1} l_w + c_0 c(T) \varepsilon^{m-1} l_w \right) \\
&+ \|\nabla w_2\|_{\Omega_\varepsilon}^2 \left( D_v + c_0 \varepsilon^{m+1} l_w + c_0 c(T) \varepsilon^{m+1} l_w \right) \\
&+ \|\varphi_1\|_{\Omega_\varepsilon}^2 \left( c_0 \varepsilon^{l-1} l_s + c_0 c(T) \varepsilon^{l-1} l_s \right) \\
&+ \|\nabla \varphi_1\|_{\Omega_\varepsilon}^2 \left( D_u + c_0 \varepsilon^{l+1} l_s + c_0 c(T) \varepsilon^{l+1} l_s \right) \\
&+ \|\varphi_2\|_{\Omega_\varepsilon}^2 \left( c_0 \varepsilon^{m-1} l_w + c_0 c(T) \varepsilon^{m-1} l_w \right) \\
&+ \|\nabla \varphi_2\|_{\Omega_\varepsilon}^2 \left( D_v + c_0 \varepsilon^{m+1} l_w + c_0 c(T) \varepsilon^{m+1} l_w \right),
\end{aligned}
$$

which is bounded for every $\varepsilon > 0$ and $T$ bounded, since $w, \varphi \in V$. $\qquad\square$

**Lemma 2.26** *The operators $\mathcal{A}_\varepsilon$ and $\mathcal{B}_\varepsilon$ defined in (2.15) and (2.16), re-*

*spectively, satisfy*

$$2(\mathcal{A}_\varepsilon(t)w)w + \lambda(\mathcal{B}_\varepsilon(t)w)w + (\mathcal{B}_\varepsilon'(t)w)w \geq \kappa\|w\|^2$$

*for every $w \in V$ and $0 \leq t \leq T$ for constants $\lambda, \kappa > 0$.*

**Proof** With $\mathcal{B}_\varepsilon$ constant it follow that $\mathcal{B}_\varepsilon'(t) = 0$. We consider

$$
\begin{aligned}
&2(\mathcal{A}_\varepsilon(t)w)w + \lambda(\mathcal{B}_\varepsilon(t)w)w \\
&= 2\int_{\Omega_\varepsilon} D_u \nabla w_1 \nabla w_1 dx + 2\int_{\Omega_\varepsilon} D_v \nabla w_2 \nabla w_2 dx \\
&\quad + 2\varepsilon^l l_s \int_{\Gamma_\varepsilon^{ER}} (w_1 - F(t)(w_1))w_1 d\sigma_x + \lambda \int_{\Omega_\varepsilon} w_1 w_1 d\sigma_x \\
&\quad + 2\varepsilon^m l_w \int_{\Gamma_\varepsilon^{ER}} (w_2 - G(t)(w_1, w_2))w_2 d\sigma_x + \lambda \int_{\Omega_\varepsilon} w_2 w_2 d\sigma_x \\
&= 2D_u \|\nabla w_1\|_{\Omega_\varepsilon}^2 + 2D_v \|\nabla w_2\|_{\Omega_\varepsilon}^2 + 2\varepsilon^l l_s \|w_1\|_{\Gamma_\varepsilon^{ER}}^2 \\
&\quad - 2\varepsilon^l l_s \int_{\Gamma_\varepsilon^{ER}} F(t)(w_1)w_1 d\sigma_x + 2\varepsilon^m l_w \|w_2\|_{\Gamma_\varepsilon^{ER}}^2 \\
&\quad - 2\varepsilon^m l_w \int_{\Gamma_\varepsilon^{ER}} G(t)(w_1, w_2)w_2 d\sigma_x + \lambda\|w_1\|_{\Omega_\varepsilon}^2 + \lambda\|w_2\|_{\Omega_\varepsilon}^2.
\end{aligned}
$$

With lemma 2.24 we deduce

$$
\begin{aligned}
&2(\mathcal{A}_\varepsilon(t)w)w + \lambda(\mathcal{B}_\varepsilon(t)w)w \\
&\geq 2D_u \|\nabla w_1\|_{\Omega_\varepsilon}^2 + 2D_v \|\nabla w_2\|_{\Omega_\varepsilon}^2 + 2\varepsilon^l l_s \|w_1\|_{\Gamma_\varepsilon^{ER}}^2 - 2c(T)\varepsilon^l l_s \|w_1\|_{\Gamma_\varepsilon^{ER}}^2 \\
&\quad + 2\varepsilon^m l_w \|w_2\|_{\Gamma_\varepsilon^{ER}}^2 - c(T)\varepsilon^m l_w \|w_1\|_{\Gamma_\varepsilon^{ER}}^2 - c(T)\varepsilon^m l_w \|w_2\|_{\Gamma_\varepsilon^{ER}}^2 \\
&\quad - 2c(T)\varepsilon^m l_w \|w_2\|_{\Gamma_\varepsilon^{ER}} + \lambda\|w_1\|_{\Omega_\varepsilon}^2 + \lambda\|w_2\|_{\Omega_\varepsilon}^2.
\end{aligned}
$$

Dropping some positive terms and applying the trace inequality yields

$$
\begin{aligned}
&2(\mathcal{A}_\varepsilon(t)w)w + \lambda(\mathcal{B}_\varepsilon(t)w)w \\
&\geq \|w_1\|_{\Omega_\varepsilon}^2 \left(\lambda - 2c_0 c(T)\varepsilon^{l-1} l_s - c_0 c(T)\varepsilon^{m-1} l_w\right) \\
&\quad + \|\nabla w_1\|_{\Omega_\varepsilon}^2 \left(2D_u - 2c_0 c(T)\varepsilon^{l+1} l_s - c_0 c(T)\varepsilon^{m+1} l_w\right) \\
&\quad + \|w_2\|_{\Omega_\varepsilon}^2 \left(\lambda - c_0 c(T)\varepsilon^{m-1} l_w - 2c_0 c(T)\varepsilon^{m-1} l_w\right) \\
&\quad + \|\nabla w_2\|_{\Omega_\varepsilon}^2 \left(2D_v - 2c_0 c(T)\varepsilon^{m+1} l_w - c_0 c(T)\varepsilon^{m+1} l_w\right).
\end{aligned}
$$

Because $l, m \geq 0$, the factors at the norm of gradients of $w_1$, $w_2$ are positive for $\varepsilon$ small enough. If $l, m \leq 1$ one chooses $\lambda$ big enough such that the factors at the norm of $w_1$, $w_2$ are positive even for small $\varepsilon > 0$. Then we merge the factors to a constant $\kappa$ and conclude that

$$2(\mathcal{A}_\varepsilon(t)w)w + \lambda(\mathcal{B}_\varepsilon(t)w)w \geq \kappa \|w\|_V^2,$$

which completes the proof. $\qquad\qquad\qquad\qquad\qquad\qquad\qquad$ $\square$

Now we are ready to prove the existence of a solution.

**Theorem 2.27** *The partial differential equations* (2.11) *has at least one solution* $(u_\varepsilon, v_\varepsilon)$ *in* $\mathcal{V}_C(\Omega_\varepsilon) \times \mathcal{V}_N(\Omega_\varepsilon)$.

**Proof** Using theorem 2.23 we find a solution $(\bar{u}_\varepsilon, v_\varepsilon)$ of the problem (2.14) in $\mathcal{V}_{C0}(\Omega_\varepsilon) \times \mathcal{V}_N(\Omega_\varepsilon)$ with initial condition $(\bar{u}_\varepsilon(0), v_\varepsilon(0)) = (u_I - u_B, v_I)$, because the linear forms $\mathcal{A}_\varepsilon(t)$ and $\mathcal{B}_\varepsilon(t)$, defined in (2.15) and (2.16), satisfy the necessary conditions (proven in lemma 2.25 and lemma 2.26). Furthermore, the right-hand side in equation (2.14) is in $L^2([0, T], V')$ and the initial conditions $(u_I - u_B, v_I)$ are an element of $W$.
Setting $u_\varepsilon = \bar{u}_\varepsilon + u_B$, we find a solution $(u_\varepsilon, v_\varepsilon) \in \mathcal{V}_C(\Omega_\varepsilon) \times \mathcal{V}_N(\Omega_\varepsilon)$. $\qquad\qquad\qquad\qquad\qquad\qquad\qquad\qquad\qquad\qquad$ $\square$

**Corollary 2.28** *With equations* (2.8) *and* (2.9) *there also exist solutions for* $s_\varepsilon$ *and* $w_\varepsilon$.

**Remark 2.29** With applying proposition 2.15 we know that $\partial_t u_\varepsilon$, $\partial_t v_\varepsilon \in L^2([0, T], H^1(\Omega_\varepsilon)')$ and $\partial_t s_\varepsilon$, $\partial_t w_\varepsilon \in L^2([0, T], H^1(\Gamma_\varepsilon^{ER})')$.

# 2.6 Identification of the two-scale limit

In this section we derive the limit equation for the system of equations (2.6). At first, we consider the limit of the binding terms for different values for $l$ and $m$. Then we find the limit of the whole system.

**Limit of the binding terms**

To find the limit equations for $u_\varepsilon$, $v_\varepsilon$, $s_\varepsilon$ and $w_\varepsilon$ we need to distinguish different cases for $l$ and $m$ in the binding terms. The limit derivation for $v_\varepsilon$ and $w_\varepsilon$ is analogous to the limit derivation for $u_\varepsilon$ and $s_\varepsilon$, respectively. Hence, we only consider $u_\varepsilon$ and $s_\varepsilon$ here.

In every case we use that $\varepsilon \|u_\varepsilon\|^2_{\Gamma^{ER}_\varepsilon}$ and $\varepsilon\|s_\varepsilon\|^2_{\Gamma^{ER}_\varepsilon}$ are bounded, see corollary 2.18 and lemma 2.17. Thus, with theorem 2.10 we find $u_0$ and $s_0$ such that $u_\varepsilon$ two-scale converges to $u_0$ and $s_\varepsilon$ two-scale converges to $s_0$.

Further, we know that $\varepsilon^l\|u_\varepsilon - s_\varepsilon\|^2_{\Gamma^{ER}_\varepsilon}$ is bounded, see lemma 2.17, and we also want to use theorem 2.10. Therefore, we consider the following cases:

- For $l > 1$ and with lemma 2.17 we obtain for $\varepsilon$ tending to zero that

$$\varepsilon^l\|u_\varepsilon - s_\varepsilon\|^2_{\Gamma^{ER}_\varepsilon} = \underbrace{\varepsilon^{l-1}}_{\to 0}\underbrace{\varepsilon\|u_\varepsilon - s_\varepsilon\|^2_{\Gamma^{ER}_\varepsilon}}_{\text{bounded}} \xrightarrow{\varepsilon\to 0} 0. \tag{2.17}$$

- $l = 1$ is the standard case and we find using the definition of two-scale convergence lemma 2.9 that

$$\varepsilon\int_{\Gamma^{ER}_\varepsilon}(u_\varepsilon - s_\varepsilon)\varphi_\varepsilon d\sigma_x$$
$$\xrightarrow{\varepsilon\to 0} \int_\Omega\int_\Gamma(u_0(x,y,t) - s_0(x,y,t))\varphi_0(x,y)d\sigma_y dx. \tag{2.18}$$

- For $0 \leq l < 1$ and with lemma 2.17 we get for $\varepsilon \to 0$ that

$$\underbrace{\varepsilon^l\|u_\varepsilon - s_\varepsilon\|^2_{\Gamma^{ER}_\varepsilon}}_{\text{bounded}} = \underbrace{\varepsilon^{l-1}}_{\to\infty}\underbrace{\varepsilon\|u_\varepsilon - s_\varepsilon\|^2_{\Gamma^{ER}_\varepsilon}}_{\Rightarrow\ \to 0}. \tag{2.19}$$

Thus the right-hand side remains bounded, it must hold that $\lim_{\varepsilon\to\infty}\varepsilon\|u_\varepsilon - s_\varepsilon\|^2_{\Gamma^{ER}_\varepsilon} = \|u_0 - s_0\|^2_{\Omega\times\Gamma} = 0$, where we use the particular limit derivations for $u_\varepsilon$ and $s_\varepsilon$. Hence, $u_0 = s_0$ almost everywhere on $\Omega \times \Gamma$.

Now, knowing the limits of the binding terms, we continue with the limit derivation of the whole equations.

To be able to use definition 2.2 the domain $\Omega_\varepsilon$ should not depend on $\varepsilon$, so we define the characteristic function $\chi : \Omega \to \mathbb{R}$ by $\chi(x/\varepsilon) = 1$ for $x \in \Omega_\varepsilon$ and 0 otherwise. We obtain the limit equations for $u_\varepsilon$, $v_\varepsilon$, $s_\varepsilon$ and $w_\varepsilon$ by using theorem 2.3, lemma 2.10 and proposition 2.5. In lemma 2.17 we checked that the conditions are fulfilled. As test functions $\varphi_\varepsilon \in C^\infty(\Omega, C_\#^\infty(Y))$ we choose functions of the form

$$\varphi_\varepsilon \left( x, \frac{x}{\varepsilon} \right) = \varphi_0(x) + \varepsilon \varphi_1 \left( x, \frac{x}{\varepsilon} \right)$$

with $(\varphi_0, \varphi_1) \in C^\infty(\Omega) \times C^\infty(\Omega, C_\#^\infty(Y))$.

## The case $l > 1, m > 1$.

We start with the equation $u_\varepsilon$ and test with the admissible test function $\varphi_\varepsilon$,

$$\int_\Omega \chi \left( \frac{x}{\varepsilon} \right) \partial_t u_\varepsilon (x,t) \, \varphi_\varepsilon \left( x, \frac{x}{\varepsilon} \right) dx$$

$$+ D_u \int_\Omega \chi \left( \frac{x}{\varepsilon} \right) \nabla u_\varepsilon (x,t) \, \nabla \varphi_\varepsilon \left( x, \frac{x}{\varepsilon} \right) dx$$

$$+ \varepsilon^l l_s \int_{\Gamma_\varepsilon^{\mathrm{ER}}} \left( u_\varepsilon (x,t) - s_\varepsilon (x,t) \right) \varphi_\varepsilon \left( x, \frac{x}{\varepsilon} \right) d\sigma_x = 0.$$

For $\varepsilon \to 0$ we obtain the limit equation

$$\int_\Omega \int_{Y^*} \partial_t u_0(x,t) \varphi_0(x) dy dx + D_u \int_\Omega \int_{Y^*} [\nabla_x u_0(x,t) + \nabla_y u_1(x,y,t)]$$

$$[\nabla_x \varphi_0(x) + \nabla_y \varphi_1(x,y)] dy dx = 0 \quad (2.20)$$

for all admissible test functions $(\varphi_0, \varphi_1) \in V_{C0}(\Omega) \times V(\Omega, Y)$, where $u_0 \in V_C(\Omega)$ is independent of $y$ and $u_1 \in V(\Omega, Y) = L^2([0,T] \times \Omega, H_\#^1(Y))$.

Analogously we take the equation for $v_\varepsilon$,

$$\int_\Omega \chi\left(\frac{x}{\varepsilon}\right) \partial_t v_\varepsilon\left(x,t\right) \varphi_\varepsilon\left(x,\frac{x}{\varepsilon}\right) dx$$

$$+ D_v \int_\Omega \chi\left(\frac{x}{\varepsilon}\right) \nabla v_\varepsilon\left(x,t\right) \nabla \varphi_\varepsilon\left(x,\frac{x}{\varepsilon}\right) dx$$

$$+ \varepsilon^m l_w \int_{\Gamma_\varepsilon^{ER}} \left(v_\varepsilon\left(x,t\right) - w_\varepsilon\left(x,t\right)\right) \varphi_\varepsilon\left(x,\frac{x}{\varepsilon}\right) d\sigma_x = 0.$$

For $\varepsilon \to 0$ we get the limit equation

$$\int_\Omega \int_{Y^*} \partial_t v_0(x,t) \varphi_0(x) dy dx$$

$$+ D_v \int_\Omega \int_{Y^*} [\nabla_x v_0(x,t) + \nabla_y v_1(x,y,t)]$$

$$[\nabla_x \varphi_0(x) + \nabla_y \varphi_1(x,y)] dy dx = 0$$

for all admissible test functions $(\varphi_0, \varphi_1) \in V_N(\Omega) \times V(\Omega, Y)$, where $v_0 \in V_N(\Omega)$ is independent of $y$ and $v_1 \in \mathcal{V}(\Omega, Y)$.

Now we determine the limit equations for $s_\varepsilon$ and $w_\varepsilon$,

$$\varepsilon \int_{\Gamma_\varepsilon^{ER}} \partial_t s_\varepsilon\left(x,t\right) \varphi_\varepsilon\left(x,\frac{x}{\varepsilon}\right) d\sigma_x$$

$$+ \varepsilon \int_{\Gamma_\varepsilon^{ER}} D_s \varepsilon \nabla_\Gamma s_\varepsilon\left(x,t\right) \varepsilon \nabla_\Gamma \varphi_\varepsilon\left(x,\frac{x}{\varepsilon}\right) d\sigma_x$$

$$+ \varepsilon \int_{\Gamma_\varepsilon^{ER}} f s_\varepsilon\left(x,t\right) \varphi_\varepsilon\left(x,\frac{x}{\varepsilon}\right) d\sigma_x$$

$$- \varepsilon^l \int_{\Gamma_\varepsilon^{ER}} l_s\left(u_\varepsilon\left(x,t\right) - s_\varepsilon\left(x,t\right)\right) \varphi_\varepsilon\left(x,\frac{x}{\varepsilon}\right) d\sigma_x = 0.$$

We find the limit function $s_0 \in \mathcal{V}(\Omega, \Gamma)$ satisfying for $\varepsilon \to 0$

$$\int_\Omega \int_\Gamma \partial_t s_0(x,y,t) \varphi_0(x,y) d\sigma_y dx$$

$$+ \int_\Omega \int_\Gamma D_s \nabla_\Gamma s_0(x,y,t) \nabla_\Gamma \varphi_0(x,y) d\sigma_y dx$$

$$+ \int_\Omega \int_\Gamma f s_0(x,y,t) \varphi_0(x,y) d\sigma_y dx = 0$$

for all $\varphi_0 \in V(\Omega, \Gamma)$.

Lastly we consider the equation for $w_\varepsilon$,

$$\varepsilon \int_{\Gamma_\varepsilon^{ER}} \partial_t w_\varepsilon(x,t) \, \varphi_\varepsilon \left(x, \frac{x}{\varepsilon}\right) \mathrm{d}\sigma_x$$

$$+ \varepsilon \int_{\Gamma_\varepsilon^{ER}} D_w \varepsilon \nabla_\Gamma w_\varepsilon(x,t) \, \varepsilon \nabla_\Gamma \varphi_\varepsilon \left(x, \frac{x}{\varepsilon}\right) \mathrm{d}\sigma_x$$

$$- \varepsilon \int_{\Gamma_\varepsilon^{ER}} f s_\varepsilon(x,t) \, \varphi_\varepsilon \left(x, \frac{x}{\varepsilon}\right) \mathrm{d}\sigma_x$$

$$- \varepsilon \int_{\Gamma_\varepsilon^{ER}} l_w (v_\varepsilon(x,t) - w_\varepsilon(x,t)) \varphi_\varepsilon \left(x, \frac{x}{\varepsilon}\right) \mathrm{d}\sigma_x = 0.$$

We find the limit function $w_0 \in \mathcal{V}(\Omega, \Gamma)$ satisfying

$$\int_\Omega \int_\Gamma \partial_t w_0(x,y,t) \varphi_0(x,y) \mathrm{d}\sigma_y \mathrm{d}x$$

$$+ \int_\Omega \int_\Gamma D_w \nabla_\Gamma w_0(x,y,t) \nabla_\Gamma \varphi_0(x,y) \mathrm{d}\sigma_y \mathrm{d}x$$

$$- \int_\Omega \int_\Gamma f s_0(x,y,t) \varphi_0(x,y) \mathrm{d}\sigma_y \mathrm{d}x = 0$$

for all $\varphi_0 \in V(\Omega, \Gamma)$.

**Identification of $u_1(x,y,t)$ and $v_1(x,y,t)$**
We take the limit equation for $u_0$ in (2.20), and set $\varphi_0 = 0$. Then we obtain

$$D_u \int_\Omega \int_{Y^*} [\nabla_x u_0(x,t) + \nabla_y u_1(x,y,t)] \nabla_y \varphi_1(x,y) \mathrm{d}y \mathrm{d}x = 0 \quad (2.21)$$

for all $\varphi_1 \in L^2(\Omega, H_\#^1(Y))$. We set $u_1(x,y,t) = \sum_{j=1}^n \partial_{x_j} u_0(x,t) \mu_j(t,y)$ for some functions $\mu_j \in L^2([0,T], H_\#^1(Y))$ and $j = 1, \ldots, n$. Hence, it follows $\nabla_y u_1(x,y,t) = \sum_{j=1}^n \partial_{x_j} u_0(x,t) \nabla_y \mu_j(t,y)$. Also $\nabla_x u_0(x,t)$ can be expressed by using the sum of the partial derivatives. Hence,

$$D_u \int_\Omega \int_{Y^*} \left[ \sum_{j=1}^n \partial_{x_j} u_0(x,t) e_j + \sum_{j=1}^n \partial_{x_j} u_0(x,t) \nabla_y \mu_j(y,t) \right]$$

$$\nabla_y \varphi_1(x,t) \mathrm{d}y \mathrm{d}x = 0$$

and

$$\int_\Omega \int_{Y^*} \sum_{j=1}^n \partial_{x_j} u_0(x,t) D_u[e_j + \nabla_y \mu_j(y,t)] \nabla_y \varphi_1(x,y) dy dx = 0$$

for all $\varphi_1 \in L^2(\Omega, H^1_\#(Y))$. Integration by parts leads to

$$\int_\Omega \sum_{j=1}^n \partial_{x_j} u_0(x,t) \int_{\partial Y^*} D_u[e_j + \nabla_y \mu_j(y,t)] \cdot n \varphi_1(x,y) d\sigma_y dx$$

$$- \int_\Omega \sum_{j=1}^n \partial_{x_j} u_0(x,t) \int_{Y^*} \nabla_y \cdot \left( D_u[e_j + \nabla_y \mu_j(y,t)] \right) \varphi_1(x,y) dy dx = 0.$$

This leads to the strong formulation, the *cell problem*

$$\begin{aligned}
\nabla_y \cdot D_u(e_j + \nabla_y \mu_j) &= 0 && \text{in } Y^* \\
D_u(e_j + \nabla_y \mu_j) \cdot n &= 0 && \text{on } \partial Y^*,
\end{aligned} \tag{2.22}$$

where $\mu_j$ must be $Y$-periodic for all $j = 1, \ldots, n$.

The solution of the cell problem contains information about the behavior in the microscopic view. As we described in section 2.1, in the end of homogenization we will find two partial differential equations to be solved: one on the homogeneous domain $\Omega$ and one on the unit cell $Y^*$.

The cell problem needs to be solved on the unit cell $Y$. Now we are going to find the diffusion tensor $P^u$ and derive the partial differential equation to be solved on the homogeneous domain $\Omega$.

Therefore, we consider the diffusion term of the limit equation for $u_0$ in (2.20). We set $\varphi_1 = 0$ and arrive at

$$\int_\Omega \int_{Y^*} \sum_{j=1}^n \partial_{x_j} u_0(x,t) D_u[e_j + \nabla_y \mu_j(y,t)] \nabla_x \varphi_0(x) dy dx.$$

Further,

$$\int_\Omega \sum_{j=1}^n \sum_{i=1}^n \partial_{x_j} u_0 \int_{Y^*} D_u[\delta_{ij} + \partial_{y_i} \mu_j] dy \, \partial_{x_i} \varphi_0 dx = \int_\Omega P^u \nabla u_0 \nabla \varphi_0 dx$$

48

where $P^u$ is defined as

$$P^u_{ij} = \int_{Y^*} D_u[\delta_{ij} + \partial_{y_i}\mu_j]\mathrm{d}y. \qquad (2.23)$$

For $v_1$ we assume that there are functions $\eta_j \in L^2([0,T], H^1_\#(Y))$ for $j = 1,\ldots,n$ such
that $v_1 = \sum_{j=1}^{N} \partial_{x_j} v_0(x,t)\eta_j(y,t)$. Then, with similar steps we find the cell problem given by

$$\begin{aligned}
\nabla_y \cdot D_v(e_j + \nabla_y\eta_j) &= 0 \quad \text{in } Y^* \\
D_v(e_j + \nabla_y\eta_j) \cdot n &= 0 \quad \text{on } \partial Y^*
\end{aligned} \qquad (2.24)$$

and $\eta_j$ $Y$-periodic for all $j = 1,\ldots,N$. We define the diffusion tensor $P^v$ by

$$P^v = \int_{Y^*} D_v[\delta_{ij} + \partial_{y_i}\eta_j]\mathrm{d}y.$$

**Macroscopic system of equations**
We summarize our results with the weak macroscopic system of equations and use that $u_0$ and $v_0$ are independent of $y$. Let $(u_0, v_0, s_0, w_0) \in \mathcal{V}_C(\Omega) \times \mathcal{V}_N(\Omega) \times \mathcal{V}(\Omega, \Gamma)^2$ such that

$$\begin{aligned}
|Y^*|(\partial_t u_0, \varphi_1)_\Omega + (P^u \nabla u_0, \nabla\varphi_1)_\Omega &= 0 \\
|Y^*|(\partial_t v_0, \varphi_2)_\Omega + (P^v \nabla v_0, \nabla\varphi_2)_\Omega &= 0 \\
(\partial_t s_0, \psi)_{\Omega\times\Gamma} + (D_s \nabla_\Gamma s_0, \nabla_\Gamma\psi)_{\Omega\times\Gamma} + f(s_0, \psi)_{\Omega\times\Gamma} &= 0 \\
(\partial_t w_0, \psi)_{\Omega\times\Gamma} + (D_w \nabla_\Gamma w_0, \nabla_\Gamma\psi)_{\Omega\times\Gamma} - f(s_0, \psi)_{\Omega\times\Gamma} &= 0
\end{aligned} \qquad (2.25)$$

for all $(\varphi_1, \varphi_2, \psi) \in V_{C0}(\Omega) \times V_N(\Omega) \times V(\Omega, \Gamma)$.

We see that for $l > 1$ or $m > 1$ either no BP molecules bind or unbind to the membrane of the ER, or no DE molecules bind or unbind to the membrane of the ER, respectively. Hence, no metabolism from BP to DE molecules takes place.
Because we know from biological examinations that there are metabolisms in real live, we conclude that this model is not a good approximation to the reality for $l > 1$ or $m > 1$.
We do not consider this case any longer.

49

## The case $l = 1, m = 1$.

Here, we distinguish two cases. At first, we consider the limit derivation of the systems of equation (2.6). Secondly, we derive the limit equation for the abbreviated system of equations (2.11). But until now, it is an open question how to derive the limits $\lim_{\varepsilon \to 0} F(u_\varepsilon)$ and $\lim_{\varepsilon \to 0} G(u_\varepsilon, v_\varepsilon)$, as defined in (2.8) and (2.9), in the sense of two-scale convergence.

We are going to determine the limits for $F(u_\varepsilon)$ and $G(u_\varepsilon, v_\varepsilon)$ indirectly: With the systems (2.11) and (2.6) being equivalent, also the corresponding limit equations must be equivalent. In the limit we are going to identify these two limit equations with each other and are able to determine $\lim_{\varepsilon \to 0} F(u_\varepsilon)$ and $\lim_{\varepsilon \to 0} G(u_\varepsilon, v_\varepsilon)$.

For the limit derivations we may adopt the two-scale limit terms from the case $l, m > 1$ for the time-derivatives, the diffusion terms and the linear terms. The limit equations only differ in the binding terms, calculated in (2.18).

**Identification of the two-scale limit of the model** (2.6)

We start with the equation for $u_\varepsilon$ and obtain for $\varepsilon \to 0$

$$\int_\Omega \int_{Y^*} \partial_t u_0(x,t) \varphi_0(x) dy dx$$

$$+ D_u \int_\Omega \int_{Y^*} [\nabla_x u_0(x,t) + \nabla_y u_1(x,y,t)][\nabla_x \varphi_0(x) + \nabla_y \varphi_1(x,y)] dy dx$$

$$+ \int_\Omega \int_\Gamma l_s(u_0(x,t) - s_0(x,y,t)) \varphi_0(x) d\sigma_y dx = 0$$

for all admissible test functions $(\varphi_0, \varphi_1) \in V_{C0}(\Omega) \times V(\Omega, Y)$, where $u_0 \in L^2([0,T], H^1(\Omega))$ and $u_1 \in L^2([0,T] \times \Omega, H^1_\#(Y))$.

For the equation with $v_\varepsilon$ we find

$$\int_\Omega \int_{Y^*} \partial_t v_0(x,t) \varphi_0(x) dy dx$$

$$+ D_v \int_\Omega \int_{Y^*} [\nabla_x v_0(x,t) + \nabla_y v_1(x,y,t)][\nabla_x \varphi_0(x) + \nabla_y \varphi_1(x,y)] dy dx$$

$$+ \int_\Omega \int_\Gamma l_w(v_0(x,t) - w_0(x,y,t)) \varphi_0(x) d\sigma_y dx = 0$$

for all $(\varphi_0, \varphi_1) \in V_N(\Omega) \times V(\Omega, Y)$, where $v_0 \in L^2([0,T], H^1(\Omega))$

and $v_1 \in L^2([0,T] \times \Omega, H^1_\#(Y))$.

We find the limit function $s_0 \in L^2([0,T] \times \Omega, H^1_\#(Y))$ satisfying the limit equation

$$\int_\Omega \int_\Gamma \partial_t s_0(x,y,t)\varphi_0(x,y)d\sigma_y dx$$
$$+ \int_\Omega \int_\Gamma D_s \nabla_\Gamma s_0(x,y,t)\nabla_\Gamma \varphi_0(x,y)d\sigma_y dx$$
$$+ \int_\Omega \int_\Gamma f s_0(x,y,t)\varphi_0(x,y)d\sigma_y dx$$
$$- \int_\Omega \int_\Gamma l_s(u_0(x,y,t) - s_0(x,y,t))\varphi_0(x,y)d\sigma_y dx = 0$$

for all admissible test functions $\varphi_0 \in V(\Omega, \Gamma)$. From the limit derivation above we see that $u_0$ is independent of $y$. Thus, we can simplify the terms $\int_\Gamma l_s u_0(x,t)d\sigma_y$ to $|\Gamma|l_s u_0(x,t)$, where $|\Gamma|$ means the Lebesgue measure of $\Gamma$.

Analogously we obtain for the equation for $w_\varepsilon \in L^2([0,T] \times \Omega, H^1_\#(Y))$ and $\varepsilon \to 0$

$$\int_\Omega \int_\Gamma \partial_t w_0(x,y,t)\varphi_0(x,y)d\sigma_y dx$$
$$+ \int_\Omega \int_\Gamma D_w \nabla_\Gamma w_0(x,y,t)\nabla_\Gamma \varphi_0(x,y)d\sigma_y dx$$
$$- \int_\Omega \int_\Gamma f s_0(x,y,t)\varphi_0(x,y)d\sigma_y dx$$
$$- \int_\Omega \int_\Gamma l_w(v_0(x,y,t) - w_0(x,y,t))\varphi_0(x,y)d\sigma_y dx = 0$$

for all admissible test functions $\varphi_0 \in V(\Omega, \Gamma)$.

We find the same cell problem (2.22) as in the case $l, m > 1$ and the derivation of the diffusion tensors $P^u$ and $P^v$ is equivalent to (2.23).

**Macroscopic system of equations**
We summarize our results with the macroscopic weak formulation. Let $(u_0, v_0, s_0, w_0) \in \mathcal{V}_C(\Omega) \times \mathcal{V}_N(\Omega) \times \mathcal{V}(\Omega, \Gamma)^2$ with

$(u_0(x,0), v_0(x,0), s_0(x,0), w_0(x,0)) = (u_I, v_I, s_I, w_I)$ such that

$$|Y^*|(\partial_t u_0, \varphi_1)_\Omega + (P^u \nabla u_0, \nabla \varphi_1)_\Omega$$
$$= -l_s|\Gamma|(u_0, \varphi_1)_\Omega + l_s(s_0, \varphi_1)_{\Omega \times \Gamma}$$
$$|Y^*|(\partial_t v_0, \varphi_2)_\Omega + (P^v \nabla v_0, \nabla \varphi_2)_\Omega$$
$$= -l_w|\Gamma|(v_0, \varphi_2)_\Omega + l_w(w_0, \varphi_2)_{\Omega \times \Gamma}$$
$$(\partial_t s_0, \psi)_{\Omega \times \Gamma} + (D_s \nabla_\Gamma s_0, \nabla_\Gamma \psi)_{\Omega \times \Gamma}$$
$$= -(f + l_s)(s_0, \psi)_{\Omega \times \Gamma} + l_s|\Gamma|(u_0, \psi)_\Omega$$
$$(\partial_t w_0, \psi)_{\Omega \times \Gamma} + (D_w \nabla_\Gamma w_0, \nabla_\Gamma \psi)_{\Omega \times \Gamma}$$
$$= +(fs_0 + l_w w_0, \psi)_{\Omega \times \Gamma} - l_w|\Gamma|(v_0, \psi)_\Omega$$

$$(2.26)$$

for all $(\varphi_1, \varphi_2, \psi) \in V_{C0}(\Omega) \times V_N(\Omega) \times V(\Omega, \Gamma)$.

The strong formulation is

$$
\begin{aligned}
|Y^*|\partial_t u_0 - \nabla \cdot (P^u \nabla u_0) + l_s|\Gamma|u_0 &= l_s \int_\Gamma s_0 d\sigma_y && \text{in } \Omega \\
u_0 &= u_{\text{Boundary}} && \text{on } \Gamma_C \\
-P^u \nabla u_0 \cdot n &= 0 && \text{on } \Gamma_N
\end{aligned}
$$

$$
\begin{aligned}
|Y^*|\partial_t v_0 - \nabla \cdot (P^v \nabla v_0) + l_w|\Gamma|v_0 &= l_w \int_\Gamma w_0 d\sigma_y && \text{in } \Omega \\
-P^v \nabla v_0 \cdot n &= 0 && \text{on } \Gamma_C \\
v_0 &= 0 && \text{on } \Gamma_N
\end{aligned}
$$

$$
\begin{aligned}
\partial_t s_0 - D_s \Delta_\Gamma s_0 + (f + l_s)s_0 &= l_s u_0 && \text{on } \Omega \times \Gamma \\
\partial_t w_0 - D_w \Delta_\Gamma w_0 - fs_0 + l_w w_0 &= l_w v_0 && \text{on } \Omega \times \Gamma.
\end{aligned}
$$

$$(2.27)$$

Here again there is an analytical solution for $s_0$ and $w_0$ on the manifold $\Gamma$, because $s_0$ and $w_0$ are solutions of well-known inhomogeneous heat equations. As in section 2.4 using theorem 2.19, we obtain functions $F_0(u_0) = s_0$ and $G_0(u_0, v_0) = w_0$ with

$$F_0(u_0) = e^{(D_s \Delta_\Gamma - f - l_s)t} s_I(x)$$
$$+ e^{-(f+l_s)t} \int_0^t e^{D_s \Delta_\Gamma(t-s)} l_s u_0(s, x) e^{(f+l_s)s} ds = s_0 \quad (2.28)$$

and

$$G_0(u_0, v_0) = e^{(D_w \Delta_\Gamma - l_w)t} w_I(x)$$
$$+ e^{-l_w t} \int_0^t e^{D_w(t-s)\Delta_\Gamma} (l_w v_0(s, x) + f F_0(u_0)(s, x)) e^{l_w s} \mathrm{d}s = w_0.$$

$$(2.29)$$

The abbreviated limit equation of (2.26) is

$$|Y^*|(\partial_t u_0, \varphi_1)_\Omega + (P^u \nabla u_0, \nabla \varphi_1)_\Omega$$
$$= - l_s |\Gamma|(u_0, \varphi_1)_\Omega + l_s (F_0(u_0), \varphi_1)_{\Omega \times \Gamma}$$
$$|Y^*|(\partial_t v_0, \varphi_2)_\Omega + (P^v \nabla v_0, \nabla \varphi_2)_\Omega$$
$$= - l_w |\Gamma|(v_0, \varphi_2)_\Omega + l_w (G_0(u_0, v_0), \varphi_2)_{\Omega \times \Gamma}$$

$$(2.30)$$

for all $(\varphi_1, \varphi_2) \in V_{C0}(\Omega) \times V_N(\Omega)$.


**Identification of the two-scale limit of the abbreviated model (2.11)**
We may adopt the limit terms from the case $l, m > 1$ for the time-derivative, the diffusion term and the linear term. The following equation

$$\int_\Omega \chi\left(\frac{x}{\varepsilon}\right) \partial_t u_\varepsilon \varphi_\varepsilon \mathrm{d}x + \int_\Omega \chi\left(\frac{x}{\varepsilon}\right) \nabla u_\varepsilon \nabla \varphi_\varepsilon \mathrm{d}x$$
$$+ \varepsilon \int_{\Gamma_\varepsilon^{ER}} l_s (u_\varepsilon - F(u_\varepsilon)) \varphi_\varepsilon \mathrm{d}\sigma_x = 0$$

yields for $\varepsilon$ tending to zero

$$|Y^*| \int_\Omega \partial_t u_0 \varphi_0 \mathrm{d}x + \int_\Omega P^u \nabla u_0 \nabla \varphi_0 \mathrm{d}x + |\Gamma| l_s \int_\Omega u_0 \varphi_0 \mathrm{d}x$$
$$- l_s \lim_{\varepsilon \to 0} \varepsilon \int_{\Gamma_\varepsilon^{ER}} F(u_\varepsilon) \varphi_\varepsilon \mathrm{d}\sigma_x = 0,$$

for all admissible test functions $(\varphi_0, \varphi_1) \in V_{C0}(\Omega) \times V(\Omega, Y)$, where $u_0 \in L^2([0, T], H^1(\Omega))$ and $u_1 \in L^2([0, T] \times \Omega, H^1_\#(Y))$. As mentioned before, it is not known how to find the limit $\lim_{\varepsilon \to 0} \varepsilon \int_{\Gamma_\varepsilon^{ER}} F(u_\varepsilon) \varphi_\varepsilon \mathrm{d}\sigma_x$ directly.

By comparing this limit model with model (2.30) we find that

$$\lim_{\varepsilon \to 0} \varepsilon \int_{\Gamma_\varepsilon^{ER}} F(u_\varepsilon) \varphi_\varepsilon d\sigma_x = \int_\Omega \int_\Gamma F_0(u_0) \varphi_0 d\sigma_y dx$$

with $F_0$ as in (2.28).

We perform analogous steps for the equation

$$\int_\Omega \chi\left(\frac{x}{\varepsilon}\right) \partial_t v_\varepsilon \varphi_\varepsilon dx + \int_\Omega \chi\left(\frac{x}{\varepsilon}\right) \nabla v_\varepsilon \nabla \varphi_\varepsilon dx$$
$$+ \varepsilon \int_{\Gamma_\varepsilon^{ER}} l_w \left(v_\varepsilon - G(u_\varepsilon, v_\varepsilon)\right) \varphi_\varepsilon d\sigma_x = 0$$

and find

$$\lim_{\varepsilon \to 0} \varepsilon \int_{\Gamma_\varepsilon^{ER}} G(u_\varepsilon, v_\varepsilon) \varphi_\varepsilon d\sigma_x = \int_\Omega \int_\Gamma G_0(u_0, v_0) \varphi_0 d\sigma_y dx$$

with $G_0$ as in (2.29).

This case $l = m = 1$ is the most relevant one from the biological point of view. The binding process of the molecules to the membrane of the ER has regular speed and hence, we obtain four coupled limit equations for $u_0, v_0, s_0, w_0$.

In this context we were also able to find a two-scale limit for the $\varepsilon$-dependent operators $F$ and $G$, (2.8) and (2.9).

## The case $l < 1, m < 1$.

We are going to use the binding limit term (2.19) and recall that $u_0 = s_0$ on $\Gamma$ in the limit for $\varepsilon$ tending to zero. Since the domain of $u_0$ is $[0, T] \times \Omega \times Y^*$, the domain of $s_0$ is $[0, T] \times \Omega \times \Gamma$ and $\Gamma \subset Y^*$, the function $s_0$ must be equal to $u_0$ in its whole domain.

This means that in the limit, the solution $u_0$ will have to satisfy the limit equations for $u_0$ and for $s_0$. To be able to relate these limit equations for $u_0$ and for $s_0$, we add the equations for $u_\varepsilon$ and

$s_\varepsilon$ before the limit derivation and test with $\varphi_\varepsilon$.

$$\int_\Omega \chi\left(\frac{x}{\varepsilon}\right) \partial_t u_\varepsilon(x,t)\, \varphi_\varepsilon\left(x,\frac{x}{\varepsilon}\right) dx + \varepsilon \int_{\Gamma_\varepsilon^{ER}} \partial_t s_\varepsilon(x,t)\, \varphi_\varepsilon\left(x,\frac{x}{\varepsilon}\right) d\sigma_x$$

$$+ D_u \int_\Omega \chi\left(\frac{x}{\varepsilon}\right) \nabla u_\varepsilon(x,t)\, \nabla \varphi_\varepsilon\left(x,\frac{x}{\varepsilon}\right) dx$$

$$+ \varepsilon \int_{\Gamma_\varepsilon^{ER}} D_s \varepsilon \nabla_\Gamma s_\varepsilon(x,t)\, \varepsilon \nabla_\Gamma \varphi_\varepsilon\left(x,\frac{x}{\varepsilon}\right) d\sigma_x$$

$$+ \varepsilon \int_{\Gamma_\varepsilon^{ER}} f s_\varepsilon(x,t)\, \varphi_\varepsilon\left(x,\frac{x}{\varepsilon}\right) d\sigma_x = 0.$$

For $\varepsilon$ tending to zero we find

$$\int_\Omega \int_{Y^*} \partial_t u_0(x,t)\varphi_0(x)dydx + \int_\Omega \int_\Gamma \partial_t s_0(x,y,t)\varphi_0(x)d\sigma_y dx$$

$$+ D_u \int_\Omega \int_{Y^*} [\nabla_x u_0(x,t) + \nabla_y u_1(x,y,t)][\nabla_x \varphi_0(x) + \nabla_y \varphi_1(x,y)]dydx$$

$$+ \int_\Omega \int_\Gamma D_s \nabla_\Gamma s_0(x,y,t)\nabla_\Gamma \varphi_0(x)d\sigma_y dx$$

$$+ \int_\Omega \int_\Gamma f s_0(x,y,t)\varphi_0(x)d\sigma_y dx = 0$$

for all $(\varphi_0,\varphi_1) \in V_{C0}(\Omega) \times V(\Omega,Y)$, where $u_1 \in L^2([0,T] \times \Omega, H^1_\#(Y))$ and $u_0 \in L^2([0,T], H^1(\Omega))$ and $s_0 \in L^2([0,T] \times \Omega, H^1_\#(Y))$.
Now we set $u_0 = s_0$ on $\Gamma$ and obtain that

$$\int_\Omega (|Y^*| + |\Gamma|)\partial_t u_0(x,t)\varphi_0(x)dx$$

$$+ D_u \int_\Omega \int_{Y^*} [\nabla_x u_0(x,t) + \nabla_y u_1(x,y,t)][\nabla_x \varphi_0(x) + \nabla_y \varphi_1(x,y)]dydx$$

$$+ \int_\Omega \int_\Gamma D_s \underbrace{\nabla_\Gamma u_0(x,t)}_{=0} \nabla_\Gamma \varphi_0(x)d\sigma_y dx + \int_\Omega |\Gamma| f u_0(x,t)\varphi_0(x)dx = 0$$

where $\nabla_\Gamma u_0 = 0$ because $u_0$ is independent of $y$. Analogously we find for $v_0$ and $w_0$ the limit equation

$$\int_\Omega (|Y^*| + |\Gamma|)\partial_t v_0(x,t)\varphi_0(x)dx$$

$$+ D_u \int_\Omega \int_{Y^*} [\nabla_x v_0(x,t) + \nabla_y v_1(x,y,t)][\nabla_x \varphi_0(x) + \nabla_y \varphi_1(x,y)]dydx$$

$$+ \int_\Omega \int_\Gamma D_w \underbrace{\nabla_\Gamma v_0(x,t)}_{=0} \nabla_\Gamma \varphi_0(x)d\sigma_y dx - \int_\Omega |\Gamma| f u_0(x,t)\varphi_0(x)dx = 0$$

for all $(\varphi_0, \varphi_1) \in V_N(\Omega) \times V(\Omega, Y)$, where $v_1 \in L^2([0,T] \times \Omega, H^1_\#(Y))$ and $v_0 \in L^2([0,T], H^1(\Omega))$.

Again we refer to the cell problem (2.22) how the diffusion tensors $P^u$ and $P^v$ (2.23) can be derived. Hence, we arrive at the macroscopic system of equations.

**Macroscopic system of equations**
The weak formulation of the macroscopic system of equations is given by $(u_0, v_0) \in V_C(\Omega) \times V_N(\Omega)$ such that

$$(|Y^*| + |\Gamma|)(\partial_t u_0, \varphi_1)_\Omega + (P^u \nabla u_0, \nabla \varphi_1)_\Omega + |\Gamma| f(u_0, \varphi_1)_\Omega = 0$$

$$(|Y^*| + |\Gamma|)(\partial_t v_0, \varphi_2)_\Omega + (P^v \nabla v_0, \nabla \varphi_2)_\Omega - |\Gamma| f(u_0, \varphi_2)_\Omega = 0$$

$$(2.31)$$

for all $(\varphi_1, \varphi_2) \in V_{C0}(\Omega) \times V_N(\Omega)$.

The strong formulation is given by

$$
\begin{aligned}
(|Y^*| + |\Gamma|)\partial_t u_0 - \nabla \cdot (P^u \nabla u_0) + f|\Gamma|u_0 &= 0 && \text{in } \Omega \\
u_0 &= u_{\text{Boundary}} && \text{on } \Gamma_C \\
-P^u \nabla u_0 \cdot n &= 0 && \text{on } \Gamma_N
\end{aligned}
$$

$$
\begin{aligned}
(|Y^*| + |\Gamma|)\partial_t v_0 - \nabla \cdot (P^v \nabla v_0) - f|\Gamma|u_0 &= 0 && \text{in } \Omega \\
-P^v \nabla v_0 \cdot n &= 0 && \text{on } \Gamma_C \\
v_0 &= 0 && \text{on } \Gamma_N.
\end{aligned}
$$

The third case with $l, m < 1$ is also interesting. We see that BP molecules directly transform into DE molecules. Hence, the metabolism takes place very quickly. In the next section we also will show uniqueness of the solution of system (2.31).

Nevertheless, from the biological point of view the most important case is for $l = m = 1$. We will illustrate this case with numerical simulations in section 2.8.

## 2.7 Uniqueness of the limit model

In this section we are going to show that the solutions of systems (2.25) (case $l > 1$, $m > 1$), (2.26) (case $l = m = 1$) and the solution of system (2.31) (case $l < 1$, $m < 1$) are unique.

**Theorem 2.30 (Uniqueness for $l > 1$, $m > 1$)**
*There exists at most one solution of the problem (2.25).*

**Proof** To show uniqueness of the solution $s_0$ we assume the existence of two solutions $s_1$ and $s_2$. We test the difference of these two solutions with $\varphi = s_1 - s_2$, integrate from 0 to $t$, and find for the third equation of system (2.25) that

$$\frac{1}{2}\|s_1 - s_2\|^2_{\Omega \times \Gamma} + D_s\|\nabla_\Gamma(s_1 - s_2)\|^2_{\Omega \times \Gamma, t} + f\|s_1 - s_2\|^2_{\Omega \times \Gamma, t} = 0.$$

Hence, $s_1 = s_2$ almost everywhere on $\Omega \times \Gamma$.

Uniqueness of the diffusion tensors $P^u$ and $P^v$ will be shown in the proof of the next theorem 2.31. Then the differential equation for the functions $u_0$, $v_0$ and $w_0$ are standard heat equations with homogeneous or inhomogeneous right-hand sides, respectively, for which uniqueness can be obtained in a standard way, see [21]. □

**Theorem 2.31 (Uniqueness for $l = m = 1$)**
*There is at most one solution of the problem (2.26).*

**Proof** We assume there are two solutions $(u_1, v_1, s_1, w_1)$ and $(u_2, v_2, s_2, w_2)$ and prove that these solutions must be equal. Uniqueness of the problem (2.26) also includes uniqueness of the diffusion tensor $P^u$. Therefore, we first prove that there exists at most one solution of the cell problem (2.22). We take two solutions $\mu_j^1, \mu_j^2$ and subtract the corresponding equations

$$\int_Y (\nabla_y \mu_j^1(y,t) + e_j)\nabla_y \varphi \, dy - \int_Y (\nabla_y \mu_j^2(y,t) + e_j)\nabla_y \varphi \, dy = 0.$$

Hence,

$$\int_Y (\nabla_y(\mu_j^1(y,t) - \mu_j^2(y,t)))\nabla_y \varphi \, dx = 0.$$

Testing with $\varphi = \mu_j^1 - \mu_j^2$ leads to

$$\int_Y (\nabla_y(\mu_j^1(y,t) - \mu_j^2(y,t)))^2 dx = \|\nabla_y(\mu_j^1(y,t) - \mu_j^2(y,t))\|_Y^2 = 0$$

and we conclude $\nabla_y \mu_j^1(y,t) = \nabla_y \mu_j^2(y,t)$ which yields $\partial_{y_i}\mu_j^1(y,t) = \partial_{y_i}\mu_j^2(y,t)$ for $i,j = 1,2$ and for almost every $y \in Y$. This yields

$$P_{ij}^u = \int_Y D_u(\delta_{ij} + \partial_{y_i}\mu_j^1(t,y))dy = \int_Y D_u(\delta_{ij} + \partial_{y_i}\mu_j^2(t,y))dy$$

for $i,j = 1,2$ with $P^u$ positive definite and symmetric. Hence, $P^u = P^{u_1} = P^{u_2}$. Analogously we find $P^v = P^{v_1} = P^{v_2}$.

Now we subtract the equations for $u_2$ from the equation for $u_1$ and test it with $\varphi = u_1 - u_2$, which yields

$$|Y^*|(\partial_t u_1 - \partial_t u_2, u_1 - u_2)_\Omega + (P^u \nabla u_1 - P^u \nabla u_2, \nabla u_1 - \nabla u_2)_\Omega$$
$$+ l_s|\Gamma|(u_1 - u_2, u_1 - u_2)_\Omega = l_s\left(\int_\Gamma (s_1 - s_2)d\sigma_x, u_1 - u_2\right)_\Omega.$$

Integration from 0 to $t$ and the binomial theorem yield

$$\frac{1}{2}\|u_1 - u_2\|_\Omega^2 + \|\sqrt{P^u}\nabla(u_1 - u_2)\|_{\Omega,t}^2 + l_s|\Gamma|\|u_1 - u_2\|_{\Omega,t}^2$$
$$\leq l_s\frac{1}{2}\|s_1 - s_2\|_{\Omega \times \Gamma, t}^2 + l_s\frac{1}{2}\|u_1 - u_2\|_{\Omega,t}^2 + \underbrace{\frac{1}{2}\|u_1(0) - u_2(0)\|_\Omega^2}_{=0}.$$

Similarly we find

$$\frac{1}{2}\|s_1 - s_2\|_{\Omega \times \Gamma}^2 + D_s \|\nabla_\Gamma (s_1 - s_2)\|_{\Omega \times \Gamma, t}^2$$
$$+ (f + l_s)\|s_1 - s_2\|_{\Omega \times \Gamma, t}^2 \leq \frac{1}{2}|\Gamma| l_s \|u_1 - u_2\|_{\Omega, t}^2$$
$$+ \frac{1}{2} l_s \|s_1 - s_2\|_{\Omega \times \Gamma, t}^2 + \underbrace{\frac{1}{2}\|s_1(0) - s_2(0)\|_{\Omega \times \Gamma}^2}_{=0}$$

and

$$\frac{1}{2}\|v_1 - v_2\|_\Omega^2 + \|\sqrt{P^v}\nabla(v_1 - v_2)\|_{\Omega, t}^2$$
$$+ l_w|\Gamma|\|v_1 - v_2\|_{\Omega, t}^2 \leq l_w \frac{1}{2}\|w_1 - w_2\|_{\Omega \times \Gamma, t}^2$$
$$+ l_w \frac{1}{2}\|v_1 - v_2\|_{\Omega, t}^2 + \underbrace{\frac{1}{2}\|v_1(0) - v_2(0)\|_\Omega^2}_{=0}$$

and

$$\frac{1}{2}\|w_1 - w_2\|_{\Omega \times \Gamma}^2 + D_w \|\nabla_\Gamma (w_1 - w_2)\|_{\Omega \times \Gamma, t}^2 + l_w \|w_1 - w_2\|_{\Omega \times \Gamma, t}^2$$
$$\leq \frac{1}{2}|\Gamma| l_s \|v_1 - v_2\|_{\Omega, t}^2 + \frac{1}{2}(l_w + f)\|w_1 - w_2\|_{\Omega \times \Gamma, t}^2$$
$$+ \frac{1}{2} f \|s_1 - s_2\|_{\Omega \times \Gamma, t}^2 + \underbrace{\frac{1}{2}\|w_1(0) - w_2(0)\|_{\Omega \times \Gamma}^2}_{=0}.$$

We add the four results to deduce

$$\|u_1 - u_2\|_\Omega^2 + \|v_1 - v_2\|_\Omega^2 + \|s_1 - s_2\|_{\Omega \times \Gamma}^2 + \|w_1 - w_2\|_{\Omega \times \Gamma}^2$$
$$\leq c_1 \left( \|u_1 - u_2\|_{\Omega, t}^2 + \|v_1 - v_2\|_{\Omega, t}^2 \right.$$
$$\left. + \|s_1 - s_2\|_{\Omega \times \Gamma, t}^2 + \|w_1 - w_2\|_{\Omega \times \Gamma, t}^2 \right)$$

for a $c_1 > 0$ depending on $l_s, l_w, f$ and $|\Gamma|$. With Gronwall's lemma 2.12 it follows that

$$\|u_1 - u_2\|_\Omega^2 + \|v_1 - v_2\|_\Omega^2 + \|s_1 - s_2\|_{\Omega \times \Gamma}^2 + \|w_1 - w_2\|_{\Omega \times \Gamma}^2 = 0$$

for almost every $t \in [0, T]$ and hence $(u_1, v_1, s_1, w_1) = (u_2, v_2, s_2, w_2)$ almost everywhere. $\square$

**Theorem 2.32 (Uniqueness for $l < 1$, $m < 1$)**
*There is at most one solution of the problem* (2.31).

**Proof** In the proof of theorem 2.31 we already saw that the diffusion tensors $P^u$ and $P^v$ are unique.
It remains to show that there is at most one solution $u_0$ and $v_0$ solving (2.31). We assume two solution $(u_1, v_1)$ and $(u_2, v_2)$ and are going to show that they must be equal.
Therefore we subtract the equation for $(u_1, v_1)$ and for $(u_2, v_2)$ and test it with $\varphi = u_1 - u_2$ and $\varphi = v_1 - v_2$, respectively.

$$(|Y^*| + |\Gamma|)(\partial_t u_1 - \partial_t u_2, u_1 - u_2)_\Omega + (P^u \nabla(u_1 - u_2), \nabla(u_1 - u_2))_\Omega$$
$$+ |\Gamma| f(u_1 - u_2, u_1 - u_2)_\Omega = 0.$$

Integration from 0 to $t$ leads to

$$(|Y^*| + |\Gamma|)\frac{1}{2}\|u_1 - u_2\|_\Omega^2$$
$$+ \|\sqrt{P^u}\nabla(u_1 - u_2)\|_{\Omega,t}^2 + |\Gamma| f \|u_1 - u_2\|_{\Omega,t}^2 = 0.$$

Hence, we find that $u_1 = u_2$ almost everywhere in $\Omega$ and for almost every $t \in [0, T]$.
Similarly, we find that

$$(|Y^*| + |\Gamma|)(\partial_t v_1 - \partial_t v_2, v_1 - v_2)_\Omega + (P^v \nabla(v_1 - v_2), \nabla(v_1 - v_2))_\Omega$$
$$- |\Gamma| f(u_1 - u_2, v_1 - v_2)_\Omega = 0.$$

Integration from 0 to $t$ leads to

$$(|Y^*| + |\Gamma|)\frac{1}{2}\|v_1 - v_2\|_\Omega^2 + \|\sqrt{P^v}\nabla(v_1 - v_2)\|_{\Omega,t}^2$$
$$= |\Gamma| f \underbrace{\|u_1 - u_2\|_{\Omega,t}^2}_{=0} + |\Gamma| f \|v_1 - v_2\|_{\Omega,t}^2.$$

With Gronwall's lemma we deduce

$$(|Y^*| + |\Gamma|)\frac{1}{2}\|v_1 - v_2\|_\Omega^2 + \|\sqrt{P^v}\nabla(v_1 - v_2)\|_{\Omega,t}^2 \leq 0$$

and we obtain that $v_1 = v_2$ almost everywhere in $\Omega$ and for almost every $t \in [0, T]$. This completes the proof. $\qquad\qquad\qquad\square$

We conclude this chapter with results from simulations of the system (2.27) discussed above.

## 2.8 Simulations of the linear carcinogenesis model

We are going to use simulations to illustrate the model (2.27). For convenience we write $(u, v, s, w)$ for the solution $(u_0, v_0, s_0, w_0)$. We choose the dimension $n$ of the domain $\Omega$ to be 2 and $\Omega = B(0,0)_{r=5\mu m} \setminus B(0,0)_{r=1.5\mu m}$. Further, $Y = [0,1]^2$ is the unit cell, $Y^* = Y \setminus B(\frac{1}{2}, \frac{1}{2})_{r=0.45}$ and $\Gamma = \{(x, y) \mid (x - \frac{1}{2})^2 + (y - \frac{1}{2})^2 = 0.45^2\} \subset Y$ are a characteristical part of the lumen of the ER and the surface of the ER, respectively.

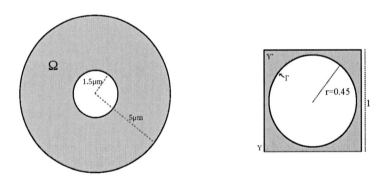

**Remark 2.33** Since the domain and initial condition are radially symmetric and because of the form of the partial differential equation, the solution remains radially symmetric, a fact we will exploit by doing simulations in just one spatial dimension, the radius of the cell.

We use polar coordinates, hence the solution depends on the radius, but not on the angle.

**Remark 2.34** The following numerical simulations are for illustrative purpose only. Accurate biological parameters may be difficult to obtain, and the chosen parameters represent our best guesses.

Hence we will not give any qualitative or quantitative biological interpretation or implication of the simulation results. Rather, we want to demonstrate that all the simulations can be done and that the numerical results agree with our analysis above.

The exemplary parameters we choose in this chapter are the following.

**Parameters**

| Notation | Value | Unit | Description |
|----------|-------|------|-------------|
| $u_I$ | 0 | $\mu M$ | Initial value for BP in cytosol |
| $v_I$ | 0 | $\mu M$ | Initial value for DE in cytosol |
| $s_I$ | 0 | $\mu M$ | Initial value for BP on ER |
| $w_I$ | 0 | $\mu M$ | Initial value for DE on ER |
| $D_u$ | 1 | $\mu m/s$ | Diff. coefficient for BP in cytosol |
| $D_v$ | 1 | $\mu m/s$ | Diff. coefficient for DE in cytosol |
| $D_s$ | 0.05 | $\mu m/s$ | Diff. coefficient for BP on ER |
| $D_w$ | 0.05 | $\mu m/s$ | Diff. coefficient for DE on ER |
| $l_s$ | 0.1 | $1/s$ | Binding rate to ER for BP |
| $l_w$ | 0.1 | $1/s$ | Binding rate to ER for DE |
| $f$ | 0.01 | $1/s$ | Metabolism rate from BP to DE |
| $u_{Boundary}$ | 1 | $\mu M$ | BP concentr. in intercellular space |

Before we solve the equation numerically, we need to calculate $P^u$, $P^v$, $|Y^*|$, $|\Gamma|$, $\int_\Gamma s d\sigma_y$ and $\int_\Gamma w d\sigma_y$. We start with the two latter integrals.

**Computation of $l_s \int_\Gamma s d\sigma_y$ and $l_w \int_\Gamma w d\sigma_y$**
The functions $s$ and $w$ solve the partial differential equation

$$\partial_t s - D_s \Delta_\Gamma s + (f + l_s)s = l_s u \qquad \text{on } \Omega \times \Gamma$$

and

$$\partial_t w - D_w \Delta_\Gamma w - fs + l_w w = l_w v \qquad \text{on } \Omega \times \Gamma.$$

for every discretization point for $u$, $v$ on $\Omega$. Because $s_I$, $w_I$ and $u$, $v$ are independent of the variable $y \in \Gamma$ and the circle $\Gamma$ has no boundaries where some boundary conditions could be defined, the functions $s$ and $w$ will also be $y$-independent. Hence, the spatial derivatives $\Delta_\Gamma s$ and $\Delta_\Gamma w$ are zero.

This simplifies the equation to the ordinary differential equations

$$\partial_t s + (f + l_s)s = l_s u \qquad \text{on } \Omega \times \Gamma$$

and

$$\partial_t w - fs + l_w w = l_w v \qquad \text{on } \Omega \times \Gamma.$$

The analytical solution of these differential equations is given by

$$s(x,t) = e^{-(f+l_s)t}\left(s_I - \frac{l_s u(x)}{f + l_s}\right) + \frac{l_s u(x)}{f + l_s}$$

and

$$w(x,t) = e^{-l_w t}\left(w_I - \frac{l_w v(x) + fs(x)}{l_w}\right) + \frac{l_w v(x) + fs(x)}{l_w}.$$

Integration over $\Gamma$ is easily done by multiplying with $2\pi r$. Hence, we obtain

$$l_s \int_\Gamma s \, \mathrm{d}\sigma_y = 2\pi r l_s \left(e^{-(f+l_s)t}\left(s_I - \frac{l_s u}{f + l_s}\right) + \frac{l_s u}{f + l_s}\right), \quad (2.32)$$

$$l_w \int_\Gamma w \, \mathrm{d}\sigma_y = 2\pi r l_w \left(e^{-l_w t}\left(w_I - \frac{l_w v + fs}{l_w}\right) + \frac{l_w v + fs}{l_w}\right). \tag{2.33}$$

**Computation of $P^u$, $P^v$, $|Y|$ and $|\Gamma|$**

Since $D_u$ and $D_v$ are constant, the matrices $P^u$ and $P^v$ are given by

$$P^u = (P^u_{ij})_{ij} \qquad \text{with} \qquad P^u_{ij} = D_u \int_Y (\delta_{ij} + \partial_{y_i} \mu_j(y)) \mathrm{d}y,$$

$$P^v = (P^v_{ij})_{ij} \qquad \text{with} \qquad P^v_{ij} = D_v \int_Y (\delta_{ij} + \partial_{y_i} \mu_j(y)) \mathrm{d}y$$

with

$$\nabla \cdot (e_j + \nabla \mu_j) = 0 \quad \text{in } Y^*$$
$$(e_j + \nabla \mu_j) \cdot n = 0 \quad \text{on } \partial Y^* = \Gamma$$

and $\mu_j$ $Y$ periodic for $i, j = 1, 2$.
Using the shape of $Y^*$ as described above, this PDE cannot be solved analytically. The numerical solver COMSOL is handy to solve classical partial differential equations and calculations of the cell problem in COMSOL yield

$$P^u = D_u \begin{pmatrix} 0.19649 & 0 \\ 0 & 0.19649 \end{pmatrix}, \quad P^v = D_v \begin{pmatrix} 0.19649 & 0 \\ 0 & 0.19649 \end{pmatrix}.$$

Further,

$$|Y^*| = 1 - \pi r^2 = 1 - \pi(0.45)^2 = 0.363827$$
$$|\Gamma| = 2\pi r = 2\pi \cdot 0.45 = 2.82743.$$

Then the equation we want to solve is

$$0.3638 \partial_t u - 0.0982 \Delta u + 0.28274 u = 0.1 \int_\Gamma s d\sigma_y,$$
$$0.3638 \partial_t v - 0.0982 \Delta v + 0.28274 v = 0.1 \int_\Gamma w d\sigma_y,$$

where we calculate $\int_\Gamma s d\sigma_y$ and $\int_\Gamma w d\sigma_y$ for every small time adaptive step. The time interval is chosen to be $[0, T] = [0s, 300s] = [0\text{min}, 5\text{min}]$.
For the numerical calculations of the macroscopic problem we are going to use MATLAB. For every adaptive time step $\tau$ and for every spatial discretization point in $\Omega$, at first the integrals $\int_\Gamma s d\sigma_y$ and $\int_\Gamma w d\sigma_y$ are calculated. Only then, calculations for $u$ and $v$ in $\Omega$ are continued for the time step $\tau$. This means, computations on $\Gamma$ and on $\Omega$ alternate.
This unusual procedure is hard to implement in COMSOL and thus, we switch to the more flexible program MATLAB. We use semi-discretization in space and the finite element method with

linear Lagrange elemtents to solve the boundary value problem. The spatial one-dimensional interval is given by $[1.5, 5]$ and the spatial mesh is fixed with mesh size 0.07.

In the unit cell we use the analytical solutions for $s$ and $w$ given in (2.32). In the time variable the MATLAB ode solver *ode15s* is used, which uses Gear's method to solve the ordinary differential equations.

**Results and Simulations**

We see simulations made with MATLAB. On the $x$-axis we see the radius of the cell, on the $y$-axis we see the time and on the $z$-axis the concentration of the molecules in $\mu M$ is visualized.

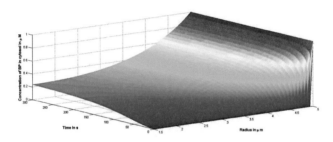

Figure 2.1: The solution $u$, the concentration of BP molecules in the cytosol.

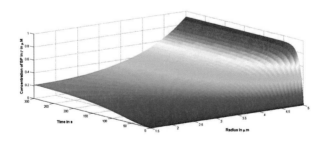

Figure 2.2: The solution $s$, the concentration of BP molecules bound to the membrane of the ER.

65

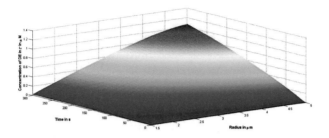

Figure 2.3: The solution $v$, the concentration of DE molecules in the cytosol.

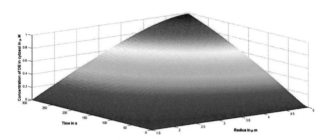

Figure 2.4: The solution $w$, the concentration of DE molecules bound to the membrane of the ER.

On the first figure 2.1 we see the solution $u$, the concentration of BP molecules in cytosol. We recognize the Dirichlet boundary condition $u_{\text{Boundary}} = 1\mu M$ at the plasma membrane (in the figure on the right upper side, where radius $r = 5\mu m$) and the initial condition $u_I = 0\mu M$ (in the figure on the right lower side, where $t = 0s$). The concentration increases with time and for $t = 300s$ the concentration of BP molecules is $\approx 0.225\mu M$ at the nuclear membrane.

On the surface of the ER the concentration of the BP molecules is very similar to the BP molecules in cytosol, see figure 2.2. This is, because the dominating term in the dynamics is the binding term $l_s(u - s)$ to the ER, which evens the concentrations. We recognize the initial condition $s_I = 0\mu M$ for $t = 0s$. The concentration increases in time and reaches its maximal values $\approx 0.9\mu M$ at the plasma membrane and $\approx 0.21\mu M$ at the nuclear membrane for $t = 300s$.

Next in figure 2.3, we see the solution $w$, the concentration of DE molecules bound to the surface of the ER. The BP molecules just transformed into these molecules. Because the changing rate $f$ is small compared to the binding rates $l_s$ and $l_w$, the influence of the boundary conditions of $v$, the DE concentration in cytosol, at the nuclear membrane $v = 0\mu M$ for $r = 1.5\mu m$ is clearly visible and there the DE concentration on ER is much lower than at the plasma membrane.
At the plasma membrane, where we have a no-flux boundary condition for $v$, the metabolism from BP to DE molecules is strongest. For time $t = 300s$ the maximal DE concentration on the ER is $\approx 1.08\mu M$.

In the last figure 2.4, where we see the solution $v$, the concentration of DE molecules in cytosol, the shape of the graph looks very similar to the one of the function $w$. We notice the initial condition $v_I = 0\mu M$ and also the boundary condition at the nuclear membrane $v = 0\mu M$. The DE concentration has its maximum of $\approx 0.98\mu M$ at the plasma membrane for $t = 300s$.

# THREE

# NONLINEAR CARCINOGENESIS MODEL

## Introduction

In this chapter we extend the linear carcinogenesis model from chapter 2 with nonlinear terms. Therefore mathematical tools to handle the nonlinearities are necessary.
These tools are presented in section 3.1. Mainly we need to show strong convergence for the functions and continuity of the functionals.

In section 3.2 we introduce another way to derive the homogeneous limit model, the periodic unfolding method. We explain the strategy of the periodic unfolding operator $\mathcal{T}_\varepsilon$ and the boundary periodic unfolding operator $\mathcal{T}_\varepsilon^b$. The periodic unfolding method was developed by D. Cioranescu, P. Donato, R. Zaki and others, see [13, 11, 10, 17]. The first basic idea was found by T. Arbogast, J. Douglas and U. Hornung in [5].
We are able to show new results for the operator $\mathcal{T}_\varepsilon^b$ concerning function products (lemma 3.18), diffusion on Riemannian manifolds (lemmas 3.19, 3.20, 3.21, 3.22, 3.27) and trace inequalities (lemma 3.24).

We continue with modeling the nonlinear carcinogenesis model in section 3.3. We include cleaning mechanisms that reduce the number of BP and DE molecules in the cytosol. The metabolism on the surface of the endoplasmic reticulum becomes nonlinear and will be bounded by a natural threshold. Further, we assume a finite number of receptors living on the membrane of the endoplasmic reticulum the BP and DE molecules can bind to.

Again we provide the system of equations with suitable $\varepsilon$ exponents.

In section 3.4 we prove that the equations satisfy the necessary conditions to apply two-scale convergence and the periodic unfolding method.

We show existence of a solution for every $\varepsilon > 0$ in section 3.5. We use Schauder's fixed-point theorem and Carathéodory's existence theorem for the proof.

In section 3.6, we derive the homogeneous limit model for the system of equations. At first we consider the limits of the nonlinear terms, afterwards we derive the limit for $\varepsilon$ tending to zero using two-scale convergence or the periodic unfolding operator.

We show that there exists at most one solution of the limit model in section 3.7.

In the last section of this chapter we do numerical simulations of the homogeneous limit model in section 3.8. We use corresponding parameters to the numerical simulations in the linear case to be able to compare the results.

## 3.1 Nonlinear homogenization

We are going to extend the theory of periodic homogenization to classes of partial differential equations with nonlinear terms.

It is important to distinguish if the nonlinear terms live in the domain $\Omega_\varepsilon$ itself or on its periodic boundary $\Gamma_\varepsilon$.

In the first case, functions which occur in the nonlinear terms are extended from the domain $\Omega_\varepsilon$ to the domain $\Omega$. Then, these function are defined on a domain that is not changing with $\varepsilon$

and we use Lion-Aubin's lemma 3.5 and lemma 3.4 to prove that these function converge strongly in $L^2([0,T], L^2(\Omega))$. It remains to show that the nonlinear operators occuring in the nonlinear terms are continuous and then, the limit for $\varepsilon$ tending to zero can be derived.

In the second case, functions defined on $\Gamma_\varepsilon$ cannot be extended to $\Omega$ so easily and we use another idea to prove strong convergence, the periodic unfolding method, introduced in the next section 3.2. But still, the nonlinear operators must be continuous in this case as well.

We need to extend functions $u_\varepsilon \in L^2([0,T], L^2(\Omega_\varepsilon))$ to functions that live on the whole, $\varepsilon$-independent domain $\Omega$, since most of the theorems, we are going to use now, only work on such domains. Therefore we cite the following extension theorem (see [41]).

As usual, let $\Omega \subset \mathbb{R}^n$, $Y = [0,1]^n$ and $\Omega_\varepsilon := \bigcup_{k \in \mathbb{Z}^n} \varepsilon(k+Y) \cap \Omega$.

**Theorem 3.1** *A function $u_\varepsilon \in H^1(\Omega_\varepsilon)$ can be extended to a function $\tilde{u}_\varepsilon$ defined on all of $\Omega$ in such a way that it holds that*

$$\|\tilde{u}_\varepsilon\|^2_{H^1(\Omega)} \leq C \|u_\varepsilon\|^2_{H^1(\Omega_\varepsilon)} \tag{3.1}$$

*for some constant $C > 0$. For $u_\varepsilon \in L^2(\Omega_\varepsilon)$, the extension $\tilde{u}_\varepsilon$ further satisfies*

$$\|\tilde{u}_\varepsilon\|^2_\Omega \leq C \|u_\varepsilon\|^2_{\Omega_\varepsilon}.$$

*for some constant $C > 0$.*

For convenience we will denote the extended function $\tilde{u}_\varepsilon$ again with $u_\varepsilon$.

**Remark 3.2** For the extension from theorem 3.1, analogous estimations to (3.1) hold for the $L^\infty$-norm, if the $L^\infty$-norm of the original functions is bounded.

**Remark 3.3** If the function to be extended depends on additional variables, $u_\varepsilon \in L^2([0,T], H^1(\Omega_\varepsilon))$ for example, one can still consider its extension in the sense of theorem 3.1 since the extension operator is linear and $u_\varepsilon(t, \cdot) \in H^1(\Omega_\varepsilon)$ for almost every

$t \in [0, T]$. Thus, a linear and continuous extension operator $E : L^2([0,T], H^1(\Omega_\varepsilon)) \to L^2([0,T], H^1(\Omega))$ exists whose norm is independent of $\varepsilon$.

These two remarks and the following lemma are cited from article [49].

**Lemma 3.4** *Let $u_\varepsilon$ be a bounded sequence in $L^2([0,T], H^1(\Omega)) \cap H^1([0,T], H_0^1(\Omega)') \cap L^\infty(\Omega \times [0,T])$ and let $u_0$ be the corresponding two-scale limit. Then, there exists a subsequence of $u_\varepsilon$, which strongly converges to $u_0$ in $L^p(\Omega \times [0,T])$ for any $1 \leq p < \infty$.*

The proof of this statement is based on the lemma 3.5, the lemma of Lions-Aubin (see Showalter 1997, p. 106 [53]).

**Lemma 3.5** *(Lions-Aubin)*
*Let $B_0, B, B_1$ be Banach spaces with $B_0 \subset B \subset B_1$; assume $B_0 \hookrightarrow B$ is compact and $B \hookrightarrow B_1$ is continuous. Let $1 < p < \infty$, $1 < q < \infty$, let $B_0$ and $B_1$ be reflexive and define*

$$W := \{u \in L^p([0,T], B_0) : \partial_t u \in L^q([0,T], B_1)\}.$$

*Then the inclusion $W \hookrightarrow L^p([0,T], B)$ is compact.*

When we have strong convergence for the sequence of functions $u_\varepsilon$ to $u_0$ in $L^2([0,T], L^2(\Omega))$, it remains to show that it holds for a nonlinear function $f : \mathbb{R} \to \mathbb{R}$ with $f : u_\varepsilon(t,x) \mapsto f(u_\varepsilon(t,x))$ that $\int_{[0,T] \times \Omega} |f(u_\varepsilon(x,t)) - f(u_0(x,t))|^2 dx dt$ tends to zero for $\varepsilon$ tending to zero. For that purpose we employ Nemytskii operators and cite the following theorem (see Showalter 1997 p.48, [53]).

**Theorem 3.6** *(Nemytskii)*
*Let $1 \leq p, q < \infty$ and let $f : \mathbb{R} \to \mathbb{R}$ be a continuous function that satisfies*

$$|f(x)| \leq C|x|^{p/q}, \qquad x \in \mathbb{R}$$

*for some $C > 0$. Then the Nemytskii operator defined by*

$$F : L^p(\Omega) \to L^q(\Omega)$$
$$(F(u))(x) = f(u(x))$$

*is bounded and continuous.*

Now it immediately follows the continuity of $F$ and strong convergence of $u_\varepsilon$, with $\lim_{\varepsilon\to0} u_\varepsilon = u_0$, that

$$\lim_{\varepsilon\to0} F(u_\varepsilon) = F(u_0).$$

Now we consider a different case. Let us look at a sequence of functions $u_\varepsilon$, defined on $\Omega$, and a periodic manifold $\Gamma_\varepsilon$, as defined in (2.1).
A nonlinear operator is applied to the function $u_\varepsilon$ restricted to $\Gamma_\varepsilon$. To perform homogenization of this nonlinear term, we cite theorem 2.5 (step 2) in [16] and modify the proof for our setting.

**Lemma 3.7** *Let $h : \mathbb{R} \to \mathbb{R}$ be bounded, Lipschitz-continuous and satisfy*

$$|h(x)| \leq c|x|^{p/q}.$$

*Let $\Omega \subset \mathbb{R}^n$ be a union of squares and $\Gamma \subset Y = [0,1]^n$ be a Lipschitz boundary in $Y$, which divides $Y$ into two subsets $T \subset Y$ and $Y^* = Y\setminus\overline{T}$, such that $T$, $\Gamma$ and $Y^*$ are $Y$-periodic. Furthermore, let $u_\varepsilon$ strongly converge to $u_0$ in $L^2([0,T], L^2(\Omega))$ and $\|u_\varepsilon\|_{L^2([0,T],H^1(\Omega))}$ be bounded independent of $\varepsilon$.*
*Then it holds that*

$$\lim_{\varepsilon\to0} \varepsilon(h(u_\varepsilon), \varphi)_{L^2(\Gamma^\varepsilon),t} = \frac{|\Gamma|}{|Y|} (h(u_0), \varphi)_{\Omega,t}$$

*for any $\varphi \in C^\infty(\Omega)$.*

**Proof** We prove the assertion in two steps. In the first step we show that the auxiliary linear form $\mu_\lambda^\varepsilon$ defined in (3.5) strongly converges in $(H^1(\Omega))'$. Therefore, we mainly adopt and rearrange section 3 "Lemmes de base" from [12]. In the second step we use the form $\mu_\lambda^\varepsilon$ to complete the proof.

1.  Let the function $\lambda \in H^{1/2}(\Gamma)$ be $Y$-periodic. We define

$$C_\lambda = \frac{1}{|Y^*|} \int_\Gamma \lambda(y)d\sigma_y.$$

Let $\Psi_\lambda$ be the solution of the problem

$$
\begin{cases}
-\Delta\Psi_\lambda &=\ -C_\lambda & \text{in } Y^*, \\
\nabla\Psi_\lambda \cdot n &=\ \lambda & \text{on } \Gamma, \\
\Psi_\lambda & & Y - \text{periodic}.
\end{cases}
\tag{3.2}
$$

A solution $\Psi_\lambda$ exists, since $-\int_{Y^*} C_\lambda dy = \int_\Gamma \lambda d\sigma_y$.
Furthermore, we define the function $\Psi_\lambda^\varepsilon(x) := \Psi_\lambda(x/\varepsilon)$. Note that $\Psi_\lambda^\varepsilon$ satisfies

$$
\begin{cases}
-\Delta\Psi_\lambda^\varepsilon &=\ -\frac{1}{\varepsilon^2}C_\lambda & \text{in } \Omega_\varepsilon \\
\nabla\Psi_\lambda^\varepsilon \cdot n &=\ \frac{1}{\varepsilon}\lambda(x/\varepsilon) & \text{on } \Gamma_\varepsilon.
\end{cases}
\tag{3.3}
$$

Testing problem (3.3) with $\varphi \in H^1(\Omega_\varepsilon)$ and integrating by parts leads to

$$
\varepsilon \int_{\Gamma_\varepsilon} \lambda(x/\varepsilon)\varphi(x)d\sigma_x = \varepsilon \int_{\Omega_\varepsilon} \nabla_y\Psi_\lambda(x/\varepsilon)\nabla\varphi(x)dx
$$
$$
+ \int_{\Omega_\varepsilon} C_\lambda\varphi(x)dx - \varepsilon \int_{\partial\Omega\cap\partial\Omega_\varepsilon} \nabla_y\Psi_\lambda(x/\varepsilon) \cdot n\varphi(x)d\sigma_x, \tag{3.4}
$$

where $y = x/\varepsilon$. Here we used that $\varepsilon\nabla_x\Psi_\lambda^\varepsilon(x) = \nabla_y\Psi_\lambda(x/\varepsilon)$. We define a linear form on $H^1(\Omega)$ by

$$
\langle \mu_\lambda^\varepsilon, \varphi \rangle = \varepsilon \int_{\Gamma_\varepsilon} \lambda(x/\varepsilon)\varphi(x)d\sigma_x \qquad \forall\, \varphi \in H^1(\Omega). \tag{3.5}
$$

With equation (3.4) we find that

$$
\varepsilon \int_{\Gamma_\varepsilon} \lambda(x/\varepsilon)\varphi(x)d\sigma_x \leq \varepsilon C \|\nabla_y\Psi_\lambda\|_{L^2(Y^*)}\|\nabla\varphi\|_{L^2(\Omega_\varepsilon)}
$$
$$
+ C'C_\lambda\|\varphi\|_{L^2(\Omega_\varepsilon)} + \varepsilon C''\|\nabla_y\Psi_\lambda\|_{L^2(\partial Y^*)}\|\varphi\|_{L^2(\partial\Omega)}.
$$

In this setup, the following convergence holds

$$
\mu_\lambda^\varepsilon \to \mu_\lambda \qquad \text{strongly in } (H^1(\Omega))' \tag{3.6}
$$

with

$$
\langle \mu_\lambda, \varphi \rangle = \underbrace{\frac{1}{|Y|} \int_\Gamma \lambda(y)d\sigma_y}_{=\mu_\lambda} \int_\Omega \varphi dx.
$$

74

To prove statement (3.6), we define the following linear forms using equation (3.4)

$$\mu_\lambda^\varepsilon = \tilde{\mu}_\lambda^\varepsilon - \hat{\mu}_\lambda^\varepsilon + C_\lambda \chi_{\Omega_\varepsilon}$$

with

$$\langle \tilde{\mu}_\lambda^\varepsilon, \varphi \rangle = \varepsilon \int_{\Omega_\varepsilon} \nabla_y \Psi_\lambda(x/\varepsilon) \nabla \varphi(x) dx$$

and

$$\langle \hat{\mu}_\lambda^\varepsilon, \varphi \rangle = \varepsilon \int_{\partial\Omega \cap \partial\Omega_\varepsilon} \nabla_y \Psi_\lambda(x/\varepsilon) \cdot n \varphi(x) d\sigma_x.$$

*a)* It holds that

$$|\langle \tilde{\mu}_\lambda^\varepsilon, \varphi \rangle| \leq \varepsilon C \underbrace{\|\nabla_y \Psi_\lambda\|_{L^2(Y^*)}}_{\text{bounded, since } \Psi_\lambda \in H^1(Y^*)} \underbrace{\|\nabla \varphi\|_{L^2(\Omega_\varepsilon)}}_{\text{bounded, since } \varphi \in H^1(\Omega)}$$

Hence, $\tilde{\mu}_\lambda^\varepsilon \xrightarrow{\varepsilon \to 0} 0$ strongly in $(H^1(\Omega))'$.

*b)* To see that $\|\nabla_y \Psi\|_{L^2(\partial Y^*)}$ is bounded, the solution $\Psi$ of problem (3.2) is copied to the adjacent cells of $Y$. This means, for $K = \{k \in \mathbb{Z}^n \mid k_i \in \{-1, 0, 1\}\}$ we define the domains $\tilde{Y} := \bigcup_{k \in K}(Y^* + k)$ and $\tilde{\Gamma} := \bigcup_{k \in K}(\Gamma + k)$ and then it holds by adding over $k \in K$ that

$$\int_{\tilde{Y}} \nabla\Psi\nabla\varphi dy + \int_{\tilde{\Gamma}} \lambda\varphi d\sigma_y - \int_{\bigcup_{k\in K}(\partial Y^*+k)} \nabla\Psi \cdot n\varphi d\sigma_y = \int_{\tilde{Y}} C_\lambda dy$$

for all $\varphi \in H^1(\tilde{Y})$. Therefore, $\lambda(y + k) = \lambda(y)$ for $y \in \Gamma$ and $k \in K$. Since $\Gamma$ is smooth and $Y$-periodic, $\tilde{\Gamma}$ is also smooth. With $\lambda$ being $Y$-periodic and an element of $H^{1/2}(\Gamma)$, the extended $\lambda$ is an element of $H^{1/2}(\tilde{\Gamma})$.

If we choose $\varphi \in H_0^1(\tilde{Y})$ as test functions, with $\varphi = 0$ on the outer boundary $\partial\tilde{Y}$ and not necessary on $\tilde{\Gamma}$, it follows that $\Psi$ satisfies

$$\int_{\tilde{Y}} \nabla\Psi\nabla\varphi dy + \int_{\tilde{\Gamma}} \lambda\varphi d\sigma_y = \int_{\tilde{Y}} C_\lambda dy.$$

That is because $\varphi = 0$ on $\partial\tilde{Y}$ and, due to the periodicity of $\Psi$, the integrals over the inner boundaries of $\partial Y^* + k$ occur once with positive and once with negative sign. In the following figure we see an exemplary sketch of $\tilde{Y}$ and $\tilde{\Gamma}$ in $\mathbb{R}^2$.

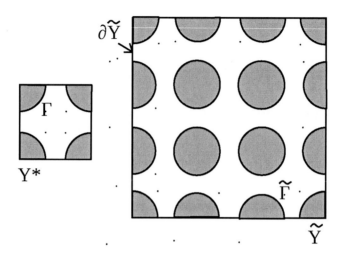

With $\tilde{\Gamma}$ being smooth, $\lambda \in H^{1/2}(\tilde{\Gamma})$ and $C_\lambda$ a constant, the solution $\Psi$ satisfies in the interior of $\tilde{Y}$

$$\|\Psi\|_{H^2(\text{int}\tilde{Y})} \leq C \left( \|C_\lambda\|_{L^2(\text{int}\tilde{Y})} + \|\lambda\|_{H^{1/2}(\tilde{\Gamma}\cap\text{int}\tilde{Y})} \right),$$

where the interior of $\tilde{Y}$ is a domain with $C^2$-boundary and has a positive distance from the outer boundary $\partial\tilde{Y}$, see [28]. This means that $\Psi$ is a $H^2$-function on $Y^*$.

Every side of the unit cell $Y$ intersected with $Y^*$ is a smooth boundary and hence, using the continuous trace operator $\gamma : H^1(Y^*) \rightarrow L^2((\partial Y^*)_i)$, one finds that $\|\nabla_y\Psi\|_{L^2((\partial Y^*)_i)} \leq c_0\|\nabla_y\Psi\|_{H^1(Y^*)}$ for every side $i = 1,\dots,2n$. Summing over all sides $i = 1,\dots,2n$ we find that $\|\nabla_y\Psi\|_{L^2(\partial Y^*)}$ is bounded. We use this fact to deduce that

$$|\langle \mathring{\mu}_\lambda^\varepsilon, \varphi \rangle| \leq \varepsilon C'' \|\nabla_y\Psi_\lambda\|_{L^2(\partial Y^*)} \underbrace{\|\varphi\|_{L^2(\partial\Omega)}}_{\text{bounded, since}\varphi\in L^2(\partial\Omega)} .$$

Hence, $\mathring{\mu}_\lambda^\varepsilon \xrightarrow{\varepsilon\to 0} 0$ strongly in $(H^1(\Omega))'$.

*c)* Furthermore, we have

$$C_\lambda \chi_{\Omega_\varepsilon} \to C_\lambda \frac{|Y^*|}{|Y|} = \mu_\lambda \qquad \text{weak}^* \text{ in } L^\infty(\Omega)$$

and hence strongly in $(H^1(\Omega))'$, which completes the proof that $\mu_\lambda^\varepsilon$ converges strongly to $\mu_\lambda$ in $H^1(\Omega)'$.

2. Before we go back to $\mu_\lambda^\varepsilon$, we deduce from theorem 3.6 that the Nemytskii operator $H$ defined by

$$H : L^p(\Omega) \to L^q(\Omega)$$
$$H(u_\varepsilon)(x,t) := h(u_\varepsilon(x,t))$$

is bounded and continuous.

Now we are going to use part 1 of the proof, where we take $\lambda = 1 \in H^{1/2}(\Gamma)$. Notice that in this case $\mu_\lambda = \mu_1 = \frac{|\Gamma|}{|Y|}$.

Moreover, if a sequence of functions $z^\varepsilon \in H^1(\Omega)$ is such that $z^\varepsilon \rightharpoonup z$ weakly in $H^1(\Omega)$, then

$$\langle \mu_\lambda^\varepsilon, z^\varepsilon \rangle \xrightarrow{\varepsilon \to 0} \mu_\lambda \int_\Omega z \, dx. \qquad (3.7)$$

With $u_\varepsilon \to u_0$ strongly in $L^2([0,T] \times \Omega)$ and continuity of $h$ we obtain

$$h(u_\varepsilon) \to h(u_0) \qquad \text{strongly in } L^2([0,T] \times \Omega).$$

Furthermore, $\|\nabla h(u_\varepsilon)\|_{L^2([0,T]\times\Omega)} = \|h'(u_\varepsilon)\nabla u_\varepsilon\|_{L^2([0,T]\times\Omega)}$ is bounded, since $h$ is Lipschitz-continuous and $\|u_\varepsilon\|_{L^2([0,T],H^1(\Omega))}$ is bounded. Hence, we deduce a weakly converging subsequence $h(u_\varepsilon)$ in $L^2([0,T], H^1(\Omega))$ and for any $\varphi \in C^\infty(\Omega)$ it holds that

$$\varphi h(u_\varepsilon) \xrightarrow{\varepsilon \to 0} \varphi h(u_0) \qquad \text{weakly in } L^2([0,T], H^1(\Omega)).$$

We deduce that $\varphi h(u_\varepsilon(t)) \rightharpoonup \varphi h(u_0(t))$ for almost every $t \in [0,T]$. Combining this with (3.7) yields for $z^\varepsilon = \varphi h(u_\varepsilon(t))$

$$\langle \mu_1^\varepsilon, \varphi h(u_\varepsilon(t)) \rangle \xrightarrow{\varepsilon \to 0} \frac{|\Gamma|}{|Y|} \int_\Omega \varphi h(u_0(t)) dx$$

for almost every $t \in [0, T]$.

Finally, we are in the position to use Lebesgue's convergence theorem and get

$$\lim_{\varepsilon \to 0} \varepsilon (h(u_\varepsilon), \varphi)_{\Gamma_\varepsilon, t} = \frac{|\Gamma|}{|Y|} (h(u_0), \varphi)_{\Omega, t}$$

which concludes the proof. □

We showed that in this setting homogenization of nonlinear terms acting on $\Omega_\varepsilon$ and on $\Gamma_\varepsilon$ are possible.

If a sequence of functions is only defined on $\Gamma_\varepsilon$, then we cannot use an extension on $\Omega$ and strong convergence as we did here. Hence, we have to switch to yet other tools.

One possibility is to use the periodic unfolding operator. We want to introduce this technique to determine homogeneous limit functions in the next section.

## 3.2 The periodic unfolding method

The periodic unfolding method is – just like the two-scale convergence – a technique to homogenize a partial differential equation. The main idea is the development of an operator $\mathcal{T}_\varepsilon$ which maps a function $\varphi_\varepsilon$ defined on a finely structured, periodic domain $\Omega_\varepsilon \in \mathbb{R}^n$ to a function $\mathcal{T}_\varepsilon(\varphi_\varepsilon)$ defined on $\Omega \times Y$. With $\Omega \subset \mathbb{R}^n$ being homogeneous and $Y = [0,1]^n$ the unit cell, the domain of the function $\mathcal{T}_\varepsilon(\varphi_\varepsilon)$ is independent of $\varepsilon$ and hence, we are able to use well-known convergence results from functional analysis.

The periodic unfolding method was developed by D. Cioranescu, P. Donato, R. Zaki and others, see [13, 11, 10, 17]; the first basic idea was found by T. Arbogast, J. Douglas and U. Hornung in [5]. In this section, we briefly describe the setting and important results.

Let

$$\Omega_\varepsilon = \bigcup_{k \in \mathbb{Z}^n} \varepsilon(k + Y^*) \subset \mathbb{R}^n \text{ and } \Gamma_\varepsilon = \bigcup_{k \in \mathbb{Z}^n} \varepsilon(k + \Gamma) \subset \mathbb{R}^n \quad (3.8)$$

be a periodic domains with unit cell $Y = [0,1]^n$, $Y^* \subset Y$, and smooth manifolds $\Gamma \subset Y$ and $\Gamma_\varepsilon$, respectively. Further, let $\Omega \subset \mathbb{R}^n$.

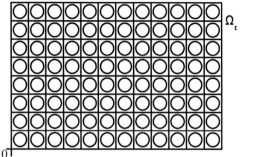
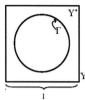

For every $z \in \mathbb{R}^n$ we define $[z]_Y$ as the bottom-left corner of the $\varepsilon$-sized square, where $z$ is located. Then we define

$$\{z\}_Y := z - [z]_Y \in \varepsilon Y^*,$$

such that

$$[z]_Y + \{z\}_Y = z \in \Omega_\varepsilon.$$

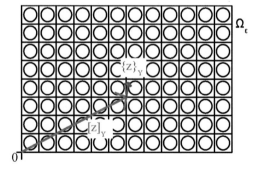
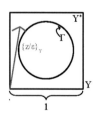

By scaling with $1/\varepsilon$ it follows that $\left\{\frac{z}{\varepsilon}\right\} \in [0,1]^n$ and

$$z - \varepsilon \left[\frac{z}{\varepsilon}\right]_Y = \varepsilon \left\{\frac{z}{\varepsilon}\right\}_Y.$$

79

## The periodic unfolding operator $\mathcal{T}_{\varepsilon}$

We define the periodic unfolding operator $\mathcal{T}_{\varepsilon}$ as follows.

**Definition 3.8** *Let $\varphi \in L^p(\Omega_{\varepsilon})$, $p \in [1, \infty]$. For any $\varepsilon > 0$ we define*

$$\mathcal{T}_{\varepsilon} \; : \; L^p(\Omega_{\varepsilon}) \to L^p(\Omega \times Y^*)$$

$$[\mathcal{T}_{\varepsilon}(\varphi)](x, y) := \varphi\left(\varepsilon\left[\frac{x}{\varepsilon}\right]_Y + \varepsilon y\right).$$

$\mathcal{T}_{\varepsilon}(\varphi)$ lives on the fixed domain $\Omega \times Y^*$ even for variating $\varepsilon$ and we do not need extensions from $\Omega_{\varepsilon}$ to $\Omega$ any more and may use convergence results from functional analysis.

Let $\Omega \subset \mathbb{R}^n$ be a bounded domain and $\Omega_{\varepsilon}$ as described in (3.8).

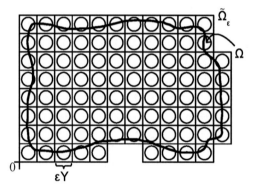

Then we define

$$\Lambda_{\varepsilon} := \{\zeta \in \mathbb{Z}^n \mid \varepsilon(\zeta + Y) \cap \Omega \neq \varnothing\} \text{ and } \tilde{\Omega}_{\varepsilon} := \text{int}\left(\bigcup_{\zeta \in \Lambda_{\varepsilon}} \varepsilon(\zeta + Y)\right).$$

In this setting we are able to obtain nice properties of the periodic unfolding operator $\mathcal{T}_{\varepsilon}$. We find this information in the paper [13]. We want to cite some of the important statements.

**Lemma 3.9** *For the periodic unfolding operator $\mathcal{T}_{\varepsilon}$ as defined in 3.8 the following assumptions hold true.*

1. $\mathcal{T}_\varepsilon$ is linear.

2. $\mathcal{T}_\varepsilon(\varphi)\left(x, \left\{\frac{x}{\varepsilon}\right\}_Y\right) = \varphi(x)$ for all $\varphi \in L^p(\Omega_\varepsilon)$.

3. $\mathcal{T}_\varepsilon(\varphi\psi) = \mathcal{T}_\varepsilon(\varphi)\mathcal{T}_\varepsilon(\psi)$ for all $\varphi, \psi \in L^p(\Omega_\varepsilon)$.

4. Let $\varphi$ in $L^p(Y^*)$ be an $Y$-periodic function. Set $\varphi_\varepsilon(x) = \varphi\left(\frac{x}{\varepsilon}\right)$. Then
$$\mathcal{T}_\varepsilon(\varphi_\varepsilon)(x,y) = \varphi(y).$$

5. For all $\varphi \in L^1(\Omega_\varepsilon)$ it holds that
$$\int_{\Omega_\varepsilon} \varphi dx = \frac{1}{|Y|} \int_{\tilde{\Omega}_\varepsilon \times Y^*} \mathcal{T}_\varepsilon(\varphi) dx dy.$$

6. If $\varphi \in L^2(\Omega_\varepsilon)$, then $\mathcal{T}_\varepsilon(\varphi) \in L^2(\tilde{\Omega}_\varepsilon \times Y^*)$.

7. If $\varphi \in L^2(\Omega_\varepsilon)$, then
$$\|\mathcal{T}_\varepsilon(\varphi)\|_{L^2(\tilde{\Omega}_\varepsilon \times Y^*)} = \sqrt{|Y|}\|\varphi\|_{L^2(\Omega_\varepsilon)}.$$

8. $\nabla_y \mathcal{T}_\varepsilon(\varphi)(x,y) = \varepsilon \mathcal{T}_\varepsilon(\nabla_x \varphi)(x,y)$.

9. If $\varphi \in H^1(\Omega_\varepsilon)$, then $\mathcal{T}_\varepsilon(\varphi) \in L^2(\tilde{\Omega}_\varepsilon, H^1(Y^*))$.

10. $\|\nabla_y \mathcal{T}_\varepsilon(\varphi)\|_{L^2(\tilde{\Omega}_\varepsilon \times Y^*)} = \varepsilon\sqrt{|Y|}\|\nabla_x \varphi\|_{L^2(\Omega_\varepsilon)}$.

The following propositions summarize some results about convergence of functions in various settings. In this section we use the notation $\tilde{\varphi} = \varphi$ in $\Omega$ and $\tilde{\varphi} = 0$ otherwise for a function $\varphi$.

**Proposition 3.10** Let $\varphi \in L^2(\Omega)$ and $Y^* \subset Y$. Then,

1. $\mathcal{T}_\varepsilon(\varphi) \to \tilde{\varphi}$ strongly in $L^2(\mathbb{R}^n \times Y)$.

2. $\varphi\chi^\varepsilon \to \theta\varphi$ weakly in $L^2(\Omega)$ where $\chi^\varepsilon$ is the characteristic function for $\bigcup_{k \in \mathbb{Z}^n} \varepsilon(k + Y^*)$ and $\theta := \frac{|Y^*|}{|Y|}$.

3. Let $\varphi_\varepsilon$ be in $L^2(\Omega)$ such that $\varphi_\varepsilon \to \varphi$ strongly in $L^2(\Omega)$. Then
$$\mathcal{T}_\varepsilon(\varphi_\varepsilon) \to \tilde{\varphi} \quad \text{strongly in } L^2(\mathbb{R}^n \times Y).$$

**Proposition 3.11** *For every* $\varepsilon > 0$ *let* $\varphi_\varepsilon$ *be in* $L^2(\Omega_\varepsilon)$, *such that* $\mathcal{T}_\varepsilon(\varphi_\varepsilon) \rightharpoonup \hat{\varphi}$ *weakly in* $L^2(\mathbb{R}^n \times Y^*)$. *Then*

$$\tilde{\varphi}_\varepsilon \rightharpoonup \frac{1}{|Y|} \int_{Y^*} \hat{\varphi}(\cdot, y) dy \quad \text{weakly in } L^2(\mathbb{R}^n).$$

The following proposition 3.12 and theorem 3.13 are the analogous statements in the periodic unfolding method to proposition 2.5 in the context of two-scale convergence.

**Proposition 3.12** *For every* $\varepsilon > 0$ *let* $\varphi_\varepsilon$ *be in* $L^2(\Omega_\varepsilon)$, *with*

$$\|\varphi_\varepsilon\|_{L^2(\Omega_\varepsilon)} \leq C,$$

$$\varepsilon \|\nabla_x \varphi_\varepsilon\|_{(L^2(\Omega_\varepsilon))^n} \leq C$$

*for* $C > 0$ *independent of* $\varepsilon$. *Then there exists* $\hat{\varphi} \in L^2(\mathbb{R}^n, H^1(Y^*))$ *such that, up to subsequences,*

$$\mathcal{T}_\varepsilon(\varphi_\varepsilon) \rightharpoonup \hat{\varphi} \quad \text{weakly in } L^2(\mathbb{R}^n, H^1(Y^*)),$$

*and*

$$\varepsilon \mathcal{T}_\varepsilon(\nabla_x \varphi_\varepsilon) \rightharpoonup \nabla_y \hat{\varphi} \quad \text{weakly in } L^2(\mathbb{R}^n \times Y^*),$$

*where*

$$y \mapsto \hat{\varphi}(\cdot, y) \quad \in L^2(\mathbb{R}^n, H^1_\#(Y^*)).$$

**Theorem 3.13** *For every* $\varepsilon > 0$ *let* $\varphi_\varepsilon$ *be in* $H^1(\Omega_\varepsilon)$, *with* $\|\varphi_\varepsilon\|_{H^1(\Omega_\varepsilon)}$ *bounded independently of* $\varepsilon$. *Then there exists* $\varphi \in H^1(\Omega)$ *and* $\hat{\varphi} \in L^2(\Omega, H^1_\#(Y^*))$ *such that, up to a subsequence,*

$$\mathcal{T}_\varepsilon(\varphi_\varepsilon) \rightharpoonup \varphi \quad \text{weakly in } L^2_{loc}(\Omega, H^1_\#(Y^*)),$$

$$\mathcal{T}_\varepsilon(\nabla_x \varphi_\varepsilon) \rightharpoonup \nabla_x \varphi + \nabla_y \hat{\varphi} \quad \text{weakly in } L^2_{loc}(\Omega, L^2(Y^*)).$$

With this information about the periodic unfolding operator we can calculate the homogeneous model of partial differential equations defined on $\Omega_\varepsilon$. If we have equations regarding the periodic boundary $\Gamma_\varepsilon$, as in the case (3.8), we need another operator working on that boundary.

# The boundary periodic unfolding operator $\mathcal{T}_{\varepsilon}^{b}$

We recall that $\Gamma \subset Y$ and $\Gamma_{\varepsilon} = \bigcup_{k \in \mathbb{Z}^n} \varepsilon(k + \Gamma)$ are smooth manifolds.

**Definition 3.14** *Let $\varphi \in L^p(\Gamma_{\varepsilon})$, $p \in [1, \infty]$. Then the boundary periodic unfolding operator $\mathcal{T}_{\varepsilon}^{b}$ is defined as*

$$\mathcal{T}_{\varepsilon}^{b} : L^p(\Gamma_{\varepsilon}) \to L^p(\mathbb{R}^n \times \Gamma)$$

$$\mathcal{T}_{\varepsilon}^{b}(\varphi)(x,y) = \varphi\left(\varepsilon \left[\frac{x}{\varepsilon}\right] + \varepsilon y\right)$$

*for all $x \in \mathbb{R}^n$ and $y \in \Gamma$.*

Again we cite some important properties of the boundary operator from the article [13].

**Lemma 3.15** *For the periodic unfolding operator $\mathcal{T}_{\varepsilon}^{b}$ as defined in definition 3.14 the following assumptions hold true.*

1. *$\mathcal{T}_{\varepsilon}^{b}$ is linear.*

2. *$\mathcal{T}_{\varepsilon}^{b}(\varphi)\left(x, \left\{\frac{x}{\varepsilon}\right\}_Y\right) = \varphi(x)$ for all $\varphi \in L^p(\Gamma_{\varepsilon})$ and $x \in \mathbb{R}^n$.*

3. *$\mathcal{T}_{\varepsilon}^{b}(\varphi \psi) = \mathcal{T}_{\varepsilon}^{b}(\varphi) \mathcal{T}_{\varepsilon}^{b}(\psi)$ for all $\varphi, \psi \in L^p(\Gamma_{\varepsilon})$.*

4. *Let $\varphi \in L^p(\Gamma_{\varepsilon})$ be a Y-periodic function. Set $\varphi_{\varepsilon}(x) = \varphi\left(\frac{x}{\varepsilon}\right)$. Then*
   $$\mathcal{T}_{\varepsilon}^{b}(\varphi_{\varepsilon})(x,y) = \varphi(y).$$

5. *For every $\varphi \in L^1(\Gamma_{\varepsilon})$, we have the integration formula*
   $$\int_{\Gamma_{\varepsilon}} \varphi(x) d\sigma_x = \frac{1}{\varepsilon|Y|} \int_{\tilde{\Omega}_{\varepsilon} \times \Gamma} \mathcal{T}_{\varepsilon}^{b}(\varphi)(x,y) dx d\sigma_y.$$

6. *For every $\varphi \in L^2(\Gamma_{\varepsilon})$ it holds that $\mathcal{T}_{\varepsilon}^{b}(\varphi) \in L^2(\tilde{\Omega}_{\varepsilon} \times \Gamma)$.*

7. *For every $\varphi \in L^2(\Gamma_{\varepsilon})$ we have*
   $$\|\mathcal{T}_{\varepsilon}^{b}(\varphi)\|_{L^2(\tilde{\Omega}_{\varepsilon} \times \Gamma)} = \sqrt{\varepsilon|Y|} \|\varphi\|_{L^2(\Gamma_{\varepsilon})}.$$

To use the boundary periodic unfolding operator for calculating the limit equations we need some propositions (see [13]).

**Proposition 3.16**    *1. Let $g \in L^2(\Gamma)$ and $\varphi \in H^1(\Omega)$. Then we have the estimate*

$$\left| \int_{\mathbb{R}^n \times \Gamma} g(y) T_\varepsilon^b(\varphi)(x,y) dx d\sigma_y \right| \leq C(|M_\Gamma(g)| + \varepsilon) \|\nabla \varphi\|_{L^2(\Omega_\varepsilon)},$$

*with $M_\Gamma(g) = \frac{1}{|\Gamma|} \int_\Gamma g(y) d\sigma_y$.*

2. *Let $g \in L^2(\Gamma)$ be a Y-periodic function, and set $g_\varepsilon(x) = g\left(\frac{x}{\varepsilon}\right)$. Then we have for all $\varphi \in H^1(\Omega)$ that*

$$\left| \int_{\Gamma_\varepsilon} g_\varepsilon(x) \varphi(x) d\sigma_x \right| \leq \frac{C}{\varepsilon} (|M_\Gamma(g)| + \varepsilon) \|\nabla \varphi\|_{L^2(\Omega_\varepsilon)}.$$

3. *Let $g \in L^2(\Gamma)$ be a Y-periodic function, and set $g_\varepsilon(x) = g\left(\frac{x}{\varepsilon}\right)$. If $\varepsilon$ tends to 0, we have the following results.*

   (a) *If $M_\Gamma(g) \neq 0$, then it holds for all $\varphi \in H^1(\Omega)$*

$$\varepsilon \int_{\Gamma_\varepsilon} g_\varepsilon(x) \varphi(x) d\sigma_x \xrightarrow{\varepsilon \to 0} \frac{|\Gamma|}{|Y|} M_\Gamma(g) \int_\Omega \varphi(x) dx.$$

   (b) *If $M_\Gamma(g) = 0$, then it holds for all $\varphi \in H^1(\Omega)$*

$$\int_{\Gamma_\varepsilon} g_\varepsilon(x) \varphi(x) d\sigma_x \xrightarrow{\varepsilon \to 0} 0.$$

The following proposition contains convergence results in $L^2$ and $H^1$. Later we will use it in the limit derivation in the application.

**Proposition 3.17**    *1. Let $\varphi \in L^2(\Omega)$. Then, as $\varepsilon \to 0$, we have the convergence*

$$\int_{\mathbb{R}^n \times \Gamma} T_\varepsilon^b(\varphi)(x,y) dx d\sigma_y \to \int_{\mathbb{R}^n \times \Gamma} \tilde{\varphi} dx d\sigma_y.$$

2. *Let $\varphi \in L^2(\Omega)$. Then*

$$T_\varepsilon^b(\varphi) \to \tilde{\varphi} \quad \text{strongly in } L^2(\mathbb{R}^n \times \Gamma).$$

84

3. *Let $\varphi_\varepsilon$ be in $L^2(\Gamma_\varepsilon)$, such that*

$$\mathcal{T}_\varepsilon^b(\varphi_\varepsilon) \rightharpoonup \hat{\varphi} \quad \text{weakly in } L^2(\mathbb{R}^n \times \Gamma).$$

*Then it holds for all $\psi \in H^1(\Omega)$*

$$\varepsilon \int_{\Gamma_\varepsilon} \varphi_\varepsilon \psi d\sigma_x \to \frac{1}{|Y|} \int_{\mathbb{R}^n \times \Gamma} \hat{\varphi}(x, y) \psi(x) dx d\sigma_y.$$

4. *For every $\varepsilon > 0$ let $\varphi_\varepsilon$ in $H^1(\Omega_\varepsilon)$ and $\hat{\varphi} \in H^1(\Omega)$ be such that*

$$\mathcal{T}_\varepsilon^b(\varphi_\varepsilon) \rightharpoonup \hat{\varphi} \quad \text{weakly in } L^2_{loc}(\Omega, H^1(Y)).$$

*Then*

$$\mathcal{T}_\varepsilon^b(\varphi_\varepsilon) \rightharpoonup \hat{\varphi} \quad \text{weakly in } L^2_{loc}(\Omega, H^{\frac{1}{2}}(\Gamma)).$$

The proofs of these results can be found in [13, 11, 10, 17]. The next results, however, have not been known before.

**Lemma 3.18 (Limit of function product)**
*Let $u_\varepsilon$, $v_\varepsilon \in L^2(\Gamma_\varepsilon)$. Let $\mathcal{T}_\varepsilon^b(u_\varepsilon)$ converge to $u_0$ weakly in $L^2(\Omega \times \Gamma)$ and let $\mathcal{T}_\varepsilon^b(v_\varepsilon)$ converge to $v_0$ strongly in $L^2(\Omega \times \Gamma)$. Then*

$$\mathcal{T}_\varepsilon^b(u_\varepsilon) \mathcal{T}_\varepsilon^b(v_\varepsilon) \rightharpoonup u_0 v_0$$

*weakly in $L^2(\Omega \times \Gamma)$.*

**Proof** The weak convergence for $u_\varepsilon$ implies for all $\varphi \in C^\infty(\Omega \times \Gamma)$ that

$$\lim_{\varepsilon \to 0} \int_{\Omega \times \Gamma} (\mathcal{T}_\varepsilon^b(u_\varepsilon) - u_0) \varphi d\sigma_y dx = 0.$$

The statement is equivalent, if we use test functions living in $L^2(\Omega \times \Gamma)$.

Then it yields for test functions $\varphi \in C^\infty(\Omega \times \Gamma)$

$$\lim_{\varepsilon \to 0} \int_{\Omega \times \Gamma} (\mathcal{T}_\varepsilon^b(u_\varepsilon)\mathcal{T}_\varepsilon^b(v_\varepsilon) - u_0 v_0) \varphi \, d\sigma_y dx$$

$$= \lim_{\varepsilon \to 0} \int_{\Omega \times \Gamma} (\mathcal{T}_\varepsilon^b(u_\varepsilon)\mathcal{T}_\varepsilon^b(v_\varepsilon)$$

$$- \mathcal{T}_\varepsilon^b(u_\varepsilon)v_0 + \mathcal{T}_\varepsilon^b(u_\varepsilon)v_0 - u_0 v_0) \varphi \, d\sigma_y dx$$

$$= \lim_{\varepsilon \to 0} \int_{\Omega \times \Gamma} \mathcal{T}_\varepsilon^b(u_\varepsilon)(\mathcal{T}_\varepsilon^b(v_\varepsilon) - v_0) \varphi \, d\sigma_y dx$$

$$+ \lim_{\varepsilon \to 0} \int_{\Omega \times \Gamma} (\mathcal{T}_\varepsilon^b(u_\varepsilon) - u_0) \underbrace{v_0 \varphi}_{\tilde{\varphi}} \, d\sigma_y dx$$

$$= \lim_{\varepsilon \to 0} \int_{\Omega \times \Gamma} \mathcal{T}_\varepsilon^b(u_\varepsilon)(\mathcal{T}_\varepsilon^b(v_\varepsilon) - v_0) \varphi \, d\sigma_y dx.$$

Here we use that $\tilde{\varphi} := v_0 \varphi \in L^2(\Omega \times \Gamma)$ can be used as test function as well. Now we continue with the absolute values of the limits.

$$\lim_{\varepsilon \to 0} \left| \int_{\Omega \times \Gamma} (\mathcal{T}_\varepsilon^b(u_\varepsilon)\mathcal{T}_\varepsilon^b(v_\varepsilon) - u_0 v_0) \varphi \, d\sigma_y dx \right|$$

$$= \lim_{\varepsilon \to 0} \left| \int_{\Omega \times \Gamma} \mathcal{T}_\varepsilon^b(u_\varepsilon)(\mathcal{T}_\varepsilon^b(v_\varepsilon) - v_0) \varphi \, d\sigma_y dx \right|$$

$$\leq \lim_{\varepsilon \to 0} \underbrace{\|\mathcal{T}_\varepsilon^b(u_\varepsilon)\varphi\|_{L^2(\Omega \times \Gamma)}}_{\text{bounded}} \underbrace{\|\mathcal{T}_\varepsilon^b(v_\varepsilon) - v_0\|_{L^2(\Omega \times \Gamma)}}_{\to 0}$$

$$= 0$$

for every $\varphi \in C^\infty(\Omega \times \Gamma)$ and the assertion holds true. $\square$

Now we regard differential equations with diffusion-terms on smooth manifolds. To do so we first have to describe a suitable setting.

Let $\Gamma \subset \mathbb{R}^n$ be a $k$-dimensional compact $C^\infty$-Riemannian manifold with Riemannian metric $g$. This means, we have an atlas $\{(U_\lambda, \alpha_\lambda) | \lambda \in \Lambda\}$ of charts on $\Gamma$ such that $\Gamma = \bigcup_\lambda U_\lambda$ and

$$\alpha_\lambda : U_\lambda \to V_\lambda \subset \mathbb{R}^k, \qquad \lambda \in \Lambda.$$

Then we conclude that $\Gamma_\varepsilon = \bigcup_{\xi \in \mathbb{Z}^n} \varepsilon(\Gamma + \xi)$ also is a Riemannian manifold with atlas $\{(U_{\lambda,\xi}^\varepsilon, \alpha_{\lambda,\xi}^\varepsilon) | \lambda \in \Lambda, \xi \in \mathbb{Z}^n\}$, where $U_{\lambda,\xi}^\varepsilon :=$

$\varepsilon(U_\lambda + \xi)$. That means $\Gamma_\varepsilon = \bigcup_{\lambda, \xi} U_{\lambda, \xi}^\varepsilon$ and

$$\alpha_{\lambda, \xi}^\varepsilon : U_{\lambda, \xi}^\varepsilon \to V_\lambda$$
$$\alpha_{\lambda, \xi}^\varepsilon(p) := \alpha_\lambda \left( \frac{1}{\varepsilon} p - \xi \right) \qquad p \in U_{\lambda, \xi}^\varepsilon, \quad \forall \lambda \in \Lambda, \, \xi \in \mathbb{Z}^n, \, \varepsilon > 0.$$

Obviously we have

$$\alpha_{\lambda, \xi}^\varepsilon(p) = \alpha_\lambda \left( \left\{ \frac{p}{\varepsilon} \right\}_Y \right) = \alpha_\lambda(y_p) \qquad p \in U_{\lambda, \xi}^\varepsilon, \quad \forall \lambda \in \Lambda, \, \xi \in \mathbb{Z}^n,$$

where $y_p := \left\{ \frac{p}{\varepsilon} \right\}_Y$. For the inverse of $\alpha_\lambda$ we have

$$\alpha_\lambda^{-1} : V_\lambda \to U_\lambda$$

$$\alpha_{\lambda, \xi}^{-1, \varepsilon} : V_\lambda \to U_{\lambda, \xi}^\varepsilon$$

$$\alpha_{\lambda, \xi}^{-1, \varepsilon}(z) = \varepsilon(\alpha_\lambda^{-1}(z) + \xi) = \pi_\xi^\varepsilon(\alpha_\lambda^{-1}(z))$$

with the function $\pi_\xi$ defined as

$$\pi_\xi^\varepsilon : \Gamma \to \Gamma_\varepsilon$$
$$\pi_\xi^\varepsilon(y) := \varepsilon(y + \xi), \qquad \xi \in \mathbb{Z}^n.$$

For any function $\varphi \in L^p(\Gamma_\varepsilon)$ the relation between $\varphi_\xi$ and $\mathcal{T}_\varepsilon^b$ is given by

$$\varphi(\pi_\xi^\varepsilon(y)) = \mathcal{T}_\varepsilon^b(\varphi)(\xi, y).$$

Now lets have a look at the tangential vectors $\frac{\mathrm{d}}{\mathrm{d}x^{i,\varepsilon}}$ on $\Gamma_\varepsilon$. Let $e_i$ be the $i$th basis vector in $\mathbb{R}^k$, $z = \alpha_{\lambda, \xi}^\varepsilon(p) \in V_\lambda$ and $t \in [-\delta, \delta]$, $\delta > 0$ small. Then $z = \alpha_\lambda(y_p)$ and

$$t \mapsto z + te_i$$

is a curve in $V_\lambda$. The relationship between tangential vectors on $\Gamma_\varepsilon$ and on $\Gamma$ in the point $p$ is given by

$$\frac{\mathrm{d}}{\mathrm{d}x^{i,\varepsilon}}(p) := \frac{\mathrm{d}}{\mathrm{d}t}_{|_{t=0}} \alpha_{\lambda, \xi}^{-1, \varepsilon}(z + te_i) = \frac{\mathrm{d}}{\mathrm{d}t}_{|_{t=0}} \varepsilon \alpha_\lambda^{-1}(z + te_i) = \varepsilon \frac{\mathrm{d}}{\mathrm{d}x^i}(y_p).$$

Next we have a look at the Riemannian metrics $g_{ij}$ and $g_{ij}^\varepsilon$. We have

$$g_{ij}^\varepsilon(p) = \left\langle \frac{d}{dx^{i,\varepsilon}}(p), \frac{d}{dx^{j,\varepsilon}}(p) \right\rangle = \left\langle \varepsilon \frac{d}{dx^i}(y_p), \varepsilon \frac{d}{dx^j}(y_p) \right\rangle = \varepsilon^2 g_{ij}(y_p),$$

which yields $\qquad g^{ij,\varepsilon}(p) = \dfrac{1}{\varepsilon^2} g^{ij}(y_p)$

for $i, j = 1, \ldots, k$.

The vectors $\left\{ \frac{d}{dx^i}(y_p) \right\}_{i=1\ldots k}$ form a basis of the tangent space

$$T_{y_p}\Gamma = \{ v \in \mathbb{R}^n : \text{ there is } \varepsilon > 0$$
$$\text{and } \gamma : (-\varepsilon, \varepsilon) \to \Gamma \text{ with } \gamma(0) = y_p, \ \dot{\gamma}(0) = v \}.$$

Within this setting we want to deduce some assertions. The first one is an extension to the fact that $\nabla_y \mathcal{T}_\varepsilon(\varphi_\varepsilon) = \varepsilon \mathcal{T}_\varepsilon(\nabla_x \varphi_\varepsilon)$ for functions $\varphi_\varepsilon \in H^1(\Omega_\varepsilon)$, see [13], to functions on $H^1(\Gamma_\varepsilon)$.

**Lemma 3.19** *Let $\varphi$ be in $H^1(\Gamma_\varepsilon)$. Then*

$$\varepsilon \mathcal{T}_\varepsilon^b(\nabla_x \varphi) = \nabla_y \mathcal{T}_\varepsilon^b(\varphi).$$

**Proof** In the proof we neglect the $\lambda$ or $\xi$ dependence of the charts $\alpha_\lambda$ and $\alpha_{\lambda,\xi}^\varepsilon$. We just claim to take the appropriate chart for any subset $U_\lambda \subset \Gamma$ and $U_{\lambda,\xi}^\varepsilon \subset \Gamma_\varepsilon$, respectively. In the setting of Riemannian manifolds the gradient $\nabla_x \varphi$ is defined as

$$\nabla_x \varphi(p) = \sum_{ij} g^{ij\varepsilon}(p) \frac{\partial \varphi}{\partial x^{j,\varepsilon}}(p) \frac{d}{dx^{i,\varepsilon}}(p), \qquad (3.9)$$

with

$$\frac{\partial \varphi}{\partial x^{i,\varepsilon}}(p) := \frac{\partial(\varphi \circ \alpha^{-1,\varepsilon})}{\partial x_i}(\alpha^\varepsilon(p)).$$

Here $x_i$, $i = 1, \ldots k$ denote the components of $\mathbb{R}^k$. Applying $\mathcal{T}_\varepsilon^b$ to

$\frac{\partial \varphi}{\partial x^{j,\varepsilon}}$ leads to

$$
\begin{aligned}
\mathcal{T}_\varepsilon^b \left( \frac{\partial \varphi}{\partial x^{j,\varepsilon}} \right) (p, y_p) &= \mathcal{T}_\varepsilon^b \left( \frac{\partial (\varphi \circ \alpha^{-1,\varepsilon})}{\partial x_j} \circ \alpha^\varepsilon \right) (p, y_p) \\
&= \left( \frac{\partial (\varphi \circ \alpha^{-1,\varepsilon})}{\partial x_j} \circ \alpha^\varepsilon \right) ([p]_Y + \varepsilon y_p) \\
&= \frac{\partial (\varphi \circ \alpha^{-1,\varepsilon})}{\partial x_j} \alpha(y_p) = \frac{\partial (\varphi \circ \pi \circ \alpha^{-1})}{\partial x_j} \alpha(y_p) \\
&= \frac{\partial_y (\mathcal{T}_\varepsilon^b(\varphi) \circ \alpha^{-1})}{\partial x_j} \alpha(y_p) = \frac{\partial_y \mathcal{T}_\varepsilon^b(\varphi)}{\partial x^j} (p, y_p).
\end{aligned}
$$

Putting the pieces together we get

$$
\begin{aligned}
\mathcal{T}_\varepsilon^b (\nabla_x \varphi)(p, y_p) &= \mathcal{T}_\varepsilon^b \left( \sum_{ij} g^{ij\varepsilon} \frac{\partial \varphi}{\partial x^{j,\varepsilon}} \frac{\mathrm{d}}{\mathrm{d}x^{i,\varepsilon}} \right) (p, y_p) \\
&= \sum_{ij} \frac{1}{\varepsilon^2} g^{ij}(y_p) \mathcal{T}_\varepsilon^b \left( \frac{\partial \varphi}{\partial x^{j,\varepsilon}} \right) (p, y_p) \varepsilon \frac{\mathrm{d}}{\mathrm{d}x^i}(y_p) \\
&= \frac{1}{\varepsilon} \sum_{ij} g^{ij}(y_p) \frac{\partial_y \mathcal{T}_\varepsilon^b(\varphi)}{\partial x^j} (p, y_p) \frac{\mathrm{d}}{\mathrm{d}x^i}(y_p) \\
&= \frac{1}{\varepsilon} \nabla_y (\mathcal{T}_\varepsilon^b(\varphi))(p, y_p).
\end{aligned}
$$

We conclude the proof with

$$
\nabla_y (\mathcal{T}_\varepsilon^b(\varphi)) = \varepsilon \mathcal{T}_\varepsilon^b (\nabla_x \varphi).
$$

$\square$

The following two lemmas easily follow.

**Lemma 3.20** Let $\varphi$ be in $H^1(\Gamma_\varepsilon)$. Then

$$
\|\nabla_y \mathcal{T}_\varepsilon^b(\varphi)\|^2_{L^2(\mathbb{R}^n \times \Gamma)} = |Y| \varepsilon^3 \|\nabla_x \varphi\|^2_{L^2(\Gamma_\varepsilon)}.
$$

**Proof** With property 5 of lemma 3.15 we have

$$\|\nabla_x \varphi\|^2_{L^2(\Gamma_\varepsilon)} = \int_{\Gamma_\varepsilon} \nabla_x \varphi \nabla_x \varphi d\sigma_x$$

$$= \frac{1}{\varepsilon|Y|} \int_{\mathbb{R}^n \times \Gamma} T^b_\varepsilon(\nabla_x \varphi) T^b_\varepsilon(\nabla_x \varphi) dx d\sigma_y$$

$$= \frac{1}{\varepsilon|Y|} \int_{\mathbb{R}^n \times \Gamma} \frac{1}{\varepsilon} \nabla_y T^b_\varepsilon(\varphi) \frac{1}{\varepsilon} \nabla_y T^b_\varepsilon(\varphi) dx d\sigma_y$$

$$= \frac{1}{\varepsilon^3 |Y|} \|\nabla_y T^b_\varepsilon(\varphi)\|^2_{L^2(\mathbb{R}^n \times \Gamma)}$$

and the assertion follows. $\qquad\qquad\qquad\qquad\qquad\qquad\qquad\square$

**Lemma 3.21** *Let* $\varphi_\varepsilon \in H^1(\Gamma_\varepsilon)$, *then* $T^b_\varepsilon(\varphi_\varepsilon) \in L^2(\Omega, H^1(\Gamma))$.

**Proof** Since $\varphi_\varepsilon \in H^1(\Gamma_\varepsilon)$, it holds that

$$\varepsilon\|\varphi_\varepsilon\|^2_{L^2(\Gamma_\varepsilon)} + \varepsilon\|\nabla_x \varphi_\varepsilon\|^2_{L^2(\Gamma_\varepsilon)} \leq C(\varepsilon)$$

for a $C(\varepsilon) > 0$. Because $\nabla_y T^b_\varepsilon(\varphi_\varepsilon) = \varepsilon T^b_\varepsilon(\nabla_x \varphi_\varepsilon)$ we have for small $\varepsilon < 1$

$$\|T^b_\varepsilon(\varphi_\varepsilon)\|^2_{L^2(\Omega, H^1(\Gamma))} = \int_\Omega \int_\Gamma T^b_\varepsilon(\varphi_\varepsilon)^2 + (\nabla_y T^b_\varepsilon(\varphi_\varepsilon))^2 d\sigma_y dx$$

$$= \int_{\Omega \times \Gamma} T^b_\varepsilon(\varphi_\varepsilon)^2 d\sigma_y dx + \int_{\Omega \times \Gamma} (\nabla_y T^b_\varepsilon(\varphi_\varepsilon))^2 d\sigma_y dx$$

$$= |Y|\varepsilon \|\varphi_\varepsilon\|^2_{L^2(\Gamma_\varepsilon)} + |Y|\varepsilon^3 \|\nabla_x \varphi_\varepsilon\|^2_{L^2(\Gamma_\varepsilon)}$$

$$\leq |Y|\varepsilon \|\varphi_\varepsilon\|^2_{L^2(\Gamma_\varepsilon)} + |Y|\varepsilon \|\nabla_x \varphi_\varepsilon\|^2_{L^2(\Gamma_\varepsilon)}$$

$$\leq |Y|C(\varepsilon).$$

$$\qquad\qquad\qquad\qquad\qquad\qquad\qquad\qquad\qquad\qquad\square$$

**Lemma 3.22** *Let* $\varphi_\varepsilon \in H^1(\Gamma_\varepsilon)$ *be bounded for every* $\varepsilon$ *such that*

$$\varepsilon\|\varphi_\varepsilon\|^2_{L^2(\Gamma_\varepsilon)} \leq C \quad and \quad \varepsilon^3 \|\nabla_x \varphi_\varepsilon\|^2_{L^2(\Gamma_\varepsilon)} \leq C,$$

*for* $C > 0$ *independent of* $\varepsilon$.
*Then there exists* $\hat{\varphi} \in L^2(\mathbb{R}^n, H^1(\Gamma))$ *such that, up to a subsequence,*

$$T^b_\varepsilon(\varphi_\varepsilon) \rightharpoonup \hat{\varphi} \quad weakly\ in\ L^2(\mathbb{R}^n, H^1(\Gamma))$$

and

$$\varepsilon \mathcal{T}_\varepsilon^b(\nabla_x \varphi_\varepsilon) \rightharpoonup \nabla_y \hat{\varphi} \quad \text{weakly in } L^2(\mathbb{R}^n \times \Gamma)$$

with

$$y \mapsto \hat{\varphi}(\cdot, y) \quad \in L^2(\mathbb{R}^n, H^1_\#(\Gamma)).$$

**Proof** We use the statement, that in a reflexive Banach space a bounded sequence has a weakly converging subsequence, see Satz 6.9 in [4]. Hence, we need to show, that $\mathcal{T}_\varepsilon^b(\varphi_\varepsilon)$ is bounded in $L^2(\mathbb{R}^n, H^1(\Gamma))$.
With property 7 of lemma 3.15 we get

$$\frac{1}{|Y|} \| \mathcal{T}_\varepsilon^b(\varphi_\varepsilon) \|^2_{L^2(\mathbb{R}^n \times \Gamma)} = \varepsilon \| \varphi_\varepsilon \|^2_{L^2(\Gamma_\varepsilon)} \leq C$$

and with the lemma 3.20

$$\frac{1}{|Y|} \| \nabla_y \mathcal{T}_\varepsilon^b(\varphi_\varepsilon) \|^2_{L^2(\mathbb{R}^n \times \Gamma)} = \varepsilon^3 \| \nabla_x \varphi_\varepsilon \|^2_{L^2(\Gamma_\varepsilon)} \leq C.$$

Hence, $\mathcal{T}_\varepsilon^b(\varphi_\varepsilon)$ is bounded in $L^2(\mathbb{R}^n \times \Gamma)$ and $\nabla_y \mathcal{T}_\varepsilon^b(\varphi)$ is bounded in $L^2(\mathbb{R}^n \times \Gamma)$.
It follows that $\mathcal{T}_\varepsilon^b(\varphi_\varepsilon)$ is bounded in $L^2(\mathbb{R}^n, H^1(\Gamma))$ and there exists $\hat{\varphi} \in L^2(\mathbb{R}^n, H^1(\Gamma))$ such that

$$\mathcal{T}_\varepsilon^b(\varphi_\varepsilon) \rightharpoonup \hat{\varphi} \quad \text{weakly in } L^2(\mathbb{R}^n, H^1(\Gamma)).$$

With lemma 3.19 we conclude

$$\varepsilon \mathcal{T}_\varepsilon^b(\nabla_x \varphi_\varepsilon) = \nabla_y \mathcal{T}_\varepsilon^b(\varphi_\varepsilon) \rightharpoonup \nabla_y \hat{\varphi} \quad \text{weakly in } L^2(\mathbb{R}^n \times \Gamma).$$

It is left to show that $\hat{\varphi}$ is $Y$-periodic. To that end, let $\psi \in C^\infty(\mathbb{R}^n \times \Gamma)$ be periodic in its second argument. Then, for $\xi \in \mathbb{Z}^n$

$$\int_{\mathbb{R}^n \times \Gamma} (\mathcal{T}_\varepsilon^b(\varphi_\varepsilon)(x, y + \xi) - \mathcal{T}_\varepsilon^b(\varphi_\varepsilon)(x, y)) \psi(x, y) dx d\sigma_y$$

$$= \int_{\mathbb{R}^n \times \Gamma} \left( \varphi_\varepsilon \left( \varepsilon \left[ \frac{x}{\varepsilon} \right]_Y + \varepsilon y + \varepsilon \xi \right) \right.$$
$$\left. - \varphi_\varepsilon \left( \varepsilon \left[ \frac{x}{\varepsilon} \right]_Y + \varepsilon y \right) \right) \psi(x, y) dx d\sigma_y$$

$$= \int_{\mathbb{R}^n \times \Gamma} \left( \varphi_\varepsilon \left( \varepsilon \left( \left[ \frac{x}{\varepsilon} \right]_Y + \xi \right) + \varepsilon y \right) \right.$$
$$\left. - \varphi_\varepsilon \left( \varepsilon \left[ \frac{x}{\varepsilon} \right]_Y + \varepsilon y \right) \right) \psi(x,y) \mathrm{d}x \mathrm{d}\sigma_y$$
$$= \int_{\mathbb{R}^n \times \Gamma} \varphi_\varepsilon \left( \varepsilon \left[ \frac{x}{\varepsilon} \right]_Y + \varepsilon y \right) (\psi(x - \varepsilon\xi, y) - \psi(x,y)) \mathrm{d}x \mathrm{d}\sigma_y.$$

Since $\psi(x - \varepsilon\xi, y) \to \psi(x,y)$ for $\varepsilon$ tending to zero, we finally conclude that

$$\int_{\mathbb{R}^n \times \Gamma} T_\varepsilon^b(\varphi_\varepsilon)(x, y + \xi)\psi(x,y)\mathrm{d}x \mathrm{d}\sigma_y$$
$$- \int_{\mathbb{R}^n \times \Gamma} T_\varepsilon^b(\varphi_\varepsilon)(x,y)\psi(x,y)\mathrm{d}x \mathrm{d}\sigma_y \xrightarrow{\varepsilon \to 0} 0.$$

$\square$

For Sobolev spaces with fractional index we use the following notation. For $\Omega \subset \mathbb{R}^n$ and $s = m + \lambda$ with $m \in \mathbb{N}_0$ and $0 < \lambda < 1$ we set

$$|u|_{H^s(\Omega)}^2 = \sum_{|\alpha|=m} \int_\Omega \int_\Omega \frac{|D^\alpha u(x) - D^\alpha u(y)|^2}{|x-y|^{n+2\lambda}} \mathrm{d}x\mathrm{d}y$$
$$\|u\|_{H^s(\Omega)}^2 = \sum_{0 \le \alpha \le m} \|u\|_{H^\alpha(\Omega)}^2 + |u|_{H^s(\Omega)}^2.$$

For functions on the hypersurface $\Gamma \subset \mathbb{R}^n$ we define

$$|u|_{H^s(\Gamma)}^2 = \sum_{|\alpha|=m} \int_\Gamma \int_\Gamma \frac{|D^\alpha u(x) - D^\alpha u(y)|^2}{|x-y|^{n-1+2\lambda}} \mathrm{d}\sigma_x\mathrm{d}\sigma_y$$
$$\|u\|_{H^s(\Gamma)}^2 = \sum_{0 \le \alpha \le m} \|u\|_{H^\alpha(\Gamma)}^2 + |u|_{H^s(\Gamma)}^2.$$

Before we can state this next new result, a theorem is needed. Therefore, we cite a theorem about traces (Theorem 2.2.2 Joel Feldmann, [22]).

**Theorem 3.23** *Let $1/2 < l \le 3/2$ and let $\Omega$ be a bounded subset of $\mathbb{R}^n$ with smooth boundary. There exists a unique map*

$$\gamma : H^l(\Omega) \to H^{l-1/2}(\partial\Omega)$$

$$u \mapsto u_{|\partial\Omega}$$

and constants $C_\gamma, C'_\gamma$ such that

- $\gamma$ is bounded with $\|\gamma u\|_{H^{l-1/2}(\Omega)} \leq C_\gamma \|u\|_{H^l(\Omega)}$.

- For each $f \in H^{l-1/2}(\partial\Omega)$ there is a $u \in H^l(\Omega)$ such that

$$\gamma u = f \qquad \text{and} \qquad \|u\|_{H^l(\Omega)} \leq C'_\gamma \|f\|_{H^{l-1/2}(\partial\Omega)}. \qquad (3.10)$$

That means, $\gamma$ has a bounded right inverse.

With help of theorem 3.23 we can derive a trace estimation for periodic domains. In the paper [42], a trace inequality regarding the spaces $H^{1/2}(\Gamma_\varepsilon)$ and $H^1(\Omega)$ is derived.
We are going to extend this result.

**Lemma 3.24** Let $\delta \in [0, 1/2)$. If $u \in H^{1-\delta}(\Omega_\varepsilon)$, then for the trace $\gamma(u) \in H^{1/2-\delta}(\Gamma_\varepsilon)$ the inequality

$$\varepsilon\|\gamma(u)\|^2_{L^2(\Gamma_\varepsilon)} + \varepsilon^{2-n}|\gamma(u)|^2_{H^{1/2-\delta}(\Gamma_\varepsilon)}$$
$$\leq C\left(\|u\|^2_{L^2(\Omega_\varepsilon)} + \varepsilon^{-n}|u|^2_{H^{1-\delta}(\Omega_\varepsilon)}\right),$$

for a constant $C < 0$, independent of $\varepsilon$, holds true.

**Proof** We know from theorem 3.23 that there is a constant $C > 0$ such that

$$|\gamma(u)|^2_{L^2(\Gamma)} + |\gamma(u)|^2_{H^{1/2-\delta}(\Gamma)} \leq C\left(|u|^2_{L^2(Y)} + |u|^2_{H^{1-\delta}(Y)}\right)$$

which means

$$\int_\Gamma \gamma(u)^2 d\sigma_y + \int_\Gamma \int_\Gamma \frac{|\gamma(u)(y_1) - \gamma(u)(y_2)|^2}{|y_1 - y_2|^{n-2\delta}} d\sigma_{y_1} d\sigma_{y_2}$$
$$\leq C\left(\int_Y u^2 dy + \int_Y \int_Y \frac{|u(y_1) - u(y_2)|^2}{|y_1 - y_2|^{n+1-2\delta}} dy_1 dy_2\right).$$

We scale the inequality by setting $y = x/\varepsilon$

$$\int_{\varepsilon\Gamma} \gamma(u) \left(\frac{x}{\varepsilon}\right)^2 \frac{d\sigma_x}{\varepsilon^{n-1}} + \int_{\varepsilon\Gamma} \int_{\varepsilon\Gamma} \frac{\left|\gamma(u)\left(\frac{x_1}{\varepsilon}\right) - \gamma(u)\left(\frac{x_2}{\varepsilon}\right)\right|^2}{\left|\frac{x_1-x_2}{\varepsilon}\right|^{n-2\delta}} \frac{d\sigma_{x_1}}{\varepsilon^{n-1}} \frac{d\sigma_{x_2}}{\varepsilon^{n-1}}$$

$$\leq C \left( \int_{\varepsilon Y} u\left(\frac{x}{\varepsilon}\right)^2 \frac{dx}{\varepsilon^n} + \int_{\varepsilon Y} \int_{\varepsilon Y} \frac{\left|u\left(\frac{x_1}{\varepsilon}\right) - u\left(\frac{x_2}{\varepsilon}\right)\right|^2}{\left|\frac{x_1-x_2}{\varepsilon}\right|^{n+1-2\delta}} \frac{dx_1}{\varepsilon^n} \frac{dx_2}{\varepsilon^n} \right).$$

By multiplying with $\varepsilon^n$ we obtain

$$\varepsilon \int_{\varepsilon\Gamma} \gamma(u) \left(\frac{x}{\varepsilon}\right) dx + \varepsilon^{2-n} \int_{\varepsilon\Gamma} \int_{\varepsilon\Gamma} \frac{\left|\gamma(u)\left(\frac{x_1}{\varepsilon}\right) - \gamma(u)\left(\frac{x_2}{\varepsilon}\right)\right|^2}{\left|\frac{x_1-x_2}{\varepsilon}\right|^{n-2\delta}} d\sigma_{x_1} d\sigma_{x_2}$$

$$\leq C \left( \int_{\varepsilon Y} u\left(\frac{x}{\varepsilon}\right)^2 dx + \varepsilon^{-n} \int_{\varepsilon Y} \int_{\varepsilon Y} \frac{\left|u\left(\frac{x_1}{\varepsilon}\right) - u\left(\frac{x_2}{\varepsilon}\right)\right|^2}{\left|\frac{x_1-x_2}{\varepsilon}\right|^{n+1-2\delta}} dx_1 dx_2 \right).$$

The assertion now follows by summing all $\varepsilon Y^k \subset \Omega_\varepsilon$ and $\varepsilon\Gamma^k \subset \Gamma_\varepsilon$. $\qquad \square$

For other trace inequalities see [15] and [41].
A theorem in a similar setting allows us to deduce global diffusion after homogenization on a manifold $\Gamma_\varepsilon$, given some technical assumptions. Now we prepare this theorem 3.27 with the following theorem 3.25 and lemma 3.26.
First we cite the theorem 3.25 proven by Peter Li in [38].

**Theorem 3.25** *Let $M$ be a $k$-dimensional compact $C^\infty$-Riemannian manifold. Then there exists a $C > 0$, such that*

$$\|f\|^2_{L^2(M)} \leq C \|\nabla f\|^2_{L^2(M)}$$

*for all $f \in H^1(M)$ with $\int_M f d\sigma_x = 0$.*

This means that the Poincaré inequality also holds on Riemannian manifolds. Furthermore, we show an inverse trace theorem in the following lemma 3.26.

**Lemma 3.26** *Let $\Omega \subset \mathbb{R}^n$ and $\Gamma \subset [0,1]^n = Y$ be a smooth and compact hypersurface such that $\Gamma_\varepsilon = \bigcup_{k\in\mathbb{Z}^n} \varepsilon(k + \Gamma) \cap \Omega$ is a periodic,*

*smooth and compact hypersurface. Let $f_\varepsilon \in H^1(\Gamma_\varepsilon)$.*
*Then, there exists a function $u_\varepsilon \in H^1(\Omega)$ with $u_{\varepsilon|_{\Gamma_\varepsilon}} = f_\varepsilon$ such that*

$$\|u_\varepsilon\|^2_{L^2(\Omega)} \le C_1\varepsilon\|f_\varepsilon\|^2_{L^2(\Gamma_\varepsilon)} \quad and \quad \|\nabla u_\varepsilon\|^2_{L^2(\Omega)} \le C_2\varepsilon\|\nabla_\Gamma f_\varepsilon\|^2_{L^2(\Gamma_\varepsilon)}$$

*for constants $C_1, C_2 > 0$ independent of $\varepsilon$.*

**Proof** For small but fixed $\delta > 0$ we define $Y^* = \{y + dn_y|\ y \in \Gamma,\ d \in (-\delta,\delta)\}$, where $n_y$ is the normal in the point $y \in \Gamma$. Because $\Gamma$ is smooth and compact, $n_y$ is well-defined and there is a $\delta > 0$ such that for every $z \in Y^*$ there exist unique $y \in \Gamma$ and $d \in (-\delta,\delta)$ with $y + dn_y = x$.
On the tube $Y^*$ we define a Riemannian metric by $g_{ij}$, $i,j = 1,\ldots n$, such that the tangential vectors $\frac{\mathrm{d}}{\mathrm{d}y^i}$, $i = 1,\ldots,n-1$ form a basis of the tangent space $T_y\Gamma$ and $\frac{\mathrm{d}}{\mathrm{d}y^n}$ treats the normal direction $n_y$. Because $n_y$ is orthogonal to the tangent space $T_y\Gamma$, it follows $g_{in} = g_{ni} = g^{in} = g^{ni} = 0$ for $i = 1,\ldots,n$.

Now we define $\Omega^*_\varepsilon = \bigcup_{k\in\mathbb{Z}^n} \varepsilon(k + Y^*) \subset \Omega$ and consider the scaled unit cell $\varepsilon Y$ with scaled tube $\varepsilon Y^*$. The width of $\varepsilon Y^*$ is now $2\delta\varepsilon$ and $d \in (-\varepsilon\delta, \varepsilon\delta)$.
Analogously one finds for every $x \in \Omega^*_\varepsilon$ unique $y \in \Gamma_\varepsilon$ and $d \in (-\varepsilon\delta, \varepsilon\delta)$ such that $y + dn_y = x$.
For small $\delta > 0$ one can estimate

$$|\Omega^*_\varepsilon| \le 2\varepsilon\delta c_1|\Gamma_\varepsilon|$$

for a constant $c_1 > 0$, independent of $\varepsilon$.
We define a function $\tilde{u}_\varepsilon \in H^1(\Omega^*_\varepsilon)$ by

$$\tilde{u}_\varepsilon(x) = \tilde{u}_\varepsilon(y + dn_y) = f_\varepsilon(y) \quad \text{for every } x \in \Omega^*_\varepsilon.$$

Then it holds

$$\begin{aligned}
\|\tilde{u}_\varepsilon\|^2_{L^2(\Omega^*_\varepsilon)} = \int_{\Omega^*_\varepsilon} \tilde{u}^2_\varepsilon(x)\mathrm{d}x &= \int_{\Omega^*_\varepsilon} \tilde{u}^2_\varepsilon(y + dn_y)\mathrm{d}x \\
&\le 2\varepsilon\delta c_1 \int_{\Gamma_\varepsilon} f^2_\varepsilon(y)\mathrm{d}\sigma_y \\
&= 2\delta\varepsilon c_1\|f_\varepsilon\|^2_{L^2(\Gamma_\varepsilon)}.
\end{aligned}$$

To estimate the gradient, we consider the gradient in the coordinates $\frac{d}{dy^{i,\varepsilon}}$, $i = 1, \ldots, n-1$ on $T_y \Gamma_\varepsilon$, and exploit that $f_\varepsilon(y)$ is independent of $d$,

$$\nabla_x \tilde{u}_\varepsilon(x) = \sum_{ij=1}^{n} g^{ij,\varepsilon}(x) \frac{\partial \tilde{u}_\varepsilon}{\partial y^{j,\varepsilon}}(y + dn_y) \frac{d}{dy^{i,\varepsilon}}$$

$$= \sum_{ij=1}^{n} g^{ij,\varepsilon}(x) \frac{\partial f_\varepsilon}{\partial y^{j,\varepsilon}}(y) \frac{d}{dy^{i,\varepsilon}} = \sum_{ij=1}^{n-1} g^{ij,\varepsilon}(x) \frac{\partial f_\varepsilon}{\partial y^{j,\varepsilon}}(y) \frac{d}{dy^{i,\varepsilon}}$$

for every $x \in \Omega_\varepsilon^*$ with $x = y + dn_y$, $y \in \Gamma_\varepsilon$. Since the Riemannian metric tensor $g^{ij,\varepsilon}$ is continuous and $\Gamma_\varepsilon$ compact, there exists a constant $c_2$ such that for small $\delta > 0$

$$|\nabla_x \tilde{u}_\varepsilon(x)|^2 \leq c_2 |\sum_{ij=1}^{n-1} g^{ij,\varepsilon}(y) \frac{\partial f_\varepsilon}{\partial y^{j,\varepsilon}}(y) \frac{d}{dy^{i,\varepsilon}}|^2 = c_2 |\nabla_\Gamma f_\varepsilon(y)|^2.$$

Now the norm of the gradient $u_\varepsilon$ can be estimated.

$$\|\nabla \tilde{u}_\varepsilon\|_{L^2(\Omega_\varepsilon^*)}^2 = \int_{\Omega_\varepsilon^*} |\nabla \tilde{u}_\varepsilon(x)|^2 dx = \int_{\Omega_\varepsilon^*} |\nabla_x u_\varepsilon(y + dn_y)|^2 dx$$

$$\leq c_2 \int_{\Omega_\varepsilon^*} |\nabla_\Gamma f_\varepsilon(y)|^2 dx$$

$$\leq c_1 c_2 2\varepsilon\delta \int_{\Gamma_\varepsilon} |\nabla_\Gamma f_\varepsilon(y)|^2 dy$$

$$= 2\delta\varepsilon c_1 c_2 \|\nabla_\Gamma f_\varepsilon\|_{L^2(\Gamma_\varepsilon)}^2.$$

We continue with the extension operator from the article [41] for connected sets $\Omega_\varepsilon$ and ensure an extended function $u_\varepsilon \in H^1(\Omega)$ such that

$$\|u_\varepsilon\|_{L^2(\Omega)}^2 \leq C \|\tilde{u}_\varepsilon\|_{L^2(\Omega_\varepsilon)}^2$$
$$\|\nabla u_\varepsilon\|_{L^2(\Omega)}^2 \leq C \|\nabla \tilde{u}_\varepsilon\|_{L^2(\Omega_\varepsilon)}^2$$

which completes the proof with the constants $C_1 = 2\delta c_1 C$ and $C_2 = 2\delta c_1 c_2 C$. $\qquad\square$

We are able to use these assertions in lemma 3.26 to deduce a statement about the $\varepsilon$-limit derivation of a function living on a manifold $\Gamma_\varepsilon$ which is bounded in $H^1(\Gamma_\varepsilon)$.

**Theorem 3.27** *Let $\varphi_\varepsilon \in H^1(\Gamma_\varepsilon)$ be a sequence of functions with*

$$\varepsilon\|\varphi_\varepsilon\|^2_{L^2(\Gamma_\varepsilon)} + \varepsilon\|\nabla_\Gamma\varphi_\varepsilon\|^2_{L^2(\Gamma_\varepsilon)} \leq C.$$

*Let $P_\Gamma$ be the orthogonal projection from $\mathbb{R}^n$ to the Tangent Space $T_y\Gamma$ for every $y \in \Gamma$.*
*Then two assertions hold true:*

1. *There exists a function $\varphi_0 \in H^1(\Omega)$ such that, up to a subsequence,*
$$\mathcal{T}_\varepsilon^b(\varphi_\varepsilon) \rightharpoonup \varphi_0 \quad \text{weakly in } L^2(\Omega \times \Gamma)$$
*and*
$$\varepsilon \int_{\Gamma_\varepsilon} \varphi_\varepsilon \psi d\sigma_x \to \frac{|\Gamma|}{|Y|} \int_\Omega \varphi_0 \psi dx$$
*for all $\psi \in C^\infty(\Omega)$.*

2. *There exists a $\hat{\varphi} \in L^2(\Omega, H^1_\#(\Gamma))$ such that, up to a subsequence,*
$$\mathcal{T}_\varepsilon^b(\nabla_x\varphi_\varepsilon) \rightharpoonup P_\Gamma\nabla_x\varphi_0 + \nabla_\Gamma\hat{\varphi} \quad \text{weakly in } L^2(\Omega \times \Gamma).$$

**Proof**
*1.* We use lemma 3.26 to deduce the existence of a function $\tilde{\varphi}_\varepsilon \in H^1(\Omega)$ such that $\gamma(\varphi_\varepsilon) = \tilde{\varphi}_\varepsilon|_{\Gamma_\varepsilon} = \varphi_\varepsilon$ and

$$\|\tilde{\varphi}_\varepsilon\|_{L^2(\Omega)} + \|\nabla\tilde{\varphi}_\varepsilon\|^2_{L^2(\Omega)} \leq c_1\varepsilon\left(\|\varphi_\varepsilon\|^2_{L^2(\Gamma_\varepsilon)} + \|\nabla_\Gamma\varphi_\varepsilon\|^2_{L^2(\Gamma_\varepsilon)}\right) \leq C.$$

Hence, $\tilde{\varphi}_\varepsilon$ has a weak limit function $\varphi_0$ in $H^1(\Omega)$. We calculate with theorem 3.13

$$\int_{\Omega \times Y} \mathcal{T}_\varepsilon(\tilde{\varphi}_\varepsilon)(x,y)\psi(x,y)\mathrm{d}x\mathrm{d}y \xrightarrow{\varepsilon \to 0} \int_{\Omega \times Y} \varphi_0(x)\psi(x,y)\mathrm{d}y\mathrm{d}x$$

weakly in $L^2(\Omega, H^1_\#(Y))$ for all $\psi \in C^\infty(\Omega \times Y)$.
The trace operator $\gamma_{\Omega \times \Gamma} : L^2(\Omega, H^1(Y)) \to L^2(\Omega, L^2(\Gamma))$ defined by $\gamma_{\Omega \times \Gamma}(\varphi) = \varphi|_{\Omega \times \Gamma}$ commutes with $\mathcal{T}_\varepsilon$ as follows. Let $\psi \in H^1(\Omega_\varepsilon)$, then

$$\mathcal{T}_\varepsilon^b(\gamma(\psi))(x,y) = \mathcal{T}_\varepsilon^b\left(\psi|_{\Gamma_\varepsilon}\right)(x,y) = \psi(\varepsilon\left[\frac{x}{\varepsilon}\right]_Y + \varepsilon\underbrace{y}_{\in \Gamma})$$

$$= \mathcal{T}_\varepsilon(\psi)|_{\Omega \times \Gamma}(x,y) = \gamma_{\Omega \times \Gamma}(\mathcal{T}_\varepsilon(\psi))(x,y).$$

97

Hence, it follows that

$$\int_{\Omega \times \Gamma} \mathcal{T}_\varepsilon^b(\varphi_\varepsilon)(x,y)\psi(x,y)\mathrm{d}\sigma_y\mathrm{d}x$$

$$= \int_{\Omega \times \Gamma} \mathcal{T}_\varepsilon^b(\gamma(\tilde{\varphi}_\varepsilon))(x,y)\psi(x,y)\mathrm{d}\sigma_y\mathrm{d}x$$

$$= \int_{\Omega \times \Gamma} \gamma_{\Omega \times \Gamma}(\mathcal{T}_\varepsilon(\tilde{\varphi}_\varepsilon))(x,y)\psi(x,y)\mathrm{d}\sigma_y\mathrm{d}x$$

$$\to \int_{\Omega \times \Gamma} \gamma_{\Omega \times \Gamma}(\varphi_0)(x,y)\psi(x,y)\mathrm{d}\sigma_y\mathrm{d}x$$

for all $\psi \in C^\infty(\Omega \times \Gamma)$.

This is true, because $\mathcal{T}_\varepsilon(\tilde{\varphi}_\varepsilon)$ converges weakly in $H^1_\#(Y)$ in its second variable and the trace operator is linear and continuous. The trace in the limit regards only this second variable, hence we have weak convergence in $L^2_\#(Y)$ in the second variable.

Because $\varphi_0$ is independent of $y$ this leads to

$$\int_{\Omega \times \Gamma} \mathcal{T}_\varepsilon^b(\varphi_\varepsilon)(x,y)\psi(x,y)\mathrm{d}\sigma_y\mathrm{d}x \to \int_{\Omega \times \Gamma} \varphi_0(x)\psi(x,y)\mathrm{d}\sigma_y\mathrm{d}x$$

for all $\psi \in C^\infty(\Omega \times \Gamma)$.

This yields

$$\varepsilon \int_{\Gamma_\varepsilon} \varphi_\varepsilon \psi \mathrm{d}\sigma_x \to \frac{|\Gamma|}{|Y|} \int_\Omega \varphi_0 \psi \mathrm{d}x$$

for all $\psi \in C^\infty(\Omega)$.

2. To prove the second part of the theorem, we at first need some definitions and properties.

We define for every $a \in \mathbb{R}^n$ the function $z_a : \Gamma \to \mathbb{R}$ by

$$z_a(y) = a^T \cdot y \qquad \text{for every } y \in \Gamma.$$

By using the directional derivative on $\Gamma$ we find

$$\langle \nabla_\Gamma z_a, v \rangle = dz_a(v) = \frac{\mathrm{d}}{\mathrm{d}t}\Big|_{t=0} z_a(\gamma(t))$$

for $\gamma : (-\delta, \delta) \to \Gamma, \gamma(0) = y$ and $\dot{\gamma}(0) = v$. We find that

$$\frac{\mathrm{d}}{\mathrm{d}t}\Big|_{t=0} z_a(\gamma(t)) = \frac{\mathrm{d}}{\mathrm{d}t}\Big|_{t=0} a^T \cdot \gamma(t) = a^T \cdot \dot{\gamma}(t) = a^T \cdot v$$

for all $v \in T_y\Gamma$. The only element of $T_y\Gamma$ satisfying $\langle \nabla_\Gamma z_a, v \rangle = a^T \cdot v$ for all $v \in T_y\Gamma$ is the orthogonal projection of $a$ to $T_y\Gamma$. Hence, $\nabla_\Gamma z_a = P_\Gamma a$.

We define

$$z_a^c(y) = z_a(y) - \frac{1}{|\Gamma|} \int_\Gamma z_a(y) \mathrm{d}\sigma_y.$$

Then for any vector $a \in \mathbb{R}^n$ it is true that $\int_\Gamma z_a^c(y) \mathrm{d}\sigma_y = 0$. Further we define

$$M_\varepsilon^b(\varphi_\varepsilon) = \frac{1}{|\Gamma|} \int_\Gamma \mathcal{T}_\varepsilon^b(\varphi_\varepsilon)(x, y) \mathrm{d}\sigma_y$$

and

$$Z_\varepsilon^b = \frac{1}{\varepsilon} \left( \mathcal{T}_\varepsilon^b(\varphi_\varepsilon) - M_\varepsilon^b(\varphi_\varepsilon) \right).$$

Then we deduce

$$\int_\Gamma Z_\varepsilon^b \mathrm{d}\sigma_y = 0 \qquad \text{and} \qquad \nabla_\Gamma Z_\varepsilon^b = \frac{1}{\varepsilon} \nabla_\Gamma \mathcal{T}_\varepsilon^b(\varphi_\varepsilon) = \mathcal{T}_\varepsilon^b(\nabla_x \varphi_\varepsilon),$$

since $M_\varepsilon^b(\varphi_\varepsilon)$ is independent of $y$.

Now the mainpart of the proof of part 2 starts. We consider

$$\|Z_\varepsilon^b(\varphi_\varepsilon) - z_{\nabla_x \varphi_0}^c\|_{L^2(\Gamma \times \Omega)}^2.$$

For the $y$-component the $x$-gradient $\nabla_x \varphi_0$ looks like a vector in $\mathbb{R}^n$, because $\varphi_0$ is independent of $y$. We use the Poincaré inequality on Riemannian manifolds (see lemma 3.25) to see

$$\|Z_\varepsilon^b(\varphi_\varepsilon) - z_{\nabla_x \varphi_0}^c\|_{\Omega \times \Gamma}^2 \leq \int_\Omega C \|\nabla_\Gamma Z_\varepsilon^b(\varphi_\varepsilon) - \nabla_\Gamma z_{\nabla_x \varphi_0}^c\|_\Gamma^2 \mathrm{d}x$$

$$= C \int_\Omega \|\mathcal{T}_\varepsilon^b(\nabla_x \varphi_\varepsilon) - P_\Gamma \nabla_x \varphi_0\|_\Gamma^2 \mathrm{d}x$$

$$\leq C \|\mathcal{T}_\varepsilon^b(\nabla_x \varphi_\varepsilon)\|_{\Gamma \times \Omega}^2 + C \|P_\Gamma \nabla_x \varphi_0\|_{\Omega \times \Gamma}^2$$

$$\leq C \varepsilon |Y| \|\nabla_x \varphi_\varepsilon\|_{\Gamma_\varepsilon}^2 + C |\Gamma| \underbrace{\|P_\Gamma\|}_{\leq 1} \|\nabla_x \varphi_0\|_\Omega^2 \leq \tilde{C}$$

for a constant $\tilde{C} > 0$ independent of $\varepsilon$. Hence, $Z_\varepsilon^b(\varphi_\varepsilon) - z_{\nabla_x \varphi_0}^c$ converges weakly to a function $\hat\varphi \in L^2(\Omega, H^1(\Gamma))$, up to a subsequence

$$Z_\varepsilon^b(\varphi_\varepsilon) \rightharpoonup \hat\varphi + z_{\nabla_x \varphi_0}^c,$$

$$\mathcal{T}_\varepsilon^b(\nabla_x \varphi_\varepsilon) = \nabla_\Gamma Z_\varepsilon^b(\varphi_\varepsilon) \rightharpoonup \nabla_\Gamma \hat\varphi + P_\Gamma \nabla_x \varphi_0.$$

Last we need to show that $\hat{\varphi} \in L^2(\Omega, H^1_{\#}(\Gamma))$, this means that $\tilde{\varphi}$ is $Y$-periodic in its second argument. Therefore we label the faces $[0,1]^{i-1} \times 0 \times [0,1]^{n-i}$ of the unit cell $Y$ by $\partial_i Y$ for $i = 1, \ldots, n$.

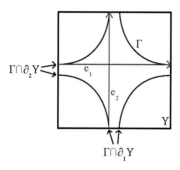

We extend the functions from $\Omega$ to $\mathbb{R}^n$ by zero and take a look at the equation

$$\int_{\mathbb{R}^n} \int_{\Gamma \cap \partial_i Y} [Z^b_\varepsilon(\varphi_\varepsilon)(x, y + e_i) - Z^b_\varepsilon(\varphi_\varepsilon)(x,y)] \psi(x,y) d\sigma_y dx$$

$$= \int_{\mathbb{R}^n} \int_{\Gamma \cap \partial_i Y} \frac{1}{\varepsilon} [T^b_\varepsilon(\varphi_\varepsilon)(x, y + e_i) - T^b_\varepsilon(\varphi_\varepsilon)(x,y)] \psi(x,y) d\sigma_y dx$$

$$= \int_{\mathbb{R}^n} \int_{\Gamma \cap \partial_i Y} \frac{1}{\varepsilon} [\varphi_\varepsilon([x]_Y + \varepsilon e_i + \varepsilon y) - \varphi_\varepsilon([x]_Y + \varepsilon y)] \psi(x,y) d\sigma_y dx$$

$$= \int_{\mathbb{R}^n} \int_{\Gamma \cap \partial_i Y} \frac{1}{\varepsilon} \varphi_\varepsilon([x]_Y + \varepsilon y) \psi(x - \varepsilon e_i, y) d\sigma_y dx$$

$$- \int_{\mathbb{R}^n} \int_{\Gamma \cap \partial_i Y} \frac{1}{\varepsilon} \varphi_\varepsilon([x]_Y + \varepsilon y) \psi(x,y) d\sigma_y dx$$

$$= \int_{\mathbb{R}^n} \int_{\Gamma \cap \partial_i Y} T^b_\varepsilon(\varphi_\varepsilon)(x,y) \frac{\psi(x - \varepsilon e_i, y) - \psi(x,y)}{\varepsilon} d\sigma_y dx$$

$$\xrightarrow{\varepsilon \to 0} \int_{\mathbb{R}^n} \int_{\Gamma \cap \partial_i Y} \gamma(\varphi_0) \left( -\frac{d\psi_0}{dx_i} \right) d\sigma_y dx$$

$$= \int_{\mathbb{R}^n} \int_{\Gamma \cap \partial_i Y} \frac{d\varphi_0}{dx_i} \psi d\sigma_y dx$$

$$= \int_{\mathbb{R}^n} \int_{\Gamma \cap \partial_i Y} e_i^T \cdot \nabla_x \varphi_0 \psi d\sigma_y dx$$

for all $\psi \in C^\infty(\mathbb{R}^n \times \Gamma)$.

Because of the extension by zero on $\mathbb{R}^n$ we conclude for the domain $\Omega$

$$\int_\Omega \int_{\Gamma \cap \partial_i Y} [Z_\varepsilon^b(\varphi_\varepsilon)(x, y + e_i) - Z_\varepsilon^b(\varphi_\varepsilon)(x, y)] \psi(x, y) d\sigma_y dx$$

$$\to \int_\Omega \int_{\Gamma \cap \partial_i Y} e_i^T \cdot \nabla_x \varphi_0 \psi d\sigma_y dx.$$

On the other hand we calculate

$$\int_\Omega \int_{\Gamma \cap \partial_i Y} [z_{\nabla_x \varphi_0}^c(y + e_i) - z_{\nabla_x \varphi_0}^c(y)] \psi(x, y) d\sigma_y dx$$

$$= \int_\Omega \int_{\Gamma \cap \partial_i Y} [\nabla_x \varphi_0 \cdot (y + e_i) - \nabla_x \varphi_0 \cdot y] \psi d\sigma_y dx$$

$$= \int_\Omega \int_{\Gamma \cap \partial_i Y} e_i^T \cdot \nabla_x \varphi_0 \psi d\sigma_y dx$$

for all $\psi \in C^\infty(\Omega \times \Gamma)$. With $Z_\varepsilon^b(\varphi_\varepsilon) \rightharpoonup \hat{\varphi} + z_{\nabla_x \varphi_0}^c$ we conclude that

$$\int_\Omega \int_{\Gamma \times \partial_i Y} [\hat{\varphi}(x, y + e_i) - \hat{\varphi}(x, y)] \psi(x, y) d\sigma_y dx = 0$$

for all $\psi \in C^\infty(\Omega \times \Gamma)$ and $i = 1, \ldots, n$. So $\hat{\varphi} \in L^2(\Omega, H_\#^1(\Gamma))$ and the proof is finished. $\square$

This theorem 3.27 can be used to derive global diffusion on a manifold for a partial differential equation using homogenization. With the following example we are going to show how to use the theorem 3.27.

**Example 3.28** Let $\Gamma \subset [0, 1]^n$ be a smooth, $n - 1$ dimensional manifold, such that $\Gamma_\varepsilon = \bigcup_{k \in \mathbb{Z}^n} \varepsilon(k + \Gamma)$ is periodic, connected and smooth. Let $\Omega \subset \mathbb{R}^n$ be bounded, $f \in C(\Omega, C_\#(Y))$ with $f_\varepsilon(x) := f(x, x/\varepsilon)$, and $D_\varepsilon(x) = D(x, x/\varepsilon)$ be an elliptic diffusion tensor on the tangent space of $T_y \Gamma_\varepsilon$, which is $\varepsilon$-periodic in its second argument and it holds that $\lim_{\varepsilon \to 0} \|D_\varepsilon\|_{\Omega_\varepsilon}^2 = \|D\|_{\Omega \times \Gamma}^2$ is bounded.

101

Further, let $u_\varepsilon$ be the solution of the partial differential equation

$$\partial_t u_\varepsilon(x,t) - \nabla_\Gamma \cdot (D_\varepsilon(x)\nabla_\Gamma u_\varepsilon(x,t)) + u_\varepsilon(x,t) = f_\varepsilon(x,t) \quad \text{on } \Gamma_\varepsilon$$
$$u_\varepsilon(x,t) = 0 \quad \text{on } \partial\Omega.$$
$$(3.11)$$

We multiply the weak formulation with $\varepsilon$ and get

$$\varepsilon \int_{\Gamma_\varepsilon} \partial_t u_\varepsilon(x,t)\psi_\varepsilon(x)\mathrm{d}\sigma_x + \varepsilon \int_{\Gamma_\varepsilon} D_\varepsilon(x)\nabla_\Gamma u_\varepsilon(x,t)\nabla_\Gamma \psi_\varepsilon(x)\mathrm{d}\sigma_x$$
$$+ \varepsilon \int_{\Gamma_\varepsilon} u_\varepsilon(x,t)\psi_\varepsilon(x)\mathrm{d}\sigma_x = \varepsilon \int_{\Gamma_\varepsilon} f_\varepsilon(x,t)\psi_\varepsilon(x)\mathrm{d}\sigma_x.$$

With standard estimations we find

$$\varepsilon\|u_\varepsilon\|^2_{L^2(\Gamma_\varepsilon)} + \varepsilon\|\sqrt{D_\varepsilon}\nabla_\Gamma u_\varepsilon\|^2_{L^2([0,t]\times\Gamma_\varepsilon)} + \varepsilon\|u_\varepsilon\|^2_{L^2([0,t]\times\Gamma_\varepsilon)} \leq C$$

and therefore the conditions to use lemma 3.27 are satisfied. Using the boundary unfolding operator leads to

$$\int_\Omega \int_\Gamma \partial_t \mathcal{T}_\varepsilon^b(u_\varepsilon)(x,y,t)\mathcal{T}_\varepsilon^b(\psi_\varepsilon)(x,y)\mathrm{d}\sigma_y\mathrm{d}x$$
$$+ \int_\Omega \int_\Gamma \mathcal{T}_\varepsilon^b(D_\varepsilon)(x,y)\mathcal{T}_\varepsilon^b(\nabla_\Gamma u_\varepsilon)(x,y,t)\mathcal{T}_\varepsilon^b(\nabla_\Gamma \varphi_\varepsilon)(x,y)\mathrm{d}\sigma_y\mathrm{d}x$$
$$+ \int_\Omega \int_\Gamma \mathcal{T}_\varepsilon^b(u_\varepsilon)(x,y,t)\mathcal{T}_\varepsilon^b(\psi_\varepsilon)(x,y)\mathrm{d}\sigma_y\mathrm{d}x$$
$$= \int_\Omega \int_\Gamma \mathcal{T}_\varepsilon^b(f_\varepsilon)(x,y,t)\mathcal{T}_\varepsilon^b(\psi_\varepsilon)(x,y)\mathrm{d}\sigma_y\mathrm{d}x.$$

Now we use theorem 3.27 and find for $\varepsilon \to 0$ and $|Y| = 1$

$$|\Gamma| \int_\Omega \partial_t u_0(x,t)\psi_0(x)\mathrm{d}x$$
$$+ \int_\Omega \int_\Gamma D(x,y)[\nabla_\Gamma \hat{u}(x,y,t) + P_\Gamma \nabla_x u_0(x)]$$
$$[\nabla_\Gamma \hat{\psi}(x,y) + P_\Gamma \nabla_x \psi_0(x)]\mathrm{d}\sigma_y\mathrm{d}x$$
$$+ |\Gamma| \int_\Omega u_0(x,t)\psi_0(x)\mathrm{d}x = \int_\Omega \int_\Gamma f_0(x,y,t)\psi_0(x)\mathrm{d}\sigma_y\mathrm{d}x.$$

To determine the cell problem we first set $\psi_0 = 0$,

$$\int_\Omega \int_\Gamma D(x,y)[\nabla_\Gamma \hat{u}(x,y,t) + P_\Gamma \nabla_x u_0(x,t)]\nabla_\Gamma \hat{\psi}(x,y)\mathrm{d}\sigma_y\mathrm{d}x = 0.$$

Let

$$\hat{u}(x,y,t) = \sum_{i=1}^{n} \partial_{x_i} u_0 \chi_i(y,t) \text{ and } \nabla_\Gamma \hat{u}(x,y,t) = \sum_{i=1}^{n} \partial_{x_i} u_0 \nabla_\Gamma \chi_i(y,t)$$

for some $\chi_i(y,t) : \Gamma \times [0,T] \to \mathbb{R}$, $i = 1,\dots,n$. Then we continue

$$\int_\Omega \int_\Gamma D(x,y) \left[ \sum_{i=1}^{n} \partial_{x_i} u_0 \nabla_\Gamma \chi_i(y,t) + \sum_{i=1}^{n} \partial_{x_i} u_0 P_\Gamma e_i \right] \nabla_\Gamma \hat{\psi}(y,t) d\sigma_y dx$$

$$= \int_\Omega \sum_{i=1}^{n} \partial_{x_i} u_0 \int_\Gamma D(x,y)(\nabla_\Gamma \chi_i(y,t) + P_\Gamma e_i) \nabla_\Gamma \hat{\psi}(x,y) d\sigma_y dx = 0$$

for all $\hat{\psi} \in C^\infty(\Omega, C^\infty_\#(\Gamma))$. Hence, the strong formulation of the cell problem is given by

$$\begin{array}{rll} -\nabla_\Gamma \cdot D(x,y)(\nabla_\Gamma \chi_i(y,t) + P_\Gamma e_i) = & 0 & \text{in } \Gamma, \\ D(x,y)(\nabla_\Gamma \chi_i(y,t) + P_\Gamma e_i) \cdot n = & 0 & \text{on } \partial\Gamma. \end{array} \tag{3.12}$$

and $\chi_i$ $Y$-periodic for all $i = 1,\dots,n$. This equation is well-defined, since $D$ maps elements of the tangent space $T_y\Gamma$ into the tangent space $T_y\Gamma$.

To find the diffusion tensor we set $\hat{\psi} = 0$. Then,

$$|\Gamma| \int_\Omega \partial_t u_0(x,t)\psi_0(x)dx$$

$$+ \int_\Omega \sum_{i=1}^{n} \partial_{x_i} u_0 \int_\Gamma D(x,y)(\nabla_\Gamma \chi_i(y,t) + P_\Gamma e_i) P_\Gamma \nabla_x \psi_0(x) d\sigma_y dx$$

$$+ |\Gamma| \int_\Omega u_0(x,t)\psi_0(x)dx = \int_\Omega \int_\Gamma f_0(x,y,t)\psi_0(x)d\sigma_y dx,$$

$$|\Gamma| \int_\Omega \partial_t u_0(x,t)\psi_0(x)dx$$

$$+ \int_\Omega \sum_{i,j=1}^{n} \partial_{x_i} u_0 \int_\Gamma \left( P_\Gamma^T D(x,y)(\nabla_\Gamma \chi_i(y,t) + P_\Gamma e_i) \right)_j d\sigma_y \partial_{x_j} \varphi_0 dx$$

$$+ |\Gamma| \int_\Omega u_0(x,t)\psi_0(x)dx = \int_\Omega \int_\Gamma f_0(x,y,t)\psi_0(x)d\sigma_y dx,$$

for all $\psi_0 \in C^\infty(\Omega)$.

In the finite dimensional case, every orthogonal Projection is symmetric $P_\Gamma^T = P_\Gamma$ and since $D(x,y)(\nabla_\Gamma \chi_i(y,t) + P_\Gamma e_i)$ is already in the tangent space $T_y\Gamma$, we drop the first $P_\Gamma$. With

$$s_{ij}(x) = \int_\Gamma (D(x,y)(\nabla_\Gamma \chi_j(y,t) + P_\Gamma e_j))_i \, d\sigma_y, \quad S = (s_{ij})_{i,j=1,\dots,n}$$

(3.13)

we write the strong formulation

$$\begin{aligned} |\Gamma|\partial_t u_0 - \nabla_x \cdot (S(x)\nabla_x u_0) + |\Gamma|u_0 &= \int_\Gamma f d\sigma_y && \text{in } \Omega \\ u_0 &= 0 && \text{on } \partial\Omega. \end{aligned}$$

(3.14)

**Remark 3.29** In section 4.1 one finds an indirect way to come to the same result as in this example 3.28 using theorem 3.27. Different from here, in section 4.1 a equation is defined on a very thin domain with width $\delta > 0$ and positive volume. Homogenization is performed using well-known results from G. Allaire in [2], and afterwards in the homogenized unit cell, the parameter $\delta$ tends to zero. The domain transforms to a smooth manifold for $\delta \to 0$ and there is global diffusion on a manifold in the limit equation. More details are found in remark 4.3.

**Remark 3.30** The time derivative $\partial_t u_\varepsilon \in L^2([0,T], H^1(\Omega_\varepsilon)')$ of functions $u_\varepsilon \in L^2([0,T] \times \Omega_\varepsilon)$ causes no trouble for the periodic unfolding method. It holds that

$$\mathcal{T}_\varepsilon(\partial_t u_\varepsilon)(t,x,y) = \partial_t u_\varepsilon \left(t, \left[\frac{x}{\varepsilon}\right]_Y + \varepsilon y\right) = \partial_t \mathcal{T}_\varepsilon(u_\varepsilon)(t,x,y).$$

Moreover, if $u_\varepsilon$ is a bounded sequence in $L^2([0,T] \times \Omega_\varepsilon)$, we know from lemma 3.9 that $\mathcal{T}_\varepsilon(u_\varepsilon)$ ist a bounded sequence in $L^2([0,T] \times \Omega \times Y)$ and there exists a weak converging subsequence with limit $u_0 \in L^2([0,T] \times \Omega \times Y)$. Testing with functions $\varphi \in C^\infty([0,T] \times$

$\Omega \times Y$) and integration by parts in the time variable yields

$$(\partial_t \mathcal{T}_\varepsilon(u_\varepsilon), \varphi)_{[0,T] \times \Omega \times Y}$$
$$= - (\mathcal{T}_\varepsilon(u_\varepsilon), \partial_t \varphi)_{[0,T] \times \Omega \times Y} - (\mathcal{T}_\varepsilon(u_\varepsilon)(0), \varphi(0))_{\Omega \times Y}$$
$$+ (\mathcal{T}_\varepsilon(u_\varepsilon)(T), \varphi(T))_{\Omega \times Y}$$
$$\overset{\varepsilon \to 0}{\to} - (u_0, \partial_t \varphi)_{[0,T] \times \Omega \times Y} - (u_0(0), \varphi(0))_{\Omega \times Y} + (u_0(T), \varphi(T))_{\Omega \times Y}$$
$$= (\partial_t u_0, \varphi)_{[0,T] \times \Omega \times Y}$$

Thus, we deduce that $\partial_t \mathcal{T}_\varepsilon(u_\varepsilon) \rightharpoonup \partial_t u_0$ if $\mathcal{T}_\varepsilon(u_\varepsilon) \rightharpoonup u_0$ in $L^2([0,T] \times \Omega \times Y)$.

## 3.3   Nonlinear carcinogenesis problem

With the tools introduced in the previous sections, the nonlinear homogenization and the periodic unfolding method, we are now prepared to improve the linear model introduced in section 2.2 by adding nonlinear terms.

Some important processes can only be formulated using nonlinearities, for example events that can not exceed a saturation value, or metabolisms that follow the law of mass action. We extend the system of differential equations (2.3) of chapter 2 with three different kinds of nonlinear terms:

1. At first we include cleaning mechanisms as we already mentioned in section 2.2. In the cytosol of a human cell there are molecules that can bind to BP or DE and render them harmless. Examples of such molecules are glutathione epoxide transferase or sulfo transferase (transferase is an enzyme), see [24]. They bind to dangerous and alien molecules and render them water-soluble. Then they can be evacuated. These cleaning enzymes are available only in limited quantities. With only a few BP or DE molecules, we assume that the cleaning is almost linear. If there are many molecules the cleaning rate will reach a threshold. The following function is suitable to describe this behavior.

$$f: \quad \mathbb{R} \to \mathbb{R}_0^+$$
$$f(x) = \begin{cases} \frac{x}{x+M} Ma & \text{for } x \geq 0 \\ 0 & \text{for } x < 0, \end{cases} \tag{3.15}$$

where $M > 0$ and $a > 0$. The function is almost linear for small $x$, because $\frac{M}{x+M} \approx 1$. Then, the number $a$ denotes the slope and the function $f$ can be approximated by

$$f(x) \approx ax \quad \text{for small } x \geq 0.$$

The greater the parameter $M > 0$ the greater is the threshold of the function $f$ for $x \to \infty$. The threshold is $Ma$. When we use this function in the nonlinear system of equations below, the variable $x$ is going to be replaced by the concentration of BP molecules in cytosol $u_\varepsilon$.
Mathematically $f$ is a nice function:

$$f \in C^\infty(\mathbb{R}^+) \quad \text{and} \quad f \geq 0 \quad \text{and} \quad f \in L^\infty(\mathbb{R}^+).$$

Using this function $f$ for the cleaning mechanisms the equation for BP molecules in the system (2.3) changes to

$$\partial_t u_\varepsilon - D_u \Delta u_\varepsilon = -f(u_\varepsilon) \quad \text{in } \Omega_\varepsilon.$$

We need the minus on the right-hand side because we have a loss in BP molecules.

We use this function also for the cleaning of DE molecules in the cytosol. We take different parameters $b > 0$ and $L > 0$ for the cleaning of DE molecules because the rate of cleaning could be different, and define

$$g : \mathbb{R} \to \mathbb{R}_0^+$$

$$g(x) = \begin{cases} \frac{x}{x+L} Lb & \text{for } x \geq 0 \\ 0 & \text{for } x < 0. \end{cases}$$

When we use this function in the nonlinear system of equations below, the variable $x$ is going to be replaced by the concentration of DE molecules in cytosol $v_\varepsilon$. Then the equation for the DE molecules in the system (2.3) becomes

$$\partial_t v_\varepsilon - D_v \Delta v_\varepsilon = -g(v_\varepsilon) \quad \text{in } \Omega_\varepsilon.$$

2. The second improvement of the original model (2.3) is a change of the way we model the transformation from BP molecules to DE molecules. In the linear equation (2.3) we assumed a linear rate. Now we consider a nonlinear term for this metabolism, because the enzymes, which induce the transformation, are also available only in limited quantities. We use the same functional form we just applied for the cleaning mechanisms.

We denote the function that describes the rate of the metabolism by

$$h : \mathbb{R} \rightarrow \mathbb{R}_0^+$$

$$h(x) = \begin{cases} \frac{x}{x+K}Kc & \text{for } x \geq 0 \\ 0 & \text{for } x < 0, \end{cases}$$

with $c > 0$ and $K > 0$. The threshold is $Kc$ as $x$ goes to $\infty$. When we use this function in the nonlinear system of equations below, the variable $x$ is going to be replaced by the concentration of BP molecules on the surface of the ER $s_\varepsilon$.

The equations on the surface of the endoplasmic reticulum in the system (2.3) transform to

$$\partial_t s_\varepsilon - \varepsilon^2 D_s \Delta_\Gamma s_\varepsilon = -h(s_\varepsilon) + l_s(u_\varepsilon - s_\varepsilon)$$
$$\partial_t w_\varepsilon - \varepsilon^2 D_w \Delta_\Gamma w_\varepsilon = h(s_\varepsilon) + l_w(v_\varepsilon - w_\varepsilon).$$

3. The last improvement is to include receptors on the surface of the endoplasmic reticulum. Molecules bind to a membrane by connecting to receptors that are attached to the membrane. BP molecules in the cytosol, denoted by $u_\varepsilon$, can only transform to BP molecules bound to the surface of the ER, denoted by $s_\varepsilon$, when they find a free receptor. The maximal number of receptors is denoted by $\overline{R}$. We define the amount of free receptors as the function $R_\varepsilon$

$$R_\varepsilon : \Gamma_\varepsilon^{ER} \times [0, T] \rightarrow [0, \overline{R}]$$

which satisfies the ordinary differential equation

$$\partial_t R_\varepsilon = -R_\varepsilon |k_u u_\varepsilon + k_v v_\varepsilon| + (\overline{R} - R_\varepsilon)|k_s s_\varepsilon + k_w w_\varepsilon| \qquad \text{on } \Gamma_\varepsilon^{ER}$$
$$R_\varepsilon(x, 0) = \overline{R}.$$

With $R_\varepsilon(x,0) = \overline{R}$ all receptors are free initially. If BP molecules $u_\varepsilon$ or DE molecules $v_\varepsilon$ bind to the surface of the endoplasmic reticulum with rate $k_u$ or $k_v$, $R_\varepsilon$ decreases. If BP molecules $s_\varepsilon$ or DE molecules $w_\varepsilon$ leave the surface of the endoplasmic reticulum with rate $l_s$ or $l_w$, then $R_\varepsilon$ increases. The factors $k_s > 0$ and $k_w > 0$ are multiples of $l_s$, $l_w$, respectively, and provide that $k_s(\overline{R} - R_\varepsilon)$ and $k_w(\overline{R} - R_\varepsilon)$ are rates.

This also leads to a different Robin boundary condition for $u_\varepsilon$ and $v_\varepsilon$ and for another equation for $s_\varepsilon$ and $w_\varepsilon$ on $\Gamma_\varepsilon^{\text{ER}}$ in the system of equations (2.3). BP and DE molecules only bind to the surface of the ER, if a free receptor meets the molecule. Following the law of mass action, this means the product between $u_\varepsilon$ and $R_\varepsilon$ or $v_\varepsilon$ and $R_\varepsilon$, respectively.

$$
\left.
\begin{aligned}
-D_u \nabla u_\varepsilon \cdot n &= \varepsilon(k_u R_\varepsilon u_\varepsilon - l_s s_\varepsilon) \\
-D_v \nabla v_\varepsilon \cdot n &= \varepsilon(k_v R_\varepsilon v_\varepsilon - l_w w_\varepsilon) \\
\partial_t s_\varepsilon - \varepsilon^2 D_s \Delta_\Gamma s_\varepsilon &= -h(s_\varepsilon) + k_u R_\varepsilon u_\varepsilon - l_s s_\varepsilon \\
\partial_t w_\varepsilon - \varepsilon^2 D_w \Delta_\Gamma w_\varepsilon &= h(s_\varepsilon) + k_v R_\varepsilon v_\varepsilon - l_w w_\varepsilon
\end{aligned}
\right\}
\quad \text{on } \Gamma_\varepsilon^{\text{ER}},
$$

for binding rates $k_u$, $k_v$, $l_s$ and $l_w > 0$. Here we remark a simplifying assumption of this formulation of the model. On the surface of the membrane the molecules diffuse by moving from one free receptor to the next free one. This means that for parts of the membrane, that are crowded with molecules, diffusion of these molecules gets harder because of the lack of free receptors. We neglect this aggregation effect in our model. Incorporating all of the above our new model becomes for $t \in [0, T]$,

$$\partial_t u_\varepsilon - D_u \Delta u_\varepsilon = - f(u_\varepsilon) \qquad \text{in } \Omega_\varepsilon$$

$$-D_u \nabla u_\varepsilon \cdot n = \varepsilon(k_u R_\varepsilon u_\varepsilon - l_s s_\varepsilon) \qquad \text{on } \Gamma_\varepsilon^{ER}$$

$$u_\varepsilon = u_{\text{Boundary}} \qquad \text{on } \Gamma_C$$

$$-D_u \nabla u_\varepsilon \cdot n = 0 \qquad \text{on } \Gamma_N$$

$$\partial_t v_\varepsilon - D_v \Delta v_\varepsilon = - g(v_\varepsilon) \qquad \text{in } \Omega_\varepsilon$$

$$-D_v \nabla v_\varepsilon \cdot n = \varepsilon(k_v R_\varepsilon v_\varepsilon - l_w w_\varepsilon) \qquad \text{on } \Gamma_\varepsilon^{ER}$$

$$-D_v \nabla v_\varepsilon \cdot n = 0 \qquad \text{on } \Gamma_C$$

$$v_\varepsilon = 0 \qquad \text{on } \Gamma_N$$

$$\partial_t s_\varepsilon - \varepsilon^2 D_s \Delta_\Gamma s_\varepsilon = - h(s_\varepsilon) + k_u R_\varepsilon u_\varepsilon - l_s s_\varepsilon \qquad \text{on } \Gamma_\varepsilon^{ER}$$

$$\partial_t w_\varepsilon - \varepsilon^2 D_w \Delta_\Gamma w_\varepsilon = h(s_\varepsilon) + k_v R_\varepsilon v_\varepsilon - l_w w_\varepsilon \qquad \text{on } \Gamma_\varepsilon^{ER}$$

$$\partial_t R_\varepsilon + R_\varepsilon |k_u u_\varepsilon + k_v v_\varepsilon| = (\overline{R} - R_\varepsilon)|k_s s_\varepsilon + k_w w_\varepsilon| \qquad \text{on } \Gamma_\varepsilon^{ER}$$

$$\tag{3.16}$$

with initial values $(u_I, v_I, s_I, w_I, R_I) = (u_\varepsilon(0), v_\varepsilon(0), s_\varepsilon(0), w_\varepsilon(0), \overline{R})$ smooth, bounded and nonnegative.

For the weak formulation we take the function spaces $\mathcal{V}_N(\Omega_\varepsilon)$, $\mathcal{V}_C(\Omega_\varepsilon)$, $\mathcal{V}(\Gamma_\varepsilon^{ER})$, $\mathcal{V}(\Omega, Y)$, $\mathcal{V}(\Omega, \Gamma)$ and the scalar products $(\cdot, \cdot)_{\Omega_\varepsilon}$, $(\cdot, \cdot)_{\Omega_\varepsilon, t}$ and $\langle \cdot, \cdot \rangle_{\Gamma_\varepsilon}$ as defined in (2.4). Further we claim the function $R_\varepsilon$ to be in $\mathcal{V}_R(\Gamma_\varepsilon^{ER}) := \{u \in L^2([0, T], L^2(\Gamma_\varepsilon)) | \partial_t u \in L^2([0, T] \times \Gamma_\varepsilon^{ER})\}$. For the test functions we need the spaces $V_{C0}(\Omega_\varepsilon)$, $V_N(\Omega_\varepsilon)$, $V(\Gamma_\varepsilon^{ER})$, $V(\Omega, Y)$ and $V(\Omega, \Gamma)$, as defined in (2.5).

Let $u_\varepsilon \in \mathcal{V}_C(\Omega_\varepsilon)$, $v_\varepsilon \in \mathcal{V}_N(\Omega_\varepsilon)$, $s_\varepsilon, w_\varepsilon \in \mathcal{V}(\Gamma_\varepsilon^{ER})$ and $R_\varepsilon \in \mathcal{V}_R(\Gamma_\varepsilon^{ER})$ such that

$$(\partial_t u_\varepsilon, \varphi_1)_{\Omega_\varepsilon} + D_u (\nabla u_\varepsilon, \nabla \varphi_1)_{\Omega_\varepsilon}$$
$$= -\varepsilon \langle k_u u_\varepsilon R_\varepsilon - l_s s_\varepsilon, \varphi_1 \rangle_{\Gamma_\varepsilon^{ER}} - (f(u_\varepsilon), \varphi_1)_{\Omega_\varepsilon}$$

$$(\partial_t v_\varepsilon, \varphi_2)_{\Omega_\varepsilon} + D_v (\nabla v_\varepsilon, \nabla \varphi_2)_{\Omega_\varepsilon}$$
$$= -\varepsilon \langle k_v v_\varepsilon R_\varepsilon - l_w w_\varepsilon, \varphi_2 \rangle_{\Gamma_\varepsilon^{ER}} - (g(v_\varepsilon), \varphi_2)_{\Omega_\varepsilon}$$

$$\langle \partial_t s_\varepsilon, \psi \rangle_{\Gamma_\varepsilon^{ER}} + D_s \varepsilon^2 \langle \nabla_\Gamma s_\varepsilon, \nabla_\Gamma \psi \rangle_{\Gamma_\varepsilon^{ER}}$$
$$= \langle k_u u_\varepsilon R_\varepsilon - l_s s_\varepsilon, \psi \rangle_{\Gamma_\varepsilon^{ER}} - \langle h(s_\varepsilon), \psi \rangle_{\Gamma_\varepsilon^{ER}}$$
$$\langle \partial_t w_\varepsilon, \psi \rangle_{\Gamma_\varepsilon^{ER}} + D_w \varepsilon^2 \langle \nabla_\Gamma w_\varepsilon, \nabla_\Gamma \psi \rangle_{\Gamma_\varepsilon^{ER}}$$
$$= \langle k_v v_\varepsilon R_\varepsilon - l_w w_\varepsilon, \psi \rangle_{\Gamma_\varepsilon^{ER}} + \langle h(s_\varepsilon), \psi \rangle_{\Gamma_\varepsilon^{ER}} \tag{3.17}$$
$$\langle \partial_t R_\varepsilon, \psi \rangle_{\Gamma_\varepsilon^{ER}} + \langle R_\varepsilon | k_u u_\varepsilon + k_v v_\varepsilon |, \psi \rangle_{\Gamma_\varepsilon^{ER}}$$
$$= \langle (\overline{R} - R_\varepsilon) | k_s s_\varepsilon + k_w w_\varepsilon |, \psi \rangle_{\Gamma_\varepsilon^{ER}}$$

for all $(\varphi_1, \varphi_2, \psi) \in V_{C0}(\Omega_\varepsilon) \times V_N(\Omega_\varepsilon) \times V(\Gamma_\varepsilon^{ER})$.

## 3.4 Estimations for the nonlinear carcinogenesis model

In this section we are going to prove that the conditions for using proposition 2.5, theorem 2.10 and lemma 3.22 are satisfied. Further, we show that the sequences of functions $(u_\varepsilon, v_\varepsilon, s_\varepsilon, w_\varepsilon, R_\varepsilon)$ converge strongly. Therefore, we additionally show that these functions are bounded in $L^\infty$ and that the time derivatives are elements of $(H_0^1)'$ independently of the variable $\varepsilon$. Strong convergence of the sequences of functions $s_\varepsilon, w_\varepsilon$ and $R_\varepsilon$ is shown by proving that they are Cauchy sequences. This idea is found in the article [17].

We start with $R_\varepsilon$ and prove that $0 \leq R_\varepsilon \leq \overline{R}$ almost everywhere.

**Lemma 3.31** *A function $R_\varepsilon$, defined in (3.17), is*

1. *nonnegative*

2. *and bounded by $\overline{R} > 0$*

*almost everywhere in $x \in \Gamma_\varepsilon^{ER}$ and $t \in [0, T]$.*

**Proof**

*1.* We test the weak formulation of $R_\varepsilon$ with $R_{\varepsilon-}$ and find

$$\langle \partial_t R_\varepsilon, R_{\varepsilon-} \rangle_{\Gamma_\varepsilon^{ER}} + \langle R_\varepsilon | k_u u_\varepsilon + k_v v_\varepsilon |, R_{\varepsilon-} \rangle_{\Gamma_\varepsilon^{ER}}$$
$$- \langle (\overline{R} - R_\varepsilon) | k_s s_\varepsilon + k_w w_\varepsilon |, R_{\varepsilon-} \rangle_{\Gamma_\varepsilon^{ER}} = 0.$$

Multiplying the equation with $-1$ leads to

$$\langle \partial_t R_{\varepsilon-}, R_{\varepsilon-} \rangle_{\Gamma_\varepsilon^{ER}} + \langle R_{\varepsilon-} |k_u u_\varepsilon + k_v v_\varepsilon|, R_{\varepsilon-} \rangle_{\Gamma_\varepsilon^{ER}}$$
$$+ \langle (\overline{R} + R_{\varepsilon-}) |k_s s_\varepsilon + k_w w_\varepsilon|, R_{\varepsilon-} \rangle_{\Gamma_\varepsilon^{ER}} = 0.$$

Integration from 0 to $t$ and with $R_\varepsilon(0) \geq 0$ yields

$$\frac{1}{2} \|R_{\varepsilon-}\|^2_{\Gamma_\varepsilon^{ER}} + \underbrace{\left\| R_{\varepsilon-} \sqrt{|k_u u_\varepsilon + k_v v_\varepsilon|} \right\|^2_{\Gamma_\varepsilon^{ER},t}}_{\geq 0}$$
$$+ \underbrace{\langle (\overline{R} + R_{\varepsilon-}) |k_s s_\varepsilon + k_w w_\varepsilon|, R_{\varepsilon-} \rangle_{\Gamma_\varepsilon^{ER},t}}_{\geq 0} = 0.$$

We deduce

$$\|R_{\varepsilon-}\|^2_{\Gamma_\varepsilon^{ER}} \leq 0$$

for almost every $t \in [0, T]$. This is equivalent to $R_\varepsilon \geq 0$ for almost every $x \in \Gamma_\varepsilon^{ER}$ and $t \in [0, T]$.

2. We test the weak formulation with $(R_\varepsilon - \overline{R})_+$ and obtain

$$\langle \partial_t R_\varepsilon, (R_\varepsilon - \overline{R})_+ \rangle_{\Gamma_\varepsilon^{ER}} + \underbrace{\langle R_\varepsilon |k_u u_\varepsilon + k_v v_\varepsilon|, (R_\varepsilon - \overline{R})_+ \rangle_{\Gamma_\varepsilon^{ER}}}_{\geq 0}$$
$$- \langle (\overline{R} - R_\varepsilon) |k_s s_\varepsilon + k_w w_\varepsilon|, (R_\varepsilon - \overline{R})_+ \rangle_{\Gamma_\varepsilon^{ER}} = 0.$$

Since $\partial_t \overline{R} = 0$, it yields

$$\langle \partial_t (R_\varepsilon - \overline{R})_+, (R_\varepsilon - \overline{R})_+ \rangle_{\Gamma_\varepsilon^{ER}}$$
$$+ \langle (R_\varepsilon - \overline{R})_+ |k_s s_\varepsilon + k_w w_\varepsilon|, (R_\varepsilon - \overline{R})_+ \rangle_{\Gamma_\varepsilon^{ER}} \leq 0.$$

Integrating from 0 to $t$ and using $R_\varepsilon(t = 0) \leq \overline{R}$ leads to

$$\frac{1}{2} \|(R_\varepsilon - \overline{R})_+\|^2_{\Gamma_\varepsilon^{ER}} + \left\| (R_\varepsilon - \overline{R})_+ \sqrt{|k_s s_\varepsilon + k_w w_\varepsilon|} \right\|^2_{\Gamma_\varepsilon^{ER},t} \leq 0.$$

We conclude that $R_\varepsilon < \overline{R}$ for almost every $x \in \Gamma_\varepsilon^{ER}$ and $t \in [0, T]$.
$\square$

Now we show positivity for the other functions, thus they have a lower bound.

**Lemma 3.32 (Positivity)**

*The functions $u_\varepsilon$, $v_\varepsilon$, $s_\varepsilon$ and $w_\varepsilon$ are nonnegative for almost every $x \in \Omega_\varepsilon$, $x \in \Gamma_\varepsilon^{ER}$, respectively, and $t \in [0, T]$.*

**Proof** We start with the equations for $u_\varepsilon$ and $s_\varepsilon$ and test the weak formulation with $u_{\varepsilon-}$ and $s_{\varepsilon-}$, respectively.

$$(\partial_t u_\varepsilon, u_{\varepsilon-})_{\Omega_\varepsilon} + D_u(\nabla u_\varepsilon, \nabla u_{\varepsilon-})_{\Omega_\varepsilon}$$
$$+ \varepsilon \langle k_u u_\varepsilon R_\varepsilon - l_s s_\varepsilon, u_{\varepsilon-} \rangle_{\Gamma_\varepsilon^{ER}} = - \underbrace{(f(u_\varepsilon), u_{\varepsilon-})_{\Omega_\varepsilon}}_{=0},$$

$$\varepsilon \langle \partial_t s_\varepsilon, s_{\varepsilon-} \rangle_{\Gamma_\varepsilon^{ER}} + D_s \varepsilon^3 \langle \nabla_\Gamma s_\varepsilon, \nabla_\Gamma s_{\varepsilon-} \rangle_{\Gamma_\varepsilon^{ER}}$$
$$- \varepsilon \langle k_u u_\varepsilon R_\varepsilon - l_s s_\varepsilon, s_{\varepsilon-} \rangle_{\Gamma_\varepsilon^{ER}} = -\varepsilon \underbrace{\langle h(s_\varepsilon), s_{\varepsilon-} \rangle_{\Gamma_\varepsilon^{ER}}}_{=0}.$$

Multiplying with $-1$ leads to

$$(\partial_t u_{\varepsilon-}, u_{\varepsilon-})_{\Omega_\varepsilon} + D_u(\nabla u_{\varepsilon-}, \nabla u_{\varepsilon-})_{\Omega_\varepsilon} + \varepsilon \langle k_u u_{\varepsilon-} R_\varepsilon, u_{\varepsilon-} \rangle_{\Gamma_\varepsilon^{ER}}$$
$$= \underbrace{-\varepsilon \langle l_s s_{\varepsilon+}, u_{\varepsilon-} \rangle_{\Gamma_\varepsilon^{ER}}}_{\leq 0} + \varepsilon \langle l_s s_{\varepsilon-}, u_{\varepsilon-} \rangle_{\Gamma_\varepsilon^{ER}},$$

$$\varepsilon \langle \partial_t s_{\varepsilon-}, s_{\varepsilon-} \rangle_{\Gamma_\varepsilon^{ER}} + D_s \varepsilon^3 \langle \nabla_\Gamma s_{\varepsilon-}, \nabla_\Gamma s_{\varepsilon-} \rangle_{\Gamma_\varepsilon^{ER}} + \varepsilon \langle l_s s_{\varepsilon-}, s_{\varepsilon-} \rangle_{\Gamma_\varepsilon^{ER}}$$
$$= \underbrace{-\varepsilon \langle k_u u_{\varepsilon+} R_\varepsilon, s_{\varepsilon-} \rangle_{\Gamma_\varepsilon^{ER}}}_{\leq 0} + \varepsilon \langle k_u u_{\varepsilon-} R_\varepsilon, s_{\varepsilon-} \rangle_{\Gamma_\varepsilon^{ER}}.$$

We add the equations and integrate from $0$ to $t$ and obtain

$$\frac{1}{2} \|u_{\varepsilon-}\|_{\Omega_\varepsilon}^2 + \frac{1}{2} \varepsilon \|s_{\varepsilon-}\|_{\Gamma_\varepsilon^{ER}}^2 + D_u \|\nabla u_{\varepsilon-}\|_{\Omega_\varepsilon,t}^2 + D_s \varepsilon^3 \|\nabla_\Gamma s_{\varepsilon-}\|_{\Gamma_\varepsilon^{ER},t}^2$$
$$+ k_u \varepsilon \|u_{\varepsilon-} \sqrt{R_\varepsilon}\|_{\Gamma_\varepsilon^{ER},t}^2 + \varepsilon l_s \|s_{\varepsilon-}\|_{\Gamma_\varepsilon^{ER},t}^2$$
$$\leq \varepsilon (l_s + k_u \overline{R}) \langle s_{\varepsilon-}, u_{\varepsilon-} \rangle_{\Gamma_\varepsilon^{ER},t}$$
$$\leq (l_s + k_u \overline{R}) \left( \varepsilon \|s_{\varepsilon-}\|_{\Gamma_\varepsilon^{ER},t}^2 + c_0 \|u_{\varepsilon-}\|_{\Omega_\varepsilon,t}^2 + \varepsilon^2 c_0 \|\nabla u_{\varepsilon-}\|_{\Omega_\varepsilon,t}^2 \right).$$

Hence,

$$
\begin{aligned}
\frac{1}{2}\|u_{\varepsilon-}\|^2_{\Omega_\varepsilon} &+ \frac{1}{2}\varepsilon\|s_{\varepsilon-}\|^2_{\Gamma_\varepsilon^{ER}} + (D_u - \varepsilon^2(l_s + k_u\overline{R})c_0)\|\nabla u_{\varepsilon-}\|^2_{\Omega_\varepsilon,t} \\
&+ D_s\varepsilon^3\|\nabla_\Gamma s_{\varepsilon-}\|^2_{\Gamma_\varepsilon^{ER},t} + k_u\varepsilon\|u_{\varepsilon-}\sqrt{R_\varepsilon}\|^2_{\Gamma_\varepsilon^{ER},t} + \varepsilon l_s\|s_{\varepsilon-}\|^2_{\Gamma_\varepsilon^{ER},t} \\
&\leq (l_s + k_u\overline{R})\left(\varepsilon\|s_{\varepsilon-}\|^2_{\Gamma_\varepsilon^{ER},t} + c_0\|u_{\varepsilon-}\|^2_{\Omega_\varepsilon,t}\right).
\end{aligned}
$$

After merging the constants, for $\varepsilon$ small and with Gronwall's lemma we deduce that $\|u_{\varepsilon-}\|^2_{\Omega_\varepsilon} + \|s_{\varepsilon-}\|^2_{\Gamma_\varepsilon^{ER}} \leq 0$ and therefore $u_\varepsilon$ and $s_\varepsilon$ are greater or equal to zero for almost every $x \in \Omega_\varepsilon$ or $x \in \Gamma_\varepsilon^{ER}$ and $t \in [0, T]$.

Then we do similar estimations for the equations for $v_\varepsilon$ and $w_\varepsilon$ and test them with $v_{\varepsilon-}$ and $w_{\varepsilon-}$, respectively.

$$
\begin{aligned}
(\partial_t v_\varepsilon, v_{\varepsilon-})_{\Omega_\varepsilon} &+ D_v(\nabla v_\varepsilon, \nabla v_{\varepsilon-})_{\Omega_\varepsilon} \\
&+ \varepsilon\langle k_v v_\varepsilon R_\varepsilon - l_w w_\varepsilon, v_{\varepsilon-}\rangle_{\Gamma_\varepsilon^{ER}} = -\underbrace{(g(v_\varepsilon), v_{\varepsilon-})_{\Omega_\varepsilon}}_{=0},
\end{aligned}
$$

$$
\begin{aligned}
\varepsilon\langle \partial_t w_\varepsilon, w_{\varepsilon-}\rangle_{\Gamma_\varepsilon^{ER}} &+ D_w\varepsilon^3\langle \nabla_\Gamma w_\varepsilon, \nabla_\Gamma w_{\varepsilon-}\rangle_{\Gamma_\varepsilon^{ER}} \\
&- \varepsilon\langle k_v v_\varepsilon R_\varepsilon - l_w w_\varepsilon, w_{\varepsilon-}\rangle_{\Gamma_\varepsilon^{ER}} = \varepsilon\langle h(s_\varepsilon), w_{\varepsilon-}\rangle_{\Gamma_\varepsilon^{ER}}.
\end{aligned}
$$

Further,

$$
\begin{aligned}
(\partial_t v_{\varepsilon-}, v_{\varepsilon-})_{\Omega_\varepsilon} &+ D_v(\nabla v_{\varepsilon-}, \nabla v_{\varepsilon-})_{\Omega_\varepsilon} \\
&+ \varepsilon\langle k_v v_{\varepsilon-} R_\varepsilon, v_{\varepsilon-}\rangle_{\Gamma_\varepsilon^{ER}} = \underbrace{-\varepsilon\langle l_w w_{\varepsilon+}, v_{\varepsilon-}\rangle_{\Gamma_\varepsilon^{ER}}}_{\leq 0} + \varepsilon\langle l_w w_{\varepsilon-}, v_{\varepsilon-}\rangle_{\Gamma_\varepsilon^{ER}},
\end{aligned}
$$

$$
\begin{aligned}
\varepsilon\langle \partial_t w_{\varepsilon-}, w_{\varepsilon-}\rangle_{\Gamma_\varepsilon^{ER}} &+ D_w\varepsilon^3\langle \nabla_\Gamma w_{\varepsilon-}, \nabla_\Gamma w_{\varepsilon-}\rangle_{\Gamma_\varepsilon^{ER}} + \varepsilon\langle l_w w_{\varepsilon-}, w_{\varepsilon-}\rangle_{\Gamma_\varepsilon^{ER}} \\
&= \underbrace{-\varepsilon\langle h(s_\varepsilon), w_{\varepsilon-}\rangle_{\Gamma_\varepsilon^{ER}} - \varepsilon\langle k_v v_{\varepsilon+} R_\varepsilon, w_{\varepsilon-}\rangle_{\Gamma_\varepsilon^{ER}}}_{\leq 0} + \varepsilon\langle k_v v_{\varepsilon-} R_\varepsilon, w_{\varepsilon-}\rangle_{\Gamma_\varepsilon^{ER}}.
\end{aligned}
$$

This leads to

$$\frac{1}{2}\|v_{\varepsilon-}\|_{\Omega_\varepsilon}^2 + \frac{1}{2}\varepsilon\|w_{\varepsilon-}\|_{\Gamma_\varepsilon^{ER}}^2 + (D_v - \varepsilon^2(l_w + k_v\overline{R})c_0)\|\nabla v_{\varepsilon-}\|_{\Omega_\varepsilon,t}^2$$

$$+ D_w\varepsilon^3\|\nabla_\Gamma w_{\varepsilon-}\|_{\Gamma_\varepsilon^{ER},t}^2 + k_v\varepsilon\|v_{\varepsilon-}\sqrt{R_\varepsilon}\|_{\Gamma_\varepsilon^{ER},t}^2 + \varepsilon l_w\|w_{\varepsilon-}\|_{\Gamma_\varepsilon^{ER},t}^2$$

$$\leq (l_w + k_v\overline{R})\left(\varepsilon\|w_{\varepsilon-}\|_{\Gamma_\varepsilon^{ER},t}^2 + c_0\|v_{\varepsilon-}\|_{\Omega_\varepsilon,t}^2\right)$$

and we also find that $v_\varepsilon$ and $w_\varepsilon$ are nonnegative for almost every $x \in \Omega_\varepsilon$ or $x \in \Gamma_\varepsilon^{ER}$ and $t \in [0,T]$. $\qquad\square$

We need the following a-priory estimates to satisfy the conditions for proposition 2.5 and lemma 3.22.

**Lemma 3.33 (Boundedness in $L^2$)**
*There exists a constant $C > 0$, independent of $\varepsilon$, such that*

$$\|u_\varepsilon\|_{\Omega_\varepsilon}^2 + \|v_\varepsilon\|_{\Omega_\varepsilon}^2 + \varepsilon\|s_\varepsilon\|_{\Gamma_\varepsilon^{ER}}^2 + \varepsilon\|w_\varepsilon\|_{\Gamma_\varepsilon^{ER}}^2$$

$$+ \|\nabla u_\varepsilon\|_{\Omega_\varepsilon,t}^2 + \|\nabla v_\varepsilon\|_{\Omega_\varepsilon,t}^2 + \varepsilon^3\|\nabla_\Gamma s_\varepsilon\|_{\Gamma_\varepsilon^{ER},t}^2 + \varepsilon^3\|\nabla_\Gamma w_\varepsilon\|_{\Gamma_\varepsilon^{ER},t}^2$$

$$+ \varepsilon\|k_u u_\varepsilon R_\varepsilon - l_s s_\varepsilon\|_{\Gamma_\varepsilon^{ER},t}^2 + \varepsilon\|k_v v_\varepsilon R_\varepsilon - l_w w_\varepsilon\|_{\Gamma_\varepsilon^{ER},t}^2 \leq C.$$

**Proof** At first we prove

$$\|u_\varepsilon\|_{\Omega_\varepsilon}^2 + \varepsilon\|s_\varepsilon\|_{\Gamma_\varepsilon^{ER}}^2 + \|\nabla u_\varepsilon\|_{\Omega_\varepsilon,t}^2 + \varepsilon\|\varepsilon\nabla_\Gamma s_\varepsilon\|_{\Gamma_\varepsilon^{ER},t}^2$$

$$+ \varepsilon\|k_u u_\varepsilon R_\varepsilon - l_s s_\varepsilon\|_{\Gamma_\varepsilon^{ER},t}^2 \leq C$$

for a $C > 0$. For that purpose we test the weak formulations for $u_\varepsilon$ and $s_\varepsilon$ with $k_u\overline{R}u_\varepsilon$ and $l_s s_\varepsilon$, respectively,

$$k_u\overline{R}(\partial_t u_\varepsilon, u_\varepsilon)_{\Omega_\varepsilon} + D_u k_u\overline{R}(\nabla u_\varepsilon, \nabla u_\varepsilon)_{\Omega_\varepsilon} + \varepsilon\langle R_\varepsilon u_\varepsilon k_u - l_s s_\varepsilon, k_u\overline{R}u_\varepsilon\rangle_{\Gamma_\varepsilon^{ER}}$$

$$+ \varepsilon l_s\langle\partial_t s_\varepsilon, s_\varepsilon\rangle_{\Gamma_\varepsilon^{ER}} + \varepsilon^3 D_s l_s\langle\nabla_\Gamma s_\varepsilon, \nabla_\Gamma s_\varepsilon\rangle_{\Gamma_\varepsilon^{ER}}$$

$$= -k_u\overline{R}(f(u_\varepsilon), u_\varepsilon)_{\Omega_\varepsilon} + \varepsilon\langle k_u R_\varepsilon u_\varepsilon - l_s s_\varepsilon, l_s s_\varepsilon\rangle_{\Gamma_\varepsilon^{ER}} - \varepsilon l_s\langle h(s_\varepsilon), s_\varepsilon\rangle_{\Gamma_\varepsilon^{ER}}.$$

With integration from 0 to $t$ we get

$$\frac{1}{2}k_u\overline{R}\|u_\varepsilon\|_{\Omega_\varepsilon}^2 + D_u k_u\overline{R}\|\nabla u_\varepsilon\|_{\Omega_\varepsilon,t}^2 + \frac{1}{2}\varepsilon l_s\|s_\varepsilon\|_{\Gamma_\varepsilon^{ER}}^2 + \varepsilon^3 D_s l_s\|\nabla_\Gamma s_\varepsilon\|_{\Gamma_\varepsilon^{ER},t}^2$$

$$+ \varepsilon\langle k_u R_\varepsilon u_\varepsilon - l_s s_\varepsilon, k_u\overline{R}u_\varepsilon - l_s s_\varepsilon\rangle_{\Gamma_\varepsilon^{ER},t} = -k_u\overline{R}(f(u_\varepsilon), u_\varepsilon)_{\Omega_\varepsilon,t}$$

$$- \varepsilon l_s\langle h(s_\varepsilon), s_\varepsilon\rangle_{\Gamma_\varepsilon^{ER},t} + \frac{1}{2}k_u\overline{R}\|u_\varepsilon(0)\|_{\Omega_\varepsilon}^2 + \frac{1}{2}\varepsilon l_s\|s_\varepsilon(0)\|_{\Gamma_\varepsilon^{ER}}^2.$$

We add $\varepsilon\langle k_u R_\varepsilon u_\varepsilon - l_s s_\varepsilon, R_\varepsilon k_u u_\varepsilon - \overline{R} k_u u_\varepsilon\rangle_{\Gamma_\varepsilon^{\mathrm{ER}},t}$ on the left- and on the right-hand side and compute

$$
\frac{1}{2} k_u \overline{R} \|u_\varepsilon\|_{\Omega_\varepsilon}^2 + D_u k_u \overline{R} \|\nabla u_\varepsilon\|_{\Omega_\varepsilon,t}^2 + \frac{1}{2}\varepsilon l_s \|s_\varepsilon\|_{\Gamma_\varepsilon^{\mathrm{ER}}}^2
$$
$$
+ \varepsilon^3 D_s l_s \|\nabla_\Gamma s_\varepsilon\|_{\Gamma_\varepsilon^{\mathrm{ER}},t}^2 + \varepsilon \|R_\varepsilon u_\varepsilon k_u - l_s s_\varepsilon\|_{\Gamma_\varepsilon^{\mathrm{ER}},t}^2
$$
$$
= -k_u \overline{R}(f(u_\varepsilon), u_\varepsilon)_{\Omega_\varepsilon,t} - \varepsilon l_s \langle h(s_\varepsilon), s_\varepsilon\rangle_{\Gamma_\varepsilon^{\mathrm{ER}},t} + \frac{1}{2} k_u \overline{R}\|u_\varepsilon(0)\|_{\Omega_\varepsilon}^2
$$
$$
+ \frac{1}{2}\varepsilon l_s \|s_\varepsilon(0)\|_{\Gamma_\varepsilon^{\mathrm{ER}}}^2 + \varepsilon\langle k_u R_\varepsilon u_\varepsilon - l_s s_\varepsilon, R_\varepsilon k_u u_\varepsilon - \overline{R} k_u u_\varepsilon\rangle_{\Gamma_\varepsilon^{\mathrm{ER}},t}
$$
$$
\leq \underbrace{\frac{1}{2} k_u \overline{R}\|u_\varepsilon(0)\|_{\Omega_\varepsilon}^2 + \frac{1}{2}\varepsilon l_s \|s_\varepsilon(0)\|_{\Gamma_\varepsilon^{\mathrm{ER}}}^2}_{=c_1}
$$
$$
+ \varepsilon\langle k_u R_\varepsilon u_\varepsilon - l_s s_\varepsilon, k_u u_\varepsilon (R_\varepsilon - \overline{R})\rangle_{\Gamma_\varepsilon^{\mathrm{ER}},t}.
$$

Using the binomial theorem gives for any $\lambda > 0$

$$
\frac{1}{2} k_u \overline{R} \|u_\varepsilon\|_{\Omega_\varepsilon}^2 + D_u k_u \overline{R} \|\nabla u_\varepsilon\|_{\Omega_\varepsilon,t}^2 + \frac{1}{2}\varepsilon l_s \|s_\varepsilon\|_{\Gamma_\varepsilon^{\mathrm{ER}}}^2
$$
$$
+ \varepsilon^3 D_s l_s \|\nabla_\Gamma s_\varepsilon\|_{\Gamma_\varepsilon^{\mathrm{ER}},t}^2 + \varepsilon \|R_\varepsilon u_\varepsilon k_u - l_s s_\varepsilon\|_{\Gamma_\varepsilon^{\mathrm{ER}},t}^2
$$
$$
\leq c_1 + \frac{1}{2\lambda}\varepsilon \|k_u R_\varepsilon u_\varepsilon - l_s s_\varepsilon\|_{\Gamma_\varepsilon^{\mathrm{ER}},t}^2
$$
$$
+ \frac{\lambda}{2} k_u^2 \overline{R}^2 c_0 \left( \|u_\varepsilon\|_{\Omega_\varepsilon,t}^2 + \varepsilon^2 \|\nabla u_\varepsilon\|_{\Omega_\varepsilon,t}^2 \right).
$$

Hence,

$$
\frac{1}{2} k_u \overline{R} \|u_\varepsilon\|_{\Omega_\varepsilon}^2 + \left( D_u k_u \overline{R} - \frac{\lambda}{2} k_u^2 \overline{R}^2 c_0 \varepsilon^2 \right) \|\nabla u_\varepsilon\|_{\Omega_\varepsilon,t}^2 + \frac{1}{2}\varepsilon l_s \|s_\varepsilon\|_{\Gamma_\varepsilon^{\mathrm{ER}}}^2
$$
$$
+ \varepsilon^3 D_s l_s \|\nabla_\Gamma s_\varepsilon\|_{\Gamma_\varepsilon^{\mathrm{ER}},t}^2 + \varepsilon\left( 1 - \frac{1}{2\lambda} \right) \|R_\varepsilon u_\varepsilon k_u - l_s s_\varepsilon\|_{\Gamma_\varepsilon^{\mathrm{ER}},t}^2
$$
$$
\leq c_1 + \frac{\lambda}{2} k_u^2 \overline{R}^2 c_0 \|u_\varepsilon\|_{\Omega_\varepsilon,t}^2.
$$

With $\lambda > \frac{1}{2}$ and $\varepsilon$ small, we can merge the constants and use Gronwall's lemma 2.12 to deduce the assertion.

To show

$$\|v_\varepsilon\|^2_{\Omega_\varepsilon} + \varepsilon\|w_\varepsilon\|^2_{\Gamma_\varepsilon^{ER}} + \|\nabla v_\varepsilon\|^2_{\Omega_\varepsilon,t} + \varepsilon^3\|\nabla_\Gamma w_\varepsilon\|^2_{\Gamma_\varepsilon^{ER},t}$$
$$+ \varepsilon\|k_v v_\varepsilon R_\varepsilon - l_w w_\varepsilon\|^2_{\Gamma_\varepsilon^{ER},t} \leq C$$

for all $\varepsilon > 0$, we make corresponding estimates and arrive at

$$\frac{1}{2}k_v\overline{R}\|v_\varepsilon\|^2_{\Omega_\varepsilon} + \left(D_v k_v\overline{R} - \frac{\lambda}{2}k_v^2\overline{R}^2 c_0\varepsilon^2\right)\|\nabla v_\varepsilon\|^2_{\Omega_\varepsilon,t} + \frac{1}{2}\varepsilon l_w\|w_\varepsilon\|^2_{\Gamma_\varepsilon^{ER}}$$
$$+ \varepsilon^3 D_w l_w\|\nabla_\Gamma w_\varepsilon\|^2_{\Gamma_\varepsilon^{ER},t} + \varepsilon\left(1 - \frac{1}{2\lambda}\right)\|R_\varepsilon v_\varepsilon k_v - l_w w_\varepsilon\|^2_{\Gamma_\varepsilon^{ER},t}$$
$$\leq \underbrace{c_1 + \varepsilon l_s\|h(s_\varepsilon)\|^2_{\Gamma_\varepsilon^{ER},t} + \varepsilon l_s\|s_\varepsilon\|^2_{\Gamma_\varepsilon^{ER},t}}_{\leq c_2} + \frac{\lambda}{2}k_v^2\overline{R}^2 c_0\|v_\varepsilon\|^2_{\Omega_\varepsilon,t}.$$

Again, with $\lambda > \frac{1}{2}$ and $\varepsilon$ small, we can merge the constants and use Gronwall's lemma to deduce the claim. $\qquad\square$

To prove strong convergence it is necessary to show that $u_\varepsilon, v_\varepsilon \in L^\infty(\Omega_\varepsilon)$ and $s_\varepsilon, w_\varepsilon \in L^\infty(\Gamma_\varepsilon^{ER})$. We make use of the fact that $R_\varepsilon \in L^\infty(\Gamma_\varepsilon^{ER})$, which we established in lemma 3.31.

### Lemma 3.34 (Boundedness in $L^\infty$)
*The functions $u_\varepsilon$, $v_\varepsilon$, $s_\varepsilon$ and $w_\varepsilon$ are bounded, independent of $\varepsilon$, for almost every $x \in \Omega_\varepsilon$ and $t \in [0,T]$, $x \in \Gamma_\varepsilon^{ER}$, respectively.*

### Proof
Let $M(t) = \max\{\|u_I\|_{L^\infty(\Omega_\varepsilon)}, \|v_I\|_{L^\infty(\Omega_\varepsilon)}, \|s_I\|_{L^\infty(\Gamma_\varepsilon^{ER})}, \|w_I\|_{L^\infty(\Gamma_\varepsilon^{ER})}\}e^{kt}$ for a constant $k \in \mathbb{R}$. The function $M$ exists because the initial conditions are bounded.
At first we prove the assertion for $u_\varepsilon$ and $s_\varepsilon$. We are testing the weak formulation for $u_\varepsilon, s_\varepsilon$ with $(\overline{R}k_u u_\varepsilon - M)_+$ and $(l_s s_\varepsilon - M)_+$,

respectively. Then we add the two equations.

$$
(\partial_t u_\varepsilon, (\overline{R}k_u u_\varepsilon - M)_+)_{\Omega_\varepsilon} + \varepsilon \langle \partial_t s_\varepsilon, (l_s s_\varepsilon - M)_+ \rangle_{\Gamma_\varepsilon^{ER}}
$$
$$
+ (D_u \nabla u_\varepsilon, \nabla (\overline{R}k_u u_\varepsilon - M)_+)_{\Omega_\varepsilon} + \varepsilon^3 \langle D_s \nabla_\Gamma s_\varepsilon, \nabla_\Gamma (l_s s_\varepsilon - M)_+ \rangle_{\Gamma_\varepsilon^{ER}}
$$
$$
+ \varepsilon \langle k_u R_\varepsilon u_\varepsilon - l_s s_\varepsilon, (\overline{R}k_u u_\varepsilon - M)_+ \rangle_{\Gamma_\varepsilon^{ER}}
$$
$$
= \underbrace{(-f(u_\varepsilon), (\overline{R}k_u u_\varepsilon - M)_+)_{\Omega_\varepsilon}}_{\leq 0} + \varepsilon \langle k_u R_\varepsilon u_\varepsilon - l_s s_\varepsilon, (l_s s_\varepsilon - M)_+ \rangle_{\Gamma_\varepsilon^{ER}}
$$
$$
+ \underbrace{\langle -h(s_\varepsilon), (l_s s_\varepsilon - M)_+ \rangle_{\Gamma_\varepsilon^{ER}}}_{\leq 0}.
$$

Regarding the time and spatial derivative of the test functions yields

$$
\frac{1}{\overline{R}k_u} (\partial_t (\overline{R}k_u u_\varepsilon - M)_+, (\overline{R}k_u u_\varepsilon - M)_+)_{\Omega_\varepsilon}
$$
$$
+ \frac{1}{l_s} \varepsilon \langle \partial_t (l_s s_\varepsilon - M)_+, (l_s s_\varepsilon - M)_+ \rangle_{\Gamma_\varepsilon^{ER}}
$$
$$
+ \frac{D_u}{\overline{R}k_u} \|\nabla (\overline{R}k_u u_\varepsilon - M)_+\|_{\Omega_\varepsilon}^2 + \frac{D_s}{l_s} \varepsilon^3 \|\nabla_\Gamma (l_s s_\varepsilon - M)_+\|_{\Gamma_\varepsilon^{ER}}^2
$$
$$
+ \varepsilon \langle k_u R_\varepsilon u_\varepsilon - l_s s_\varepsilon, (\overline{R}k_u u_\varepsilon - M)_+ - (l_s s_\varepsilon - M)_+ \rangle_{\Gamma_\varepsilon^{ER}}
$$
$$
\leq -\frac{1}{\overline{R}k_u} (Mk, (\overline{R}k_u u_\varepsilon - M)_+)_{\Omega_\varepsilon} - \frac{1}{l_s} \varepsilon \langle Mk, (l_s s_\varepsilon - M)_+ \rangle_{\Gamma_\varepsilon^{ER}}.
$$

We add $\varepsilon \langle k_u \overline{R} u_\varepsilon - k_u R_\varepsilon u_\varepsilon, (\overline{R}k_u u_\varepsilon - M)_+ - (l_s s_\varepsilon - M)_+ \rangle_{\Gamma_\varepsilon^{ER}}$ on each side of the equation and integrate from 0 to $t$.

$$
\frac{1}{2\overline{R}k_u} \|(\overline{R}k_u u_\varepsilon - M)_+\|_{\Omega_\varepsilon}^2 + \frac{1}{2l_s} \varepsilon \|(l_s s_\varepsilon - M)_+\|_{\Gamma_\varepsilon^{ER}}^2
$$
$$
+ \frac{D_u}{\overline{R}k_u} \|\nabla (\overline{R}k_u u_\varepsilon - M)_+\|_{\Omega_\varepsilon, t}^2 + \frac{D_s}{l_s} \varepsilon^3 \|\nabla_\Gamma (l_s s_\varepsilon - M)_+\|_{\Gamma_\varepsilon^{ER}, t}^2
$$
$$
+ \varepsilon \|(\overline{R}k_u u_\varepsilon - M)_+ - (l_s s_\varepsilon - M)_+\|_{\Gamma_\varepsilon^{ER}, t}^2
$$
$$
\leq \varepsilon \langle k_u \overline{R} u_\varepsilon - k_u R_\varepsilon u_\varepsilon, (\overline{R}k_u u_\varepsilon - M)_+ - (l_s s_\varepsilon - M)_+ \rangle_{\Gamma_\varepsilon^{ER}, t}
$$
$$
- \frac{1}{\overline{R}k_u} (Mk, (\overline{R}k_u u_\varepsilon - M)_+)_{\Omega_\varepsilon, t} - \frac{1}{l_s} \varepsilon \langle Mk, (l_s s_\varepsilon - M)_+ \rangle_{\Gamma_\varepsilon^{ER}, t}
$$

$$\leq \lambda \overline{R}^2 k_u^2 \varepsilon \|u_\varepsilon\|_{\Gamma_\varepsilon^{ER}}^2 + \frac{1}{2\lambda}\varepsilon\|(\overline{R}k_u u_\varepsilon - M)_+ - (l_s s_\varepsilon - M)_+\|_{\Gamma_\varepsilon^{ER},t}^2$$

$$- \frac{1}{\overline{R}k_u}(Mk, (\overline{R}k_u u_\varepsilon - M)_+)_{\Omega_\varepsilon,t} - \frac{1}{l_s}\varepsilon\langle Mk, (l_s s_\varepsilon - M)_+\rangle_{\Gamma_\varepsilon^{ER},t}.$$

Simplifying further yields

$$\frac{1}{2\overline{R}k_u}\|(\overline{R}k_u u_\varepsilon - M)_+\|_{\Omega_\varepsilon}^2 + \frac{1}{2l_s}\varepsilon\|(l_s s_\varepsilon - M)_+\|_{\Gamma_\varepsilon^{ER}}^2$$

$$+ \frac{D_u}{\overline{R}k_u}\|\nabla(\overline{R}k_u u_\varepsilon - M)_+\|_{\Omega_\varepsilon,t}^2 + \frac{D_s}{l_s}\varepsilon^3\|\nabla_\Gamma(l_s s_\varepsilon - M)_+\|_{\Gamma_\varepsilon^{ER},t}^2$$

$$+ \left(1 - \frac{1}{2\lambda}\right)\varepsilon\|(\overline{R}k_u u_\varepsilon - M)_+ - (l_s s_\varepsilon - M)_+\|_{\Gamma_\varepsilon^{ER},t}^2$$

$$\leq c_1 - \frac{1}{\overline{R}k_u}(Mk, (\overline{R}k_u u_\varepsilon - M)_+)_{\Omega_\varepsilon,t} - \frac{1}{l_s}\varepsilon\langle Mk, (l_s s_\varepsilon - M)_+\rangle_{\Gamma_\varepsilon^{ER},t}.$$

where we choose $\lambda > \frac{1}{2}$, but finite.
Now we distinguish two cases.

a) Either $\overline{R}k_u u_\varepsilon - M \leq 0$ and $l_s s_\varepsilon - M \leq 0$ almost everywhere in $\Omega_\varepsilon$ and $\Gamma_\varepsilon^{ER}$, respectively.
Then $u_\varepsilon \in L^\infty(\Omega_\varepsilon)$ and $s_\varepsilon \in L^\infty(\Gamma_\varepsilon^{ER})$ for almost every $t \in [0,T]$ and the assertion holds true.

b) Or there exists $V \subset \Omega_\varepsilon$ (not a null set) with $\overline{R}k_u u_\varepsilon - M > 0$ in $V$ or there exists $V \subset \Gamma_\varepsilon^{ER}$ (not a null set) with $l_s s_\varepsilon - M > 0$ in $V$.
Then we choose $k$ such that the right-hand side is smaller than or equal to zero and we conclude

$$\frac{1}{2\overline{R}k_u}\|(\overline{R}k_u u_\varepsilon - M)_+\|_{\Omega_\varepsilon}^2 + \frac{1}{2l_s}\varepsilon\|(l_s s_\varepsilon - M)_+\|_{\Gamma_\varepsilon^{ER}}^2$$

$$+ \frac{D_u}{\overline{R}k_u}\|\nabla(\overline{R}k_u u_\varepsilon - M)_+\|_{\Omega_\varepsilon,t}^2 + \frac{D_s}{l_s}\varepsilon^3\|\nabla_\Gamma(l_s s_\varepsilon - M)_+\|_{\Gamma_\varepsilon^{ER},t}^2$$

$$+ c_1\varepsilon\|(\overline{R}k_u u_\varepsilon - M)_+ - (l_s s_\varepsilon - M)_+\|_{\Gamma_\varepsilon^{ER},t}^2 \leq 0.$$

This yields $\overline{R}k_u u_\varepsilon - M < 0$ and $l_s s_\varepsilon - M < 0$ almost everywhere in $\Omega_\varepsilon$ and $\Gamma_\varepsilon^{ER}$, respectively, and for almost every $t \in [0,T]$.

The proof for $v_\varepsilon$ and $w_\varepsilon$ is very similar. With corresponding estimates as before we get that $\overline{Rk}_v v_\varepsilon - M \leq 0$ and $l_w w_\varepsilon - M \leq 0$ almost everywhere in $\Omega_\varepsilon$ and $\Gamma_\varepsilon^{ER}$, respectively, and for almost every $t \in [0, T]$. $\qquad\square$

**Corollary 3.35 (Boundedness in $L^\infty$)**
*The traces of the functions $u_\varepsilon$ and $v_\varepsilon$ are bounded, independent of $\varepsilon$, for almost every $x \in \Gamma_\varepsilon^{ER}$ and $t \in [0, T]$.*

**Proof** This proof is based on the proof of lemma 3.34. Furthermore, we are going to use the trace inequality

$$\varepsilon \|(\overline{Rk}_u u_\varepsilon - M)_+\|_{\Gamma_\varepsilon^{ER}}^2$$
$$\leq c_0 \left( \|(\overline{Rk}_u u_\varepsilon - M)_+\|_{\Omega_\varepsilon}^2 + \varepsilon^2 \|\nabla(\overline{Rk}_u u_\varepsilon - M)_+\|_{\Omega_\varepsilon}^2 \right).$$

From lemma 3.34 we know that $(\overline{Rk}_u u_\varepsilon - M)_+ = 0$ and $(\overline{Rk}_v v_\varepsilon - M)_+ = 0$ almost everywhere in $\Omega_\varepsilon$. It remains to show that $\|\nabla(\overline{Rk}_u u_\varepsilon - M)_+\|_{\Omega_\varepsilon}^2 = 0$ and $\|\nabla(\overline{Rk}_v v_\varepsilon - M)_+\|_{\Omega_\varepsilon}^2 = 0$ almost everywhere.

We deduce that

$$(\partial_t u_\varepsilon, \underbrace{(\overline{Rk}_u u_\varepsilon - M)_+}_{=0,\text{ a.e.}})_{\Omega_\varepsilon} + \varepsilon \langle \partial_t s_\varepsilon, \underbrace{(l_s s_\varepsilon - M)_+}_{=0,\text{ a.e.}} \rangle_{\Gamma_\varepsilon^{ER}}$$
$$+ (D_u \nabla u_\varepsilon, \nabla(\overline{Rk}_u u_\varepsilon - M)_+)_{\Omega_\varepsilon} + \varepsilon^3 \langle D_s \nabla_\Gamma s_\varepsilon, \nabla(l_s s_\varepsilon - M)_+ \rangle_{\Gamma_\varepsilon^{ER}}$$
$$+ \varepsilon \langle k_u R_\varepsilon u_\varepsilon - l_s s_\varepsilon, (\overline{Rk}_u u_\varepsilon - M)_+ \rangle_{\Gamma_\varepsilon^{ER}}$$
$$= \underbrace{(-f(u_\varepsilon), (\overline{Rk}_u u_\varepsilon - M)_+)_{\Omega_\varepsilon}}_{\leq 0} + \varepsilon \langle k_u R_\varepsilon u_\varepsilon - l_s s_\varepsilon, \underbrace{(l_s s_\varepsilon - M)_+}_{=0,\text{ a.e.}} \rangle_{\Gamma_\varepsilon^{ER}}$$
$$+ \underbrace{\langle -h(s_\varepsilon), (l_s s_\varepsilon - M)_+ \rangle_{\Gamma_\varepsilon^{ER}}}_{\leq 0}.$$

Hence,

$$\frac{D_u}{\overline{Rk}_u} \|\nabla(\overline{Rk}_u u_\varepsilon - M)_+\|_{\Omega_\varepsilon}^2 + \frac{D_s}{l_s} \varepsilon^3 \|\nabla_\Gamma(l_s s_\varepsilon - M)_+\|_{\Gamma_\varepsilon^{ER}}^2$$
$$+ \varepsilon \langle (k_u R_\varepsilon u_\varepsilon - M) - \underbrace{(l_s s_\varepsilon - M)}_{\leq 0}, (\overline{Rk}_u u_\varepsilon - M)_+ \rangle_{\Gamma_\varepsilon^{ER}} \leq 0.$$

119

This leads to

$$\frac{D_u}{\overline{R}k_u}\|\nabla(\overline{R}k_u u_\varepsilon - M)_+\|_{\Omega_\varepsilon}^2 + \frac{D_s}{l_s}\varepsilon^3\|\nabla_\Gamma(l_s s_\varepsilon - M)_+\|_{\Gamma_\varepsilon^{ER}}^2$$

$$\leq \varepsilon\langle|k_u R_\varepsilon u_\varepsilon - M|, (\overline{R}k_u u_\varepsilon - M)_+\rangle_{\Gamma_\varepsilon^{ER}}$$

$$\leq \varepsilon\langle|k_u \overline{R} u_\varepsilon - M|, (\overline{R}k_u u_\varepsilon - M)_+\rangle_{\Gamma_\varepsilon^{ER}}$$

$$= \varepsilon\|(\overline{R}k_u u_\varepsilon - M)_+\|_{\Gamma_\varepsilon^{ER}}^2$$

$$\leq c_0\left(\|(\overline{R}k_u u_\varepsilon - M)_+\|_{\Omega_\varepsilon}^2 + \varepsilon^2\|\nabla(\overline{R}k_u u_\varepsilon - M)_+\|_{\Omega_\varepsilon}^2\right).$$

Hence,

$$\underbrace{\left(\frac{D_u}{\overline{R}k_u} - c_0\varepsilon^2\right)}_{\geq 0, \text{ for small } \varepsilon}\|\nabla(\overline{R}k_u u_\varepsilon - M)_+\|_{\Omega_\varepsilon}^2 + \frac{D_s}{l_s}\varepsilon^3\|\nabla_\Gamma(l_s s_\varepsilon - M)_+\|_{\Gamma_\varepsilon^{ER}}^2$$

$$\leq c_0\|(\overline{R}k_u u_\varepsilon - M)_+\|_{\Omega_\varepsilon}^2 = 0.$$

We deduce $\|\nabla(\overline{R}k_u u_\varepsilon - M)_+\|_{\Omega_\varepsilon}^2 = 0$.
With similar estimates we get $\|\nabla(\overline{R}k_v v_\varepsilon - M)_+\|_{\Omega_\varepsilon}^2 = 0$. This completes the proof. $\square$

Now we show that the time derivatives of $u_\varepsilon$ and $v_\varepsilon$ are elements of $H_0^1(\Omega_\varepsilon)'$.

**Lemma 3.36 (Time-estimation in $(H_0^1)'$)**
*There exists a $C > 0$, independent of $\varepsilon$, such that*

$$\|\partial_t u_\varepsilon\|_{L^2([0,T],H_0^1(\Omega_\varepsilon)')} + \|\partial_t v_\varepsilon\|_{L^2([0,T],H_0^1(\Omega_\varepsilon)')} < C.$$

**Proof**
We start by writing the $H_0^1(\Omega_\varepsilon)'$-Norm in full for $\partial_t u_\varepsilon$. In the following we use that test functions $\varphi$ in $H_0^1(\Omega_\varepsilon)$ are zero on the

boundary $\Gamma_\varepsilon^{ER}$,

$$\|\partial_t u_\varepsilon\|_{H_0^1(\Omega_\varepsilon)'} = \sup_{\varphi \in H_0^1(\Omega_\varepsilon), \|\varphi\|=1} (\partial_t u_\varepsilon, \varphi)_{H_0^1(\Omega_\varepsilon)' \times H_0^1(\Omega_\varepsilon)}$$

$$= \sup_{\varphi \in H_0^1(\Omega_\varepsilon), \|\varphi\|=1} ((-D_u \nabla u_\varepsilon, \nabla \varphi)_{H_0^1(\Omega_\varepsilon)' \times H_0^1(\Omega_\varepsilon)}$$

$$\underbrace{- \varepsilon \langle k_u R u_\varepsilon - l_s s_\varepsilon, \varphi \rangle_{\Gamma_\varepsilon^{ER}}}_{=0} - (f(u_\varepsilon), \varphi)_{H_0^1(\Omega_\varepsilon)' \times H_0^1(\Omega_\varepsilon)})$$

$$\leq \sup_{\varphi \in H_0^1(\Omega_\varepsilon), \|\varphi\|=1} (c_1 \|\nabla u_\varepsilon\|_{L^2(\Omega_\varepsilon)} \|\nabla \varphi\|_{L^2(\Omega)}$$

$$+ c_2 \|f(u_\varepsilon)\|_{L^2(\Omega_\varepsilon)} \|\varphi\|_{L^2(\Omega_\varepsilon)})$$

$$\leq c_1 \left( \|\nabla u_\varepsilon\|_{\Omega_\varepsilon} + \|f(u_\varepsilon)\|_{\Omega_\varepsilon} \right).$$

Integration with respect to time yields

$$\|\partial_t u_\varepsilon\|_{L^2([0,T],H_0^1(\Omega_\varepsilon)')}^2 \leq c_1 \left( \|\nabla u_\varepsilon\|_{\Omega_\varepsilon,t}^2 + \|f(u_\varepsilon)\|_{\Omega_\varepsilon,t}^2 \right) < c_2,$$

where the boundedness holds because of lemma 3.33. The proof for $\|\partial_t v_\varepsilon\|_{L^2([0,T],H_0^1(\Omega_\varepsilon)')}$ works analogously. $\qquad \square$

Now we know that $u_\varepsilon, v_\varepsilon \in L^2([0,T], H^1(\Omega)) \cap H^1([0,T], H_0^1(\Omega)') \cap L^\infty(\Omega \times [0,T])$ and therefore, applying lemma 3.4, $u_\varepsilon, v_\varepsilon$ converge strongly to $u_0, v_0$ in $L^2([0,T], L^2(\Omega))$, respectively.

We cannot prove strong convergence of the functions $s_\varepsilon$, $w_\varepsilon$ and $R_\varepsilon$ using extensions to $\Omega$ (theorem 3.1) and applying lemma 3.4, because they are defined on the $\varepsilon$-dependent manifold $\Gamma_\varepsilon^{ER}$ which has a smaller dimension than $\Omega$. Hence, we use the boundary unfolding operator $\mathcal{T}_\varepsilon^b$, because it is already defined on a fixed domain $\Omega \times \Gamma$ and show that $\mathcal{T}_\varepsilon^b(s_\varepsilon)$, $\mathcal{T}_\varepsilon^b(w_\varepsilon)$ and $\mathcal{T}_\varepsilon^b(R_\varepsilon)$ are Cauchy-sequences. This idea is from article [17].

**Lemma 3.37** ($s_\varepsilon$, $w_\varepsilon$, $R_\varepsilon$ are Cauchy-sequences)
*For all $\delta > 0$ there exists $\tilde{\varepsilon} > 0$ such that for all $0 < \varepsilon_1, \varepsilon_2 < \tilde{\varepsilon}$ it holds that*

$$\|\mathcal{T}_{\varepsilon_1}^b(s_{\varepsilon_1}) - \mathcal{T}_{\varepsilon_2}^b(s_{\varepsilon_2})\|_{[0,T] \times \Omega \times \Gamma}^2 + \|\mathcal{T}_{\varepsilon_1}^b(w_{\varepsilon_1}) - \mathcal{T}_{\varepsilon_2}^b(w_{\varepsilon_2})\|_{[0,T] \times \Omega \times \Gamma}^2$$

$$+ \|\mathcal{T}_{\varepsilon_1}^b(R_{\varepsilon_1}) - \mathcal{T}_{\varepsilon_2}^b(R_{\varepsilon_2})\|_{[0,T] \times \Omega \times \Gamma}^2 < \delta.$$

This means that $s_\varepsilon$, $w_\varepsilon$ and $R_\varepsilon$ are Cauchy-sequences in $L^2([0,T] \times \Omega \times \Gamma)$.

**Proof** We apply part 5 of lemma 3.15 to the weak equation of $s_\varepsilon$ and find

$$(\partial_t \mathcal{T}_\varepsilon^b(s_\varepsilon), \psi)_{\Omega \times \Gamma} + D_s(\varepsilon \mathcal{T}_\varepsilon^b(\nabla_\Gamma s_\varepsilon), \nabla_\Gamma \psi)_{\Omega \times \Gamma}$$
$$= (k_u \mathcal{T}_\varepsilon^b(u_\varepsilon) \mathcal{T}_\varepsilon^b(R_\varepsilon) - l_s \mathcal{T}_\varepsilon^b(s_\varepsilon), \psi)_{\Omega \times \Gamma} - (\mathcal{T}_\varepsilon^b(h(s_\varepsilon)), \psi)_{\Omega \times \Gamma}$$

for all $\psi \in L^2(\Omega, H^1_\#(\Gamma))$.

Now we write this equation for two epsilons $\varepsilon_1$ and $\varepsilon_2$ and subtract the equations from each other. As test function $\psi$ we take $\psi = \mathcal{T}_{\varepsilon_1}^b(s_{\varepsilon_1}) - \mathcal{T}_{\varepsilon_2}^b(s_{\varepsilon_2})$.

$$(\partial_t(\mathcal{T}_{\varepsilon_1}^b(s_{\varepsilon_1}) - \mathcal{T}_{\varepsilon_2}^b(s_{\varepsilon_2})), \mathcal{T}_{\varepsilon_1}^b(s_{\varepsilon_1}) - \mathcal{T}_{\varepsilon_2}^b(s_{\varepsilon_2}))_{\Omega \times \Gamma}$$
$$+ D_s(\varepsilon_1 \mathcal{T}_{\varepsilon_1}^b(\nabla_\Gamma s_{\varepsilon_1}) - \varepsilon_2 \mathcal{T}_{\varepsilon_2}^b(\nabla_\Gamma s_{\varepsilon_2}), \varepsilon_1 \mathcal{T}_{\varepsilon_1}^b(\nabla_\Gamma s_{\varepsilon_1}) - \varepsilon_2 \mathcal{T}_{\varepsilon_2}^b(\nabla_\Gamma s_{\varepsilon_2}))_{\Omega \times \Gamma}$$
$$= (k_u \mathcal{T}_{\varepsilon_1}^b(u_{\varepsilon_1}) \mathcal{T}_{\varepsilon_1}^b(R_{\varepsilon_1}) - l_s \mathcal{T}_{\varepsilon_1}^b(s_{\varepsilon_1}) - k_u \mathcal{T}_{\varepsilon_2}^b(u_{\varepsilon_2}) \mathcal{T}_{\varepsilon_2}^b(R_{\varepsilon_2})$$
$$+ l_s \mathcal{T}_{\varepsilon_2}^b(s_{\varepsilon_2}), \mathcal{T}_{\varepsilon_1}^b(s_{\varepsilon_1}) - \mathcal{T}_{\varepsilon_2}^b(s_{\varepsilon_2}))_{\Omega \times \Gamma}$$
$$- (\mathcal{T}_{\varepsilon_1}^b(h(s_{\varepsilon_1})) - \mathcal{T}_{\varepsilon_2}^b(h(s_{\varepsilon_2})), \mathcal{T}_{\varepsilon_1}^b(s_{\varepsilon_1}) - \mathcal{T}_{\varepsilon_2}^b(s_{\varepsilon_2}))_{\Omega \times \Gamma}.$$

This leads to

$$(\partial_t(\mathcal{T}_{\varepsilon_1}^b(s_{\varepsilon_1}) - \mathcal{T}_{\varepsilon_2}^b(s_{\varepsilon_2})), \mathcal{T}_{\varepsilon_1}^b(s_{\varepsilon_1}) - \mathcal{T}_{\varepsilon_2}^b(s_{\varepsilon_2}))_{\Omega \times \Gamma}$$
$$+ D_s \|\varepsilon_1 \mathcal{T}_{\varepsilon_1}^b(\nabla_\Gamma s_{\varepsilon_1}) - \varepsilon_2 \mathcal{T}_{\varepsilon_2}^b(\nabla_\Gamma s_{\varepsilon_2})\|^2_{\Omega \times \Gamma} + l_s \|\mathcal{T}_{\varepsilon_1}^b(s_{\varepsilon_1}) - \mathcal{T}_{\varepsilon_2}^b(s_{\varepsilon_2})\|^2_{\Omega \times \Gamma}$$
$$= (k_u \mathcal{T}_{\varepsilon_1}^b(u_{\varepsilon_1}) \mathcal{T}_{\varepsilon_1}^b(R_{\varepsilon_1}) - k_u \mathcal{T}_{\varepsilon_2}^b(u_{\varepsilon_2}) \mathcal{T}_{\varepsilon_1}^b(R_{\varepsilon_1}) + k_u \mathcal{T}_{\varepsilon_2}^b(u_{\varepsilon_2}) \mathcal{T}_{\varepsilon_1}^b(R_{\varepsilon_1})$$
$$- k_u \mathcal{T}_{\varepsilon_2}^b(u_{\varepsilon_2}) \mathcal{T}_{\varepsilon_2}^b(R_{\varepsilon_2}), \mathcal{T}_{\varepsilon_1}^b(s_{\varepsilon_1}) - \mathcal{T}_{\varepsilon_2}^b(s_{\varepsilon_2}))_{\Omega \times \Gamma}$$
$$- (\mathcal{T}_{\varepsilon_1}^b(h(s_{\varepsilon_1})) - \mathcal{T}_{\varepsilon_2}^b(h(s_{\varepsilon_2})), \mathcal{T}_{\varepsilon_1}^b(s_{\varepsilon_1}) - \mathcal{T}_{\varepsilon_2}^b(s_{\varepsilon_2}))_{\Omega \times \Gamma}$$
$$\leq k_u \overline{R} \|\mathcal{T}_{\varepsilon_1}^b(u_{\varepsilon_1}) - \mathcal{T}_{\varepsilon_2}^b(u_{\varepsilon_2})\|^2_{\Omega \times \Gamma} + k_u \overline{R} \|\mathcal{T}_{\varepsilon_1}^b(s_{\varepsilon_1}) - \mathcal{T}_{\varepsilon_2}^b(s_{\varepsilon_2})\|^2_{\Omega \times \Gamma}$$
$$+ k_u \|u_\varepsilon\|_{L^\infty} \|\mathcal{T}_{\varepsilon_1}^b(R_{\varepsilon_1}) - \mathcal{T}_{\varepsilon_2}^b(R_{\varepsilon_2})\|^2_{\Omega \times \Gamma}$$
$$+ k_u \|u_\varepsilon\|_{L^\infty} \|\mathcal{T}_{\varepsilon_1}^b(s_{\varepsilon_1}) - \mathcal{T}_{\varepsilon_2}^b(s_{\varepsilon_2})\|^2_{\Omega \times \Gamma} + L_h \|\mathcal{T}_{\varepsilon_1}^b(s_{\varepsilon_1}) - \mathcal{T}_{\varepsilon_2}^b(s_{\varepsilon_2})\|^2_{\Omega \times \Gamma}.$$

Here we used that the function $h$ is Lipschitz-continuous with constant $L_h$. We integrate from 0 to $t$ and merge the constants to a single constant $c_1 > 0$.

$$\|\mathcal{T}^b_{\varepsilon_1}(s_{\varepsilon_1}) - \mathcal{T}^b_{\varepsilon_2}(s_{\varepsilon_2})\|^2_{\Omega \times \Gamma} + \|\varepsilon_1 \mathcal{T}^b_{\varepsilon_1}(\nabla_\Gamma s_{\varepsilon_1}) - \varepsilon_2 \mathcal{T}^b_{\varepsilon_2}(\nabla_\Gamma s_{\varepsilon_2})\|^2_{\Omega \times \Gamma, t}$$
$$+ \|\mathcal{T}^b_{\varepsilon_1}(s_{\varepsilon_1}) - \mathcal{T}^b_{\varepsilon_2}(s_{\varepsilon_2})\|^2_{\Omega \times \Gamma, t} \leq c_1 (\|\mathcal{T}^b_{\varepsilon_1}(s_{\varepsilon_1}) - \mathcal{T}^b_{\varepsilon_2}(s_{\varepsilon_2})\|^2_{\Omega \times \Gamma, t}$$
$$+ \|\mathcal{T}^b_{\varepsilon_1}(u_{\varepsilon_1}) - \mathcal{T}^b_{\varepsilon_2}(u_{\varepsilon_2})\|^2_{\Omega \times \Gamma, t} + \|\mathcal{T}^b_{\varepsilon_1}(R_{\varepsilon_1}) - \mathcal{T}^b_{\varepsilon_2}(R_{\varepsilon_2})\|^2_{\Omega \times \Gamma, t}).$$

With similar estimations we find

$$\|\mathcal{T}^b_{\varepsilon_1}(R_{\varepsilon_1}) - \mathcal{T}^b_{\varepsilon_2}(R_{\varepsilon_2})\|^2_{\Omega \times \Gamma} \leq c_1 \left( \|\mathcal{T}^b_{\varepsilon_1}(R_{\varepsilon_1}) - \mathcal{T}^b_{\varepsilon_2}(R_{\varepsilon_2})\|^2_{\Omega \times \Gamma, t} \right.$$
$$+ \|\mathcal{T}^b_{\varepsilon_1}(s_{\varepsilon_1}) - \mathcal{T}^b_{\varepsilon_2}(s_{\varepsilon_2})\|^2_{\Omega \times \Gamma, t} + \|\mathcal{T}^b_{\varepsilon_1}(w_{\varepsilon_1}) - \mathcal{T}^b_{\varepsilon_2}(w_{\varepsilon_2})\|^2_{\Omega \times \Gamma, t}$$
$$\left. + \|\mathcal{T}^b_{\varepsilon_1}(u_{\varepsilon_1}) - \mathcal{T}^b_{\varepsilon_2}(u_{\varepsilon_2})\|^2_{\Omega \times \Gamma, t} + \|\mathcal{T}^b_{\varepsilon_1}(v_{\varepsilon_1}) - \mathcal{T}^b_{\varepsilon_2}(v_{\varepsilon_2})\|^2_{\Omega \times \Gamma, t} \right)$$

and

$$\|\mathcal{T}^b_{\varepsilon_1}(w_{\varepsilon_1}) - \mathcal{T}^b_{\varepsilon_2}(w_{\varepsilon_2})\|^2_{\Omega \times \Gamma} + \|\varepsilon_1 \mathcal{T}^b_{\varepsilon_1}(\nabla_\Gamma w_{\varepsilon_1}) - \varepsilon_2 \mathcal{T}^b_{\varepsilon_2}(\nabla_\Gamma w_{\varepsilon_2})\|^2_{\Omega \times \Gamma, t}$$
$$+ \|\mathcal{T}^b_{\varepsilon_1}(w_{\varepsilon_1}) - \mathcal{T}^b_{\varepsilon_2}(w_{\varepsilon_2})\|^2_{\Omega \times \Gamma, t}$$
$$\leq c_1 \|\mathcal{T}^b_{\varepsilon_1}(s_{\varepsilon_1}) - \mathcal{T}^b_{\varepsilon_2}(s_{\varepsilon_2})\|^2_{\Omega \times \Gamma, t} + \|\mathcal{T}^b_{\varepsilon_1}(w_{\varepsilon_1}) - \mathcal{T}^b_{\varepsilon_2}(w_{\varepsilon_2})\|^2_{\Omega \times \Gamma, t}$$
$$+ \|\mathcal{T}^b_{\varepsilon_1}(v_{\varepsilon_1}) - \mathcal{T}^b_{\varepsilon_2}(v_{\varepsilon_2})\|^2_{\Omega \times \Gamma, t} + \|\mathcal{T}^b_{\varepsilon_1}(R_{\varepsilon_1}) - \mathcal{T}^b_{\varepsilon_2}(R_{\varepsilon_2})\|^2_{\Omega \times \Gamma, t}.$$

Adding all three inequalities and using Gronwall's lemma gives

$$\|\mathcal{T}^b_{\varepsilon_1}(s_{\varepsilon_1}) - \mathcal{T}^b_{\varepsilon_2}(s_{\varepsilon_2})\|^2_{\Omega \times \Gamma} + \|\mathcal{T}^b_{\varepsilon_1}(w_{\varepsilon_1}) - \mathcal{T}^b_{\varepsilon_2}(w_{\varepsilon_2})\|^2_{\Omega \times \Gamma}$$
$$+ \|\mathcal{T}^b_{\varepsilon_1}(R_{\varepsilon_1}) - \mathcal{T}^b_{\varepsilon_2}(R_{\varepsilon_2})\|^2_{\Omega \times \Gamma}$$
$$\leq c_1 \left( \|\mathcal{T}^b_{\varepsilon_1}(u_{\varepsilon_1}) - \mathcal{T}^b_{\varepsilon_2}(u_{\varepsilon_2})\|^2_{\Omega \times \Gamma} + \|\mathcal{T}^b_{\varepsilon_1}(v_{\varepsilon_1}) - \mathcal{T}^b_{\varepsilon_2}(v_{\varepsilon_2})\|^2_{\Omega \times \Gamma} \right)$$
$$= c_1 \left( \|\gamma_{\Omega \times \Gamma} \left( \mathcal{T}_{\varepsilon_1}(u_{\varepsilon_1}) - \mathcal{T}_{\varepsilon_2}(u_{\varepsilon_2}) \right)\|^2_{\Omega \times \Gamma} \right.$$
$$\left. + \|\gamma_{\Omega \times \Gamma} \left( \mathcal{T}_{\varepsilon_1}(v_{\varepsilon_1}) - \mathcal{T}_{\varepsilon_2}(v_{\varepsilon_2}) \right)\|^2_{\Omega \times \Gamma} \right)$$
$$\leq c_1 c_0 \left( \|\mathcal{T}_{\varepsilon_1}(u_{\varepsilon 1}) - \mathcal{T}_{\varepsilon_2}(u_{\varepsilon_2})\|^2_{\Omega \times Y} + \|\mathcal{T}_{\varepsilon_1}(v_{\varepsilon_1}) - \mathcal{T}_{\varepsilon_2}(v_{\varepsilon_2})\|^2_{\Omega \times Y} \right.$$
$$+ \|\varepsilon_1 \mathcal{T}_{\varepsilon_1}(\nabla_x u_{\varepsilon_1}) - \varepsilon_2 \mathcal{T}_{\varepsilon_2}(\nabla_x u_{\varepsilon_2})\|^2_{\Omega \times Y}$$
$$\left. + \|\varepsilon_1 \mathcal{T}_{\varepsilon_1}(\nabla_x v_{\varepsilon_1}) - \varepsilon_2 \mathcal{T}_{\varepsilon_2}(\nabla_x v_{\varepsilon_2})\|^2_{\Omega \times Y} \right)$$

where we used part 8 of lemma 3.9. With part 7 of lemma 3.9 and integration with respect to time we find

$$
\|\mathcal{T}_{\varepsilon_1}^b(u_{\varepsilon_1}) - \mathcal{T}_{\varepsilon_2}^b(u_{\varepsilon_2})\|_{\Omega \times \Gamma, t}^2 + \|\mathcal{T}_{\varepsilon_1}^b(v_{\varepsilon_1}) - \mathcal{T}_{\varepsilon_2}^b(v_{\varepsilon_2})\|_{\Omega \times \Gamma, t}^2
$$
$$
\leq c_1 |Y| (\underbrace{\|u_{\varepsilon_1} - u_{\varepsilon_2}\|_{\Omega, t}^2 + \|v_{\varepsilon_1} - v_{\varepsilon_2}\|_{\Omega, t}^2}_{< \tilde{\delta}}
$$
$$
+ \underbrace{\max\{\varepsilon_1, \varepsilon_2\}^2}_{\xrightarrow{\varepsilon \to 0} 0} (\underbrace{\|\nabla_x u_{\varepsilon_1}\|_{\Omega, t}^2 + \|\nabla_x u_{\varepsilon_2}\|_{\Omega, t}^2}_{< C, \text{ bounded}}))
$$
$$
+ \underbrace{\max\{\varepsilon_1, \varepsilon_2\}^2}_{\xrightarrow{\varepsilon \to 0} 0} (\underbrace{\|\nabla_x v_{\varepsilon_1}\|_{\Omega, t}^2 + \|\nabla_x v_{\varepsilon_2}\|_{\Omega, t}^2}_{< C, \text{ bounded}})). \quad (3.18)
$$

Because $u_\varepsilon$ and $v_\varepsilon$ converge strongly in $L^2([0, T] \times \Omega)$, there exists a $\tilde{\varepsilon} > 0$ such that the first curly bracket holds true for $\varepsilon_1, \varepsilon_2 < \tilde{\varepsilon}$. Hence, we deduce

$$
\|\mathcal{T}_{\varepsilon_1}^b(s_{\varepsilon_1}) - \mathcal{T}_{\varepsilon_2}^b(s_{\varepsilon_2})\|_{\Omega \times \Gamma, t}^2 + \|\mathcal{T}_{\varepsilon_1}^b(w_{\varepsilon_1}) - \mathcal{T}_{\varepsilon_2}^b(w_{\varepsilon_2})\|_{\Omega \times \Gamma, t}^2
$$
$$
+ \|\mathcal{T}_{\varepsilon_1}^b(R_{\varepsilon_1}) - \mathcal{T}_{\varepsilon_2}^b(R_{\varepsilon_2})\|_{\Omega \times \Gamma, t}^2 \leq c_1 (\tilde{\delta} + \tilde{\varepsilon} C) \leq \delta
$$

for $\varepsilon_1, \varepsilon_2 < \tilde{\varepsilon}$ and $\delta$ dependent on $\tilde{\varepsilon}$. This means that $s_\varepsilon$, $w_\varepsilon$ and $R_\varepsilon$ converge strongly in $L^2([0, T] \times \Omega \times \Gamma)$. $\qquad \square$

**Remark 3.38** With lemma 3.33, lemma 3.34, lemma 3.36 and the extension theorem 3.1 we deduce that $u_\varepsilon$ and $v_\varepsilon$ are elements of $L^2([0, T], H^1(\Omega)) \cap H^1([0, T], H_0^1(\Omega)') \cap L^\infty([0, T] \times \Omega)$. Using lemma 3.4 we obtain that $u_\varepsilon$, $v_\varepsilon$ converge strongly to $u_0$, $v_0$ in $L^2([0, T], L^2(\Omega))$.
This fact also implies with lemma 3.37 that $\mathcal{T}_\varepsilon^b(s_\varepsilon)$, $\mathcal{T}_\varepsilon^b(w_\varepsilon)$ and $\mathcal{T}_\varepsilon^b(R_\varepsilon)$ converge strongly to functions $s_0$, $w_0$ and $R_0$ in $L^2([0, T] \times \Omega \times \Gamma)$.

Furthermore, we need to prove that for every $\varepsilon > 0$ there exists a solution of the system of equations (3.17). We are going to show this assertion in the next section 3.5.

## 3.5 Existence of a solution

We are going to show existence of the solution $(u_\varepsilon, v_\varepsilon, s_\varepsilon, w_\varepsilon, R_\varepsilon)$ of the system of equations (3.17) for every $\varepsilon > 0$.

This will be done in two steps. At first we prove existence of the function $R_\varepsilon \in L^2([0,T] \times \Gamma_\varepsilon^{ER})$ when we assume the existence of $s_\varepsilon, w_\varepsilon, u_\varepsilon, v_\varepsilon \in L^2([0,T] \times \Gamma_\varepsilon^{ER})$. Secondly, we prove a preparing lemma to use Schauder's fixed-point theorem. Then we show by using Schauder's theorem the existence of the solutions $u_\varepsilon$, $v_\varepsilon$, $s_\varepsilon$, and $w_\varepsilon$.

Hence, we start with considering the ordinary differential equation for $R_\varepsilon$ given by

$$
\begin{aligned}
\partial_t R_\varepsilon(t,x) &= -R_\varepsilon(t,x)|k_u u_\varepsilon(t,x) + k_v v_\varepsilon(t,x)| \\
&\qquad + (\overline{R} - R_\varepsilon(t,x))|k_s s_\varepsilon(t,x) + k_w w_\varepsilon(t,x)| \\
R_\varepsilon(0,x) &= \overline{R} \in \mathbb{R}^+,
\end{aligned}
$$

(3.19)

with $k_u, k_v, k_s, k_w > 0$.

**Theorem 3.39 (Existence of $R_\varepsilon$)**
*Let $\varepsilon > 0$ and $s_\varepsilon, w_\varepsilon, u_\varepsilon, v_\varepsilon \in L^2([0,T] \times \Gamma_\varepsilon^{ER})$. Then there exists a solution $R_\varepsilon \in \{u \in L^2([0,T] \times \Gamma_\varepsilon^{ER})| \partial_t u \in L^2([0,T] \times \Gamma_\varepsilon^{ER})\}$ of the ordinary differential equation (3.19).*

**Proof** We use Carathéodory's existence theorem (see [14]). For that reason, we define for almost every $x \in \Gamma_\varepsilon^{ER}$ the function $j_x : [0,T] \times [0,\overline{R}] \to \mathbb{R}$ as

$$
j_x(t, R_\varepsilon) := -R_\varepsilon|k_u u_\varepsilon(t) + k_v v_\varepsilon(t)| + (\overline{R} - R_\varepsilon)|k_s s_\varepsilon(t) + k_w w_\varepsilon(t)|.
$$

Carathéodorie's existence theorem says: If the following conditions hold for the function $j_x$ for almost every $x \in \Gamma_\varepsilon^{ER}$, then there exists a solution $R_\varepsilon(\cdot, x) \in C([0,T])$ for almost every $x \in \Gamma_\varepsilon^{ER}$.

  a) The function $j_x$ must be defined on a rectangle $[0,T] \times [0,\overline{R}]$.

  b) The function $j_x$ has to be measurable in $t$ for all fixed $R_\varepsilon \in [0,\overline{R}]$.

  c) The function $j_x$ has to be continuous in $R_\varepsilon$ for all fixed $t \in [0,T]$.

  d) There exists a Lebesgue-integrable function $m : [0,T] \to \mathbb{R}$ such that $|j_x(t, R_\varepsilon)| \le m(t)$ for all $(t, R_\varepsilon) \in [0,T] \times [0,\overline{R}]$.

Conditions a) and c) are easily satisfied. Condition b) is true because $u_\varepsilon, v_\varepsilon, s_\varepsilon, w_\varepsilon$ are $L^2$-functions and $|\cdot|$ is continuous. For d) we use that $u_\varepsilon(x)$, $v_\varepsilon(x)$, $s_\varepsilon(x)$ and $w_\varepsilon(x)$ are elements of $L^2([0,T])$, thus Lebesgue-integrable for almost every $x \in \Gamma_\varepsilon^{ER}$. It follows that $j_x$ is Lebesgue-integrable itself and condition d) is fulfilled.

Hence, there exists a solution $R_\varepsilon(\cdot, x) \in C([0,T])$ for almost every $x \in L^2(\Gamma_\varepsilon^{ER})$.

We notice that $R_\varepsilon(t, x) \in [0, \overline{R}]$ for almost every $t, x \in [0,T] \times \Gamma_\varepsilon^{ER}$. The function $R : [L^2((0,T) \times \Gamma_\varepsilon^{ER})]^4 \rightarrow C([0,T], L^2(\Gamma_\varepsilon^{ER})) \subset L^2([0,T] \times \Gamma_\varepsilon^{ER})$ with $R(u_\varepsilon, v_\varepsilon, s_\varepsilon, w_\varepsilon) = R_\varepsilon$ is bounded and continuous.

Finally we estimate

$$\|\partial_t R_\varepsilon\|_{\Gamma_\varepsilon^{ER}, T} = \|R_\varepsilon|k_u u_\varepsilon + k_v v_\varepsilon| + (\overline{R} - R_\varepsilon)|k_s s_\varepsilon + k_w w_\varepsilon|\|_{\Gamma_\varepsilon^{ER}, T}$$
$$\leq \overline{R} k_u \|u_\varepsilon\|_{\Gamma_\varepsilon^{ER}, T} + \overline{R} k_v \|v_\varepsilon\|_{\Gamma_\varepsilon^{ER}, T}$$
$$+ \overline{R} k_s \|s_\varepsilon\|_{\Gamma_\varepsilon^{ER}, T} + \overline{R} k_w \|w_\varepsilon\|_{\Gamma_\varepsilon^{ER}, T},$$

which is bounded because $(u_\varepsilon, v_\varepsilon, s_\varepsilon, w_\varepsilon) \in L^2([0,T] \times \Gamma_\varepsilon^{ER})$.

□

To show existence of the solutions $u_\varepsilon, v_\varepsilon, s_\varepsilon$ and $w_\varepsilon$, we want to use the following fixed-point theorem from Schauder, [58].

**Theorem 3.40 (Schauder)**
*Let $X$ be a Banach space and $T : X \rightarrow X$ a continuous and compact mapping. If there exists a number $r > 0$ such that*

$$\|x\| < r \quad \text{implies} \quad \|T(x)\| < r,$$

*then there exists a $\hat{x} \in X$ with $T(\hat{x}) = \hat{x}$.*

**Lemma 3.41** *The functions $f$, $g$ and $h$ (defined in (3.15)) are continuous, bounded and satisfy the equation $|\varphi(x)| \leq C|x|^{p/q}$ for some $p = q = 2$ and a $C > 0$ for $\varphi = f$, $g$ and $h$.*

*Furthermore, the functions*

$$F, G : L^2([0,T], L^2(\Omega_\varepsilon)) \to L^2([0,T], L^2(\Omega_\varepsilon))$$
$$F(u)(t) = f(u(t))$$
$$G(v)(t) = g(v(t))$$

*and* (3.20)

$$H : L^2([0,T], L^2(\Gamma_\varepsilon^{ER})) \to L^2([0,T], L^2(\Gamma_\varepsilon^{ER}))$$
$$H(s)(t) = h(s(t))$$

*are continuous and bounded.*

**Proof** The functions $f$, $g$ and $h$ have the form

$$\varphi(x) = \begin{cases} \frac{M}{x+M} ax, & x \geq 0, \\ 0 & x < 0 \end{cases}$$

for positive constants $M$ and $a$. Continuity and boundedness are easily satisfied. Further, it holds that

$$|\varphi(x)| = \left| \frac{M}{x+M} ax \right| \leq |ax| = a|x|^{p/p}$$

with $C = a$ and $p = q = 2$.

With this result and Nemytskii's theorem 3.6 we deduce continuity and boundedness of the functions $F, G$ and $H$. $\qquad\square$

Now we come to the main theorem to show existence of a solution of the system (3.17). Theorem 3.39 and lemma 3.41 are going to be used.

**Theorem 3.42 (Existence of $u_\varepsilon$, $v_\varepsilon$, $s_\varepsilon$ and $w_\varepsilon$)**
*For every small $\varepsilon > 0$ there exists at least one solution $(u_\varepsilon, v_\varepsilon, s_\varepsilon, w_\varepsilon, R_\varepsilon) \in \mathcal{V}_C \times \mathcal{V}_N \times \mathcal{V}(\Gamma_\varepsilon^{ER})^2 \times \mathcal{V}_R(\Gamma_\varepsilon^{ER})$ of the system (3.17).*

**Proof** To prove the theorem, we use Schauders' theorem 3.40. We are going to show that there exists a solution for a small time step $[0, \tau]$. To find the solution on the whole interval $[0, T]$ the solution parts must be linked together bit by bit. We define for a $\delta \in [0, 1/2)$ the function space $V = L^2([0, \tau], H^{1-\delta}(\Omega_\varepsilon))$ and $W := L^2([0, \tau], L^2(\Gamma_\varepsilon^{ER}))$. Furthermore we define the mapping

$$T : V^2 \times W^2 \to \{u \in L^2([0, \tau], H^1(\Omega_\varepsilon)) | \partial_t u \in L^2([0, \tau], H^1(\Omega_\varepsilon)')\}^2$$
$$\times \{u \in L^2([0, \tau], H^1(\Gamma_\varepsilon^{ER})) | \partial_t u \in L^2([0, \tau], H^1(\Gamma_\varepsilon^{ER})')\}^2$$

given by

$$T(\hat{u}_\varepsilon, \hat{v}_\varepsilon, \hat{s}_\varepsilon, \hat{w}_\varepsilon) = (u_\varepsilon, v_\varepsilon, s_\varepsilon, w_\varepsilon),$$

where $(u_\varepsilon, v_\varepsilon, s_\varepsilon, w_\varepsilon)$ is given by

$$
\begin{aligned}
\partial_t u_\varepsilon - D_u \Delta u_\varepsilon &= f(\hat{u}_\varepsilon) \\
\partial_t v_\varepsilon - D_v \Delta v_\varepsilon &= g(\hat{v}_\varepsilon) \\
-D_u \nabla u_\varepsilon \cdot n &= \varepsilon(k_u u_\varepsilon R(\hat{u}_\varepsilon, \hat{v}_\varepsilon, \hat{s}_\varepsilon, \hat{w}_\varepsilon) - l_s \hat{s}_\varepsilon) \\
-D_v \nabla v_\varepsilon \cdot n &= \varepsilon(k_v v_\varepsilon R(\hat{u}_\varepsilon, \hat{v}_\varepsilon, \hat{s}_\varepsilon, \hat{w}_\varepsilon) - l_w \hat{w}_\varepsilon) \\
\partial_t s_\varepsilon - \varepsilon^2 D_s \Delta_\Gamma s_\varepsilon + l_s s_\varepsilon &= -h(\hat{s}_\varepsilon) + k_u u_\varepsilon R(\hat{u}_\varepsilon, \hat{v}_\varepsilon, \hat{s}_\varepsilon, \hat{w}_\varepsilon) \\
\partial_t w_\varepsilon - \varepsilon^2 D_w \Delta_\Gamma w_\varepsilon + l_w w_\varepsilon &= h(\hat{s}_\varepsilon) + k_v v_\varepsilon R(\hat{u}_\varepsilon, \hat{v}_\varepsilon, \hat{s}_\varepsilon, \hat{w}_\varepsilon).
\end{aligned}
\tag{3.21}
$$

The partial differential equation (3.21) is linear and has a unique solution (see [21]) and the function $T$ is continuous. With the lemma of Lions-Aubin 3.5 we know that $\{u \in L^2([0, \tau], H^1(\Omega_\varepsilon)) | \partial_t u \in L^2([0, \tau], H^1(\Omega_\varepsilon)')\}$ is compactly embedded in $V$ and $\{u \in L^2([0, \tau], H^1(\Gamma_\varepsilon^{ER})) | \partial_t u \in L^2([0, \tau], H^1(\Gamma_\varepsilon^{ER})')\}$ is compactly embedded in $W$. We deduce that the operator which maps $(\tilde{u}_\varepsilon, \tilde{v}_\varepsilon, \tilde{s}_\varepsilon, \tilde{w}_\varepsilon) \in V^2 \times W^2$ to $(u_\varepsilon, v_\varepsilon, s_\varepsilon, w_\varepsilon) \in V^2 \times W^2$ is continuous and compact.

Now, see Schauder's theorem 3.40, it is left to show that

$$\left( \|\hat{u}_\varepsilon\|_V^2 + \|\hat{v}_\varepsilon\|_V^2 + \|\hat{s}_\varepsilon\|_W^2 + \|\hat{w}_\varepsilon\|_W^2 \right) \leq r$$

implies

$$\left( \|u_\varepsilon\|_V^2 + \|v_\varepsilon\|_V^2 + \|s_\varepsilon\|_W^2 + \|w_\varepsilon\|_W^2 \right) \leq r$$

for some $r > 0$, where we may assume that the norm initial conditions are smaller than $r$.

We are testing the equation for $u_\varepsilon$ of system (3.21) with $u_\varepsilon$ and integrate from 0 to $t < \tau$,

$$\frac{1}{2}\|u_\varepsilon\|_{\Omega_\varepsilon}^2 + D_u\|\nabla u_\varepsilon\|_{\Omega_\varepsilon,t}^2$$

$$= \int_{\Omega_\varepsilon \times (0,t)} f(\hat{u}_\varepsilon) u_\varepsilon dx + \varepsilon \int_{\Gamma_\varepsilon^{ER} \times (0,t)} k_u R u_\varepsilon^2 d\sigma_x$$

$$+ \varepsilon \int_{\Gamma_\varepsilon^{ER} \times (0,t)} l_s \hat{s}_\varepsilon u_\varepsilon d\sigma_x + \underbrace{\frac{1}{2}\|u_\varepsilon(0)\|_{\Omega_\varepsilon}^2}_{\leq r}$$

$$\leq L_f \int_{\Omega_\varepsilon \times (0,t)} |\hat{u}_\varepsilon||u_\varepsilon| dx + \varepsilon \overline{R} k_u \|u_\varepsilon\|_{\Gamma_\varepsilon^{ER},t}^2$$

$$+ \varepsilon l_s \|\hat{s}_\varepsilon\|_{\Gamma_\varepsilon^{ER},t}^2 + \varepsilon l_s \|u_\varepsilon\|_{\Gamma_\varepsilon^{ER},t}^2 + r$$

$$\leq L_f \|\hat{u}_\varepsilon\|_{\Omega_\varepsilon,t}^2 + L_f \|u_\varepsilon\|_{\Omega_\varepsilon,t}^2$$

$$+ (\overline{R}k_u + l_s)\left(\|u_\varepsilon\|_{\Omega_\varepsilon,t}^2 + \varepsilon^2\|\nabla u_\varepsilon\|_{\Omega_\varepsilon,t}^2\right) + l_s\varepsilon\|\hat{s}_\varepsilon\|_{\Gamma_\varepsilon^{ER},t}^2 + r$$

$$\leq c_1\|u_\varepsilon\|_{\Omega_\varepsilon,t}^2 + \varepsilon^2 c_2\|\nabla u_\varepsilon\|_{\Omega_\varepsilon,t}^2 + c_3\|\hat{u}_\varepsilon\|_{\Omega_\varepsilon,t}^2 + c_3\|\hat{s}_\varepsilon\|_{\Gamma_\varepsilon^{ER},t}^2 + r$$

$$\leq c_1\|u_\varepsilon\|_{\Omega_\varepsilon,t}^2 + \varepsilon^2 c_2\|\nabla u_\varepsilon\|_{\Omega_\varepsilon,t}^2 + c_3 r.$$

Analogously we obtain for the equation for $v_\varepsilon$ that

$$\|v_\varepsilon\|_{\Omega_\varepsilon}^2 + D_v\|\nabla v_\varepsilon\|_{\Omega_\varepsilon,t}^2 \leq c_1\|v_\varepsilon\|_{\Omega_\varepsilon,t}^2 + \varepsilon^2 c_2\|\nabla v_\varepsilon\|_{\Omega_\varepsilon,t}^2 + c_3 r.$$

We also get for equation $s_\varepsilon$ with integration from 0 to $t < \tau$

$$\frac{1}{2}\varepsilon\|s_\varepsilon\|_{\Gamma_\varepsilon^{ER}}^2 + \varepsilon^3 D_s\|\nabla_\Gamma s_\varepsilon\|_{\Gamma_\varepsilon^{ER}}^2$$

$$= l_s\varepsilon \int_{\Gamma_\varepsilon^{ER} \times (0,t)} s_\varepsilon^2 d\sigma_x dt + \varepsilon \int_{\Gamma_\varepsilon^{ER} \times (0,t)} h(\hat{s}_\varepsilon) s_\varepsilon d\sigma_x$$

$$+ \varepsilon k_u \int_{\Gamma_\varepsilon^{ER} \times (0,t)} R u_\varepsilon s_\varepsilon d\sigma_x + \frac{1}{2}\underbrace{\|s_\varepsilon(0)\|_{\Gamma_\varepsilon^{ER}}}_{\leq r}$$

$$\leq l_s\varepsilon\|s_\varepsilon\|_{\Gamma_\varepsilon^{ER},t}^2 + \varepsilon L_h\|s_\varepsilon\|_{\Gamma_\varepsilon^{ER},t}^2 + \varepsilon L_h\|\hat{s}_\varepsilon\|_{\Gamma_\varepsilon^{ER},t}^2$$

$$+ \varepsilon k_u\overline{R}\|s_\varepsilon\|_{\Gamma_\varepsilon^{ER},t}^2 + k_u\overline{R}\varepsilon\|u_\varepsilon\|_{\Gamma_\varepsilon^{ER},t}^2 + r$$

$$\leq c_1\varepsilon\|s_\varepsilon\|_{\Gamma_\varepsilon^{ER},t}^2 + c_2\varepsilon\|\hat{s}_\varepsilon\|_{\Gamma_\varepsilon^{ER},t}^2 + c_3\|u_\varepsilon\|_{\Omega_\varepsilon,t}^2 + c_4\varepsilon^2\|\nabla u_\varepsilon\|_{\Omega_\varepsilon,t}^2 + r$$

$$\leq c_1\varepsilon\|s_\varepsilon\|_{\Gamma_\varepsilon^{ER},t}^2 + c_2\|u_\varepsilon\|_{\Omega_\varepsilon,t}^2 + c_3\varepsilon^2\|\nabla u_\varepsilon\|_{\Omega_\varepsilon,t}^2 + c_4 r$$

Analogously we obtain for equation $w_\varepsilon$ that

$$\frac{1}{2}\varepsilon\|w_\varepsilon\|^2_{\Gamma^{ER}_\varepsilon} + \varepsilon^3 D_w\|\nabla_\Gamma w_\varepsilon\|^2_{\Gamma^{ER}_\varepsilon,t}$$
$$\leq c_1\varepsilon\|w_\varepsilon\|^2_{\Gamma^{ER}_\varepsilon,t} + c_2\|v_\varepsilon\|^2_{\Omega_\varepsilon,t} + c_3\varepsilon^2\|\nabla v_\varepsilon\|^2_{\Omega_\varepsilon,t} + c_4 r.$$

We add the results above and with $\varepsilon$ small we find

$$\frac{1}{2}\left(\|u_\varepsilon\|^2_{\Omega_\varepsilon} + \|v_\varepsilon\|^2_{\Omega_\varepsilon} + \overbrace{\varepsilon\|s_\varepsilon\|^2_{\Gamma^{ER}_\varepsilon}}^{\geq 0} + \overbrace{\varepsilon\|w_\varepsilon\|^2_{\Gamma^{ER}_\varepsilon}}^{\geq 0}\right)$$
$$+ (D_u - \varepsilon^2 c_2 - \varepsilon^2 c_2)\|\nabla u_\varepsilon\|^2_{\Omega_\varepsilon,t} + (D_v - \varepsilon^2 c_2 - \varepsilon^2 c_2)\|\nabla v_\varepsilon\|^2_{\Omega_\varepsilon,t}$$
$$+ \varepsilon^3 D_s\|\nabla_\Gamma s_\varepsilon\|^2_{\Gamma^{ER}_\varepsilon,t} + \varepsilon^3 D_s\|\nabla_\Gamma w_\varepsilon\|^2_{\Gamma^{ER}_\varepsilon,t}$$
$$\leq c_1\int_0^t(\|u_\varepsilon\|^2_{\Omega_\varepsilon} + \|v_\varepsilon\|^2_{\Omega_\varepsilon} + \varepsilon\|s_\varepsilon\|^2_{\Gamma^{ER}_\varepsilon} + \varepsilon\|w_\varepsilon\|^2_{\Gamma^{ER}_\varepsilon})d\tau + 4c_4 r.$$

With Gronwall's lemma we conclude

$$\|u_\varepsilon\|^2_{\Omega_\varepsilon} + \|v_\varepsilon\|^2_{\Omega_\varepsilon} + \varepsilon\|s_\varepsilon\|^2_{\Gamma^{ER}_\varepsilon} + \varepsilon\|w_\varepsilon\|^2_{\Gamma^{ER}_\varepsilon} + \|\nabla u_\varepsilon\|^2_{\Omega_\varepsilon,t}$$
$$+ \|\nabla v_\varepsilon\|^2_{\Omega_\varepsilon,t} + \varepsilon\|\nabla_\Gamma s_\varepsilon\|^2_{\Gamma^{ER}_\varepsilon,t} + \varepsilon\|\nabla_\Gamma w_\varepsilon\|^2_{\Gamma^{ER}_\varepsilon,t} \leq c_1 r. \quad (3.22)$$

This inequality (3.22) implies by integration from 0 to $\tau$ that

$$\|u_\varepsilon\|^2_{L^2((0,\tau),H^1(\Omega_\varepsilon))} + \|v_\varepsilon\|^2_{L^2((0,\tau),H^1(\Omega_\varepsilon))}$$
$$+ \|s_\varepsilon\|^2_{L^2((0,\tau),H^1(\Gamma_\varepsilon))} + \|w_\varepsilon\|^2_{L^2((0,\tau),H^1(\Gamma_\varepsilon))} \leq c_1 r.$$

Integration from 0 to $\tau$ of inequality (3.22) gives

$$\|u_\varepsilon\|^2_{L^2((0,\tau),L^2(\Omega_\varepsilon))} + \|v_\varepsilon\|^2_{L^2((0,\tau),L^2(\Omega_\varepsilon))}$$
$$+ \varepsilon\|s_\varepsilon\|^2_{L^2((0,\tau),L^2(\Gamma_\varepsilon))} + \varepsilon\|w_\varepsilon\|^2_{L^2((0,\tau),L^2(\Gamma_\varepsilon))} \leq c_1 r\tau.$$

With the interpolation inequality (see [1])

$$\|\cdot\|_V \leq \tilde{c}\|\cdot\|^\delta_{L^2((0,\tau),L^2(\Omega_\varepsilon))}\|\cdot\|^{1-\delta}_{L^2((0,\tau),H^1(\Omega_\varepsilon))}$$

we get

$$\|u_\varepsilon\|_V^2 + \|v_\varepsilon\|_V^2 + \|s_\varepsilon\|_W^2 + \|w_\varepsilon\|_W^2 \leq \tilde{c}(c_1 r \tau)^\delta (c_1 r)^{1-\delta}$$
$$= \tilde{c} c_1 r \tau^\delta$$
$$\leq r$$

The last inequality is correct if $\tau$ is chosen smaller than $\frac{1}{(\tilde{c}c_1)^{1/\delta}}$.
Hence, the embedding of the mapping $T$ has at least one fixed-point in $\{u \in L^2([0,\tau]; H^1(\Omega_\varepsilon)) | \partial_t u \in H^1(\Omega_\varepsilon)'\}^2 \times \{u \in L^2([0,\tau]; H^1(\Gamma_\varepsilon^{ER})) | \partial_t u \in H^1(\Gamma_\varepsilon^{ER})'\}^2$. $\qquad\square$

Now we know that for every small $\varepsilon$ there exists a solution of the system of equations (3.17) and we are ready to find the $\varepsilon$-limit equation of this system.

# 3.6 Identification of the limit model for the nonlinear carcinogenesis model

To calculate the limit for $\varepsilon \to 0$ of the equations for $u_\varepsilon$, $v_\varepsilon$ and $R_\varepsilon$ we use two-scale convergence. The $\varepsilon$-limits of the equations for $s_\varepsilon$ and $w_\varepsilon$ are determined by using the periodic unfolding method with the operator $\mathcal{T}_\varepsilon^b$. First we consider the $\varepsilon$-limits of the nonlinear terms. Afterwards we derive the limit of the whole equations.

**The nonlinear terms**

- First we consider the nonlinear terms $f(u_\varepsilon)$ and $g(v_\varepsilon)$ in the equations for $u_\varepsilon$ and $v_\varepsilon$, respectively, in the system 3.17. Because $u_\varepsilon$ and $v_\varepsilon$ are bounded in $L^2([0,T], H^1(\Omega_\varepsilon))$ independently of $\varepsilon$, lemma 2.3 guarantees that they two-scale converge to limit functions $u_0$ and $v_0$, respectively
  In lemma 3.33, 3.34 and 3.36 we showed that $u_\varepsilon$ and $v_\varepsilon$ are bounded in $L^2([0,T], H^1(\Omega)) \cap H^1([0,T], H^{-1}(\Omega)) \cap L^\infty([0,T] \times \Omega)$ independently of $\varepsilon$. Then lemma 3.4 implies the existence of a subsequences, which we also denote by $u_\varepsilon$ and $v_\varepsilon$, for convenience of notation, that strongly converge to their two-scale limits (see remark 3.38). Furthermore, with theorem 3.6 the functions $F, G : L^2([0,T] \times \Omega) \to$

$L^2([0,T] \times \Omega)$ (defined in (3.20)) are bounded and continuous. Hence, it follows that

$$F(u_\varepsilon) \xrightarrow{\varepsilon \to 0} F(u_0) \quad \text{strongly in } L^2([0,T] \times \Omega)$$

and

$$G(v_\varepsilon) \xrightarrow{\varepsilon \to 0} G(v_0) \quad \text{strongly in } L^2([0,T] \times \Omega).$$

- Secondly, we calculate the limits of the nonlinear Robin-boundary terms $k_u u_\varepsilon R_\varepsilon$ and $k_v v_\varepsilon R_\varepsilon$ at the surface of the ER. With remark 3.38 we deduce that $\mathcal{T}_\varepsilon^b(R_\varepsilon)$ converges strongly to a function $R_0$ in $L^2([0,T] \times \Omega \times \Gamma)$. Proposition 3.17 implies that $\mathcal{T}_\varepsilon^b(u_\varepsilon)$, $\mathcal{T}_\varepsilon^b(v_\varepsilon)$ converge weakly to $u_0$, $v_0$ in $L^2([0,T] \times \Omega \times \Gamma)$, respectively. Then we use lemma 3.18 to deduce that

$$\lim_{\varepsilon \to 0} \int_{\Omega \times \Gamma} (\mathcal{T}_\varepsilon^b(u_\varepsilon)\mathcal{T}_\varepsilon^b(R_\varepsilon) - u_0 R_0) \varphi d\sigma_y dx = 0$$

and

$$\lim_{\varepsilon \to 0} \int_{\Omega \times \Gamma} (\mathcal{T}_\varepsilon^b(v_\varepsilon)\mathcal{T}_\varepsilon^b(R_\varepsilon) - v_0 R_0) \varphi d\sigma_y dx = 0$$

for all $\varphi \in C^\infty(\Omega \times \Gamma)$.

- Last we consider the nonlinear transformation function $h(s_\varepsilon)$ in the equations for $s_\varepsilon$ and $w_\varepsilon$. Applying lemma 3.33 yields that a subsequence of $s_\varepsilon$ has a weak limit $s_0$ in $L^2([0,T] \times \Omega \times \Gamma)$ as $\varepsilon$ tends to zero. With lemma 3.37 we know that $\mathcal{T}_\varepsilon^b(s_\varepsilon)$ has a strongly converging subsequence, denoted again by $\mathcal{T}_\varepsilon^b(s_\varepsilon)$, which converges to its weak limit $s_0$ for $\varepsilon$ tending to zero. We use theorem 3.6 and deduce that $H : L^2([0,T] \times \Omega \times \Gamma) \to L^2([0,T] \times \Omega \times \Gamma)$ with $(H(\mathcal{T}_\varepsilon^b(s_\varepsilon)))(x) = h(\mathcal{T}_\varepsilon^b(s_\varepsilon)(x))$ is bounded and continuous. Hence, we conclude that

$$H(\mathcal{T}_\varepsilon^b(s_\varepsilon)) \xrightarrow{\varepsilon \to 0} H(s_0) \quad \text{strongly in } L^2([0,T] \times \Omega \times \Gamma).$$

Now we perform the limit derivation for the equations $u_\varepsilon$, $v_\varepsilon$, $s_\varepsilon$, $w_\varepsilon$ and $R_\varepsilon$ and use the just calculated $\varepsilon$-limits of the nonlinear terms.

We test these equations with admissible test functions $\varphi_\varepsilon \in C^\infty(\Omega, C_\#^\infty(Y))$. Let $\chi$ be the characteristic function on $\Omega_\varepsilon$. As test functions $\varphi_\varepsilon \in C^\infty(\Omega, C_\#^\infty(Y))$ we choose functions of the form

$$\varphi_\varepsilon\left(x, \frac{x}{\varepsilon}\right) = \varphi_0(x) + \varepsilon\varphi_1\left(x, \frac{x}{\varepsilon}\right)$$

with $(\varphi_0, \varphi_1) \in C^\infty(\Omega) \times C^\infty(\Omega, C_\#^\infty(Y))$.

**Limit equation for $u_\varepsilon$**

We use part 5 of lemma 3.15 for the first term on $\Gamma_\varepsilon^{\mathrm{ER}}$ in the equation for $u_\varepsilon$ in system (3.17). Note also remark 3.30 for the time derivative

$$\int_\Omega \partial_t \mathcal{T}_\varepsilon(u_\varepsilon)\varphi_\varepsilon\chi\left(\frac{x}{\varepsilon}\right)\mathrm{d}x + D_u\int_\Omega \nabla u_\varepsilon \nabla\varphi_\varepsilon\chi\left(\frac{x}{\varepsilon}\right)\mathrm{d}x$$
$$+ \int_{\Omega\times\Gamma} k_u \mathcal{T}_\varepsilon^b(u_\varepsilon)\mathcal{T}_\varepsilon^b(R_\varepsilon)\varphi_\varepsilon \mathrm{d}\sigma_y \mathrm{d}x - \varepsilon\int_{\Gamma_\varepsilon} l_s s_\varepsilon \varphi_\varepsilon \mathrm{d}\sigma_y$$
$$= -\int_\Omega f(u_\varepsilon)\varphi_\varepsilon\chi\left(\frac{x}{\varepsilon}\right)\mathrm{d}x.$$

With proposition 2.5 and theorem 2.10 applied to the a-priori estimates 3.33 we find for $\varepsilon \to 0$ that

$$\int_{\Omega\times Y^*} \partial_t u_0(x,t)\varphi_0(x)\mathrm{d}x\mathrm{d}y$$
$$+ D_u\int_{\Omega\times Y^*} [\nabla_x u_0(x,t) + \nabla_y u_1(x,y,t)][\nabla_x\varphi_0(x) + \nabla_y\varphi_1(x,y)]\mathrm{d}x\mathrm{d}y$$
$$+ \int_{\Omega\times\Gamma} (k_u u_0(x,t)R_0(x,t) - l_s s_0(x,y,t))\varphi_0(x)\mathrm{d}x\mathrm{d}\sigma_y$$
$$= -\int_{\Omega\times Y^*} f(u_0(x,t))\varphi_0(x)\mathrm{d}x\mathrm{d}y \quad (3.23)$$

for all $(\varphi_0, \varphi_1) \in C^\infty(\Omega) \times C^\infty(\Omega, C_\#^\infty(Y))$, where $u_1 \in L^2(\Omega, H_\#^1(Y^*))$ and $u_0 \in H^1(\Omega)$, independent of $y \in Y^*$.

133

**Limit equation for $v_\varepsilon$**
Analogously we obtain for the equation for $v_\varepsilon$ and $\varepsilon \to 0$ that

$$\int_{\Omega \times Y^*} \partial_t v_0(x,t)\varphi_0(x)\mathrm{d}x\mathrm{d}y$$

$$+ D_v \int_{\Omega \times Y^*} [\nabla_x v_0(x,t) + \nabla_y v_1(x,y,t)][\nabla_x \varphi_0(x) + \nabla_y \varphi_1(x,y)]\mathrm{d}x\mathrm{d}y$$

$$+ \int_{\Omega \times \Gamma} (k_v v_0(x,t)R_0(x,t) - l_w w_0(x,y,t))\varphi_0(x)\mathrm{d}x\mathrm{d}\sigma_y$$

$$= -\int_{\Omega \times Y^*} g(v_0(x,t))\varphi_0(x)\mathrm{d}x\mathrm{d}y \quad (3.24)$$

for all $(\varphi_0, \varphi_1) \in C^\infty(\Omega) \times C^\infty(\Omega, C_\#^\infty(Y))$, where $v_1 \in L^2(\Omega, H_\#^1(Y^*))$ and $v_0 \in H^1(\Omega)$, independent of $y \in Y^*$.

**Limit equation for $R_\varepsilon$**
Again with theorem 2.10 for the linear terms as well as lemma 3.33 and 3.37 and lemma 3.18 for the function products we calculate the limit equation for $R_\varepsilon$.

$$\varepsilon \int_{\Omega \times \Gamma} \partial_t \mathcal{T}_\varepsilon^b(R_\varepsilon)\psi_\varepsilon \mathrm{d}x\mathrm{d}\sigma_y + \int_{\Omega \times \Gamma} \mathcal{T}_\varepsilon^b(R_\varepsilon)\left(k_u \mathcal{T}_\varepsilon^b(u_\varepsilon) + k_v \mathcal{T}_\varepsilon^b(v_\varepsilon)\right.$$

$$\left. + l_s \mathcal{T}_\varepsilon^b(s_\varepsilon) + l_w \mathcal{T}_\varepsilon^b(w_\varepsilon)\right)\psi_\varepsilon \mathrm{d}x\mathrm{d}\sigma_y$$

$$= \varepsilon \int_{\Omega \times \Gamma} \overline{R}\left(l_s \mathcal{T}_\varepsilon^b(s_\varepsilon) + l_w \mathcal{T}_\varepsilon^b(w_\varepsilon)\right)\psi_\varepsilon \mathrm{d}x\mathrm{d}\sigma_y.$$

We find for $\varepsilon \to 0$

$$\int_{\Omega \times \Gamma} \partial_t R_0(x,y,t)\psi_0(x,y)\mathrm{d}x\mathrm{d}\sigma_y$$

$$+ \int_{\Omega \times \Gamma} R_0(x,y,t)(k_u u_0(x,t) + k_v v(x,t)$$

$$+ l_s s_0(x,y,t) + l_w w_0(x,y,t))\psi_0(x,y)\mathrm{d}x\mathrm{d}\sigma_y$$

$$= \int_{\Omega \times \Gamma} \overline{R}\left(l_s s_0(x,y,t) + l_w w_0(x,y,t)\right)\psi_0(x,y)\mathrm{d}x\mathrm{d}\sigma_y$$

for all $\psi_0 \in C^\infty(\Omega, C^\infty(\Gamma))$, where $R_0 \in L^2([0,T] \times \Omega \times \Gamma)$.

**Limit equation for $s_\varepsilon$**

Calculating the limit equations for $s_\varepsilon$ and $w_\varepsilon$, we just use the boundary periodic unfolding operator $\mathcal{T}_\varepsilon^b$,

$$\varepsilon \int_{\Gamma_\varepsilon} \partial_t s_\varepsilon \psi_\varepsilon d\sigma_x + D_s \varepsilon^3 \int_{\Gamma_\varepsilon} \nabla_\Gamma s_\varepsilon \nabla_\Gamma \psi_\varepsilon d\sigma_x$$
$$= \varepsilon \int_{\Gamma_\varepsilon} (k_u u_\varepsilon R_\varepsilon - l_s s_\varepsilon) \psi_\varepsilon d\sigma_x - \varepsilon \int_{\Gamma_\varepsilon} h(s_\varepsilon) \psi_\varepsilon d\sigma_x.$$

We use property 5 and 3 of lemma 3.15 and conclude

$$\int_{\Omega \times \Gamma} \partial_t \mathcal{T}_\varepsilon^b(s_\varepsilon) \psi_\varepsilon d\sigma_y dx + D_s \int_{\Omega \times \Gamma} \varepsilon \mathcal{T}_\varepsilon^b(\nabla_\Gamma s_\varepsilon) \nabla_y \psi_\varepsilon d\sigma_y dx$$
$$= \int_{\Omega \times \Gamma} (k_u \mathcal{T}_\varepsilon^b(u_\varepsilon) \mathcal{T}_\varepsilon^b(R_\varepsilon) - l_s \mathcal{T}_\varepsilon^b(s_\varepsilon)) \psi_\varepsilon d\sigma_y dx$$
$$- \int_{\Omega \times \Gamma} \mathcal{T}_\varepsilon^b(h(s_\varepsilon)) \psi_\varepsilon d\sigma_y dx.$$

Then we use lemma 3.19 and lemma 3.22 to deduce for $\varepsilon \to 0$ that

$$\int_{\Omega \times \Gamma} \partial_t s_0 \psi_0 dx d\sigma_y + D_s \int_{\Omega \times \Gamma} \nabla_y s_0 \nabla_y \psi_0 dx d\sigma_y$$
$$= \int_{\Omega \times \Gamma} (k_u u_0 R_0 - l_s s_0) \psi_0 dx d\sigma_y - \int_{\Omega \times \Gamma} h(s_0) \psi_0 dx d\sigma_y$$

for all $\psi_0 \in C^\infty(\Omega, C_\#^\infty(\Gamma))$, where $s_0 \in L^2([0,T], L^2(\Omega, H_\#^1(\Gamma)))$.

**Limit equation for $w_\varepsilon$**

Analogously we get

$$\int_{\Omega \times \Gamma} \partial_t w_0 \psi_0 dx d\sigma_y + D_w \int_{\Omega \times \Gamma} \nabla_y w_0 \nabla_y \psi_0 dx d\sigma_y$$
$$= \int_{\Omega \times \Gamma} (k_v v_0 R_0 - l_w w_0) \psi_0 dx d\sigma_y + \int_{\Omega \times \Gamma} h(s_0) \psi_0 dx d\sigma_y$$

for all $\psi_0 \in C^\infty(\Omega, C_\#^\infty(\Gamma))$, where $w_0 \in L^2([0,T], L^2(\Omega, H_\#^1(\Gamma)))$.

**Identification of $u_1(x,y,t)$ and $v_1(x,y,t)$**

We take the limit equation (3.23) for $u_0$ set $\varphi_0 = 0$ and obtain

$$D_u \int_\Omega \int_{Y^*} [\nabla_y u_0(x,t) + \nabla_y u_1(x,y,t)] \nabla_y \varphi_1(x,y) dy dx = 0.$$

135

This is the same system to solve as in (2.21). Hence, using the notation $u_1 = \sum_j \partial_{x_j} u_0 \mu_j$ for functions $\mu_j \in H^1(Y^*)$, we obtain the same *cell problem* (2.22)

$$
\begin{aligned}
\nabla_y \cdot D_u(e_j + \nabla_y \mu_j) &= 0 && \text{in } Y^* \\
D_u(e_j + \nabla_y \mu_j) \cdot n &= 0 && \text{on } \partial Y^*
\end{aligned}
\tag{3.25}
$$

and $\mu_j$ must be $Y$-periodic for all $j = 1, \ldots, n$.

Furthermore, we obtain the same diffusion tensor $P^u$, as in (2.23) given by

$$
P_{ij}^u = \int_{Y^*} D_u[\delta_{ij} + \partial_{y_i} \mu_j]\mathrm{d}y.
$$

Analogous steps for the limit equation (3.24) for $v_0$ lead to the cell problem for $v_1 = \sum_j \partial_{x_j} v_0 \eta_j$ for functions $\eta_j \in H^1(Y^*)$,

$$
\begin{aligned}
\nabla_y \cdot D_v(e_j + \nabla_y \eta_j) &= 0 && \text{in } Y^* \\
D_v(e_j + \nabla_y \eta_j) \cdot n &= 0 && \text{on } \partial Y^*
\end{aligned}
$$

and $\eta_j$ is $Y$-periodic for all $j = 1, \ldots, N$. We define the diffusion tensor $P^v$ by

$$
P^v = \int_{Y^*} D_v[\delta_{ij} + \partial_{y_i} \eta_j]\mathrm{d}y.
$$

Now we know the $\varepsilon$-limit for every equation in the model (3.17). For convenience we denote the $\varepsilon$-limit $(u_0, v_0, s_0, w_0, R_0)$ by $(u, v, s, w, R)$. We use that $u$ and $v$ are $y$-independent and summarize the homogenized limit equation as follows.

## Weak formulation of the limit model

Let $(u, v, s, w, R) \in \mathcal{V}_C(\Omega) \times \mathcal{V}_N(\Omega) \times \mathcal{V}(\Omega, \Gamma)^2 \times \mathcal{V}_R(\Omega, \Gamma)$ such that

$$
\begin{aligned}
|Y^*|(\partial_t u, \varphi_1)_\Omega &+ (P^u \nabla u, \nabla \varphi_1)_\Omega \\
&+ ((k_u u R - l_s s), \varphi_1)_{\Omega \times \Gamma} = -|Y^*|(f(u), \varphi_1)_\Omega \\
|Y^*|(\partial_t v, \varphi_2)_\Omega &+ (P^v \nabla v, \nabla \varphi_2)_\Omega \\
&+ ((k_v v R - l_w w), \varphi_2)_{\Omega \times \Gamma} = -|Y^*|(g(v), \varphi_2)_\Omega \\
(\partial_t s, \psi)_{\Omega \times \Gamma} &+ D_s (\nabla_\Gamma s, \nabla_\Gamma \psi)_{\Omega \times \Gamma} \\
&- ((k_u u R - l_s s), \psi)_{\Omega \times \Gamma} = -(h(s), \psi)_{\Omega \times \Gamma} \\
(\partial_t w, \psi)_{\Omega \times \Gamma} &+ D_w (\nabla_\Gamma w, \nabla_\Gamma \psi)_{\Omega \times \Gamma} \\
&- ((k_v v R - l_w w), \psi)_{\Omega \times \Gamma} = (h(s), \psi)_{\Omega \times \Gamma} \\
(\partial_t R, \psi)_{\Omega \times \Gamma} &+ (R|k_u u + k_v v + l_s s + l_w w|, \psi)_{\Omega \times \Gamma} \\
&= (\overline{R}|l_s s + l_w w|, \psi)_{\Omega \times \Gamma}
\end{aligned}
\tag{3.26}
$$

for all $(\varphi_1, \varphi_2, \psi) \in V_{C0}(\Omega) \times V_N(\Omega) \times V(\Omega, \Gamma)$.

## Strong formulation of the limit model

For $t \in [0, T]$, the macroscopic strong limit formulation of the model (3.26) is given by

$$
\begin{aligned}
|Y^*|\partial_t u - \nabla(P^u \nabla u) + \int_\Gamma (k_u u R - l_s s)\mathrm{d}\sigma_y &= -|Y^*|f(u) && \text{in } \Omega \\
|Y^*|\partial_t v - \nabla(P^v \nabla v) + \int_\Gamma (k_v v R - l_w w)\mathrm{d}\sigma_y &= -|Y^*|g(v) && \text{in } \Omega \\
\partial_t s - D_s \Delta_\Gamma s - k_u u R + l_s s &= -h(s) && \text{on } \Omega \times \Gamma \\
\partial_t w - D_w \Delta_\Gamma w - k_v v R + l_w w &= h(s) && \text{on } \Omega \times \Gamma \\
\partial_t R + R|k_u u + k_v v + l_s s + l_w w| &= \overline{R}|l_s s + l_w w| && \text{on } \Omega \times \Gamma
\end{aligned}
\tag{3.27}
$$

with

$$
\begin{aligned}
u &= u_{\text{Boundary}} && \text{on } \Gamma_C \\
P^u \nabla u \cdot n &= 0 && \text{on } \Gamma_N \\
P^v \nabla v \cdot n &= 0 && \text{on } \Gamma_C \\
v &= 0 && \text{on } \Gamma_N.
\end{aligned}
$$

Before we illustrate the solution of this partial differential equation with simulations in section 3.8, we show that the solution (3.26) is unique.
This is the goal of the next section (3.26).

## 3.7  Uniqueness of the limit model

In this section we are going to show that the solution of system (3.26) is unique.

**Theorem 3.43 (Uniqueness)**
*There is at most one solution of the problem* (3.26).

**Proof** To prove uniqueness of the homogenized limit model, we need to show uniqueness of the cell problem (3.25) and the macroscopic system of equations (3.26).
We refer to the proof of theorem 2.31 to see that the cell problem (2.22), which is equal to (3.25), has a unique solution and we find a unique diffusion tensor $P^u$. Analogously we find that $P^v$ is unique.

To prove uniqueness of the macroscopic system of equations, let us suppose that there exist two solutions $(u_1, v_1, s_1, w_1, R_1)$ and $(u_2, v_2, s_2, w_2, R_2)$ of the weak problem (3.26) with the same given initial values. We want to show that $(u_1, v_1, s_1, w_1, R_1) = (u_2, v_2, s_2, w_2, R_2)$ almost everywhere. Now we take the equations for $u_1$ and $u_2$, subtract them from each other, test with $\varphi = u_1 - u_2$ and integrate from 0 to $t$,

$$|Y|\frac{1}{2}\|u_1 - u_2\|_\Omega^2 + \|\sqrt{P^u}\nabla(u_1 - u_2)\|_{\Omega,t}^2$$
$$+ k_u(u_1 R_1 - u_2 R_2, u_1 - u_2)_{\Omega \times \Gamma, t} - l_s(s_1 - s_2, u_1 - u_2)_{\Omega \times \Gamma, t}$$
$$= \underbrace{-|Y|(f(u_1) - f(u_2), u_1 - u_2)_{\Omega,t}}_{\leq 0,\ \text{since } f \text{ monotone}} \leq 0.$$

Hence,

$$|Y|\frac{1}{2}\|u_1 - u_2\|_\Omega^2 + \|\sqrt{P^u}\nabla(u_1 - u_2)\|_{\Omega,t}^2$$

$$\leq -k_u(u_1 R_1 - u_1 R_2 + u_1 R_2 - u_2 R_2, u_1 - u_2)_{\Omega\times\Gamma,t}$$
$$- l_s(s_1 - s_2, u_1 - u_2)_{\Omega\times\Gamma,t}$$

$$\leq k_u\|u\|_{L^\infty}(|R_1 - R_2|, |u_1 - u_2|)_{\Omega\times\Gamma_t}$$
$$+ k_u\overline{R}\|u_1 - u_2\|_{\Omega\times\Gamma,t}^2 + l_s(|s_1 - s_2|, |u_1 - u_2|)_{\Omega\times\Gamma,t}$$

$$\leq |\Gamma|\left(k_u\|u\|_{L^\infty} + k_u\overline{R} + l_s\right)\|u_1 - u_2\|_{\Omega,t}^2$$
$$+ k_u\|u\|_{L^\infty}\|R_1 - R_2\|_{\Omega\times\Gamma,t}^2 + l_s\|s_1 - s_2\|_{\Omega\times\Gamma,t}^2.$$

Analogously we obtain for the equation for $v$ that

$$|Y|\frac{1}{2}\|v_1 - v_2\|_\Omega^2 + \|\sqrt{P^v}\nabla(v_1 - v_2)\|_{\Omega,t}^2$$

$$\leq |\Gamma|\left(k_v\|v\|_{L^\infty} + k_v\overline{R} + l_w\right)\|v_1 - v_2\|_{\Omega,t}^2$$
$$+ k_v\|v\|_{L^\infty}\|R_1 - R_2\|_{\Omega\times\Gamma,t}^2 + l_w\|w_1 - w_2\|_{\Omega\times\Gamma,t}^2.$$

Further we continue for the equation for $s$ and we subtract the equations for $s_1$ and $s_2$. We test with the function $\psi = s_1 - s_2$ and obtain

$$\frac{1}{2}\partial_t\|s_1 - s_2\|_{\Omega\times\Gamma}^2 + D_s\|\nabla_y(s_1 - s_2)\|_{\Omega\times\Gamma}^2$$

$$+ \underbrace{(h(s_1) - h(s_2), s_1 - s_2)_{\Omega\times\Gamma}}_{\geq 0,\ \text{since } h \text{ monotone}}$$

$$= k_u(u_1 R_1 - u_2 R_2, s_1 - s_2)_{\Omega\times\Gamma} - l_s\|s_1 - s_2\|_{\Omega\times\Gamma}^2$$
$$\leq |\Gamma|\left(k_u\|u\|_{L^\infty} + k_u\overline{R} + l_s\right)\|u_1 - u_2\|_\Omega^2$$
$$+ k_u\|u\|_{L^\infty}\|R_1 - R_2\|_{\Omega\times\Gamma}^2 + l_s\|s_1 - s_2\|_{\Omega\times\Gamma}^2.$$

Integration from 0 to $t$ yields

$$\frac{1}{2}\|s_1 - s_2\|_{\Omega\times\Gamma}^2 + D_s\|\nabla_y(s_1 - s_2)\|_{\Omega\times\Gamma,t}^2$$

$$\leq |\Gamma|\left(k_u\|u\|_{L^\infty} + k_u\overline{R} + l_s\right)\|u_1 - u_2\|_{\Omega,t}^2$$
$$+ k_u\|u\|_{L^\infty}\|R_1 - R_2\|_{\Omega\times\Gamma,t}^2 + l_s\|s_1 - s_2\|_{\Omega\times\Gamma,t}^2.$$

Analogously we get

$$
\frac{1}{2}\|w_1 - w_2\|_{\Omega \times \Gamma}^2 + D_w \|\nabla_y (w_1 - w_2)\|_{\Omega \times \Gamma, t}^2
$$
$$
\leq |\Gamma| \left( k_v \|v\|_{L^\infty} + k_v \overline{R} + l_w \right) \|v_1 - v_2\|_{\Omega, t}^2 + k_v \|v\|_{L^\infty} \|R_1 - R_2\|_{\Omega \times \Gamma, t}^2
$$
$$
+ l_w \|w_1 - w_2\|_{\Omega \times \Gamma, t}^2 + L_h \|s_1 - s_2\|_{\Omega \times \Gamma, t}^2.
$$

For the equation for the receptors $R$ we obtain

$$
\frac{1}{2}\|R_1 - R_2\|_{\Omega \times \Gamma}^2 \leq (k_u \|u\|_{L^\infty} + k_v \|v\|_{L^\infty} + l_s \|s\|_{L^\infty}
$$
$$
+ l_w \|w\|_{L^\infty} + \overline{R} k_s + \overline{R} k_w) \|R_1 - R_2\|_{\Omega \times \Gamma, t}^2
$$
$$
+ \overline{R} k_s \|s_1 - s_2\|_{\Omega \times \Gamma, t}^2 + \overline{R} k_w \|w_1 - w_2\|_{\Omega \times \Gamma, t}^2.
$$

Adding the five estimates and merging the constants leads to

$$
\|u_1 - u_2\|_\Omega^2 + \|v_1 - v_2\|_\Omega^2 + \|s_1 - s_2\|_{\Omega \times \Gamma}^2 + \|w_1 - w_2\|_{\Omega \times \Gamma}^2
$$
$$
+ \|R_1 - R_2\|_{\Omega \times \Gamma}^2 \leq C(\|u_1 - u_2\|_{\Omega, t}^2 + \|v_1 - v_2\|_{\Omega, t}^2
$$
$$
+ \|s_1 - s_2\|_{\Omega \times \Gamma, t}^2 + \|w_1 - w_2\|_{\Omega \times \Gamma, t}^2 + \|R_1 - R_2\|_{\Omega \times \Gamma, t}^2)
$$

for a constant $C > 0$.
Gronwall's lemma implies

$$
\|u_1 - u_2\|_\Omega^2 + \|v_1 - v_2\|_\Omega^2 + \|s_1 - s_2\|_{\Omega \times \Gamma}^2
$$
$$
+ \|w_1 - w_2\|_{\Omega \times \Gamma}^2 + \|R_1 - R_2\|_{\Omega \times \Gamma}^2 = 0
$$

and we obtain $u_1 = u_2$ and $v_1 = v_2$ almost everywhere in $\Omega$ and $s_1 = s_2$, $w_1 = w_2$ and $R_1 = R_2$ almost everywhere in $\Omega \times \Gamma$ and almost every $t \in [0, T]$. $\qquad\square$

## 3.8 Simulations of the nonlinear carcinogenesis model

We conclude this chapter with simulations for the limit model (3.26). We take the same domains $\Omega$, $Y$ and $Y^*$ as we did in the linear case in section 2.8.

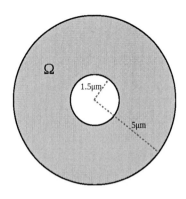
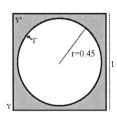

We refer to remark 2.33 to justify simulations in only one dimension, the radial component of the cell. The diffusion tensors $P^u$ and $P^v$ and the volume factors $|Y^*|$ and $|\Gamma|$ are the same as in section 2.8. The maximal time is $T = 5$min, as in the simulations for the linear case.

The following set of parameter values relate to the parameters in section 2.8. This means that, if possible, the parameters are equal. The transformation function $h$ is for small values equal to the linear transformation rate $f$. We also refer to remark 2.34 to declare that simulations only have illustrative purpose and no biological meaning.

We take the following set of exemplary parameters.

## Parameters

| Notation | Value | Unit | Description |
|---|---|---|---|
| $u_I$ | 0 | $\mu M$ | Initial value for BP in cytosol |
| $v_I$ | 0 | $\mu M$ | Initial value for DE in cytosol |
| $s_I$ | 0 | $\mu M$ | Initial value for BP on ER |
| $w_I$ | 0 | $\mu M$ | Initial value for DE on ER |
| $\overline{R}$ | 1 | $\mu M$ | Max. value for free receptors on ER |
| $R_I$ | 1 | $\mu M$ | Initial value for free receptors on ER |
| $D_u$ | 1 | $\mu m/s$ | Diff. coefficient for BP in cytosol |
| $D_v$ | 1 | $\mu m/s$ | Diff. coefficient for DE in cytosol |

| Notation | Value | Unit | Description |
|---|---|---|---|
| $D_s$ | 0.05 | $\mu m/s$ | Diff. coefficient for BP on ER |
| $D_w$ | 0.05 | $\mu m/s$ | Diff. coefficient for DE on ER |
| $l_s$ | 0.1 | $1/s$ | Binding rate to ER for BP |
| $l_w$ | 0.1 | $1/s$ | Binding rate to ER for DE |
| $u_{\text{Boundary}}$ | 1 | $\mu M$ | BP concentr. in intercellular space |
| $f(x)$ | $\frac{0.6x}{x+10}$ | $\mu M/s$ | Cleaning function for BP molecules |
| $g(x)$ | $\frac{0.6x}{x+10}$ | $\mu M/s$ | Cleaning function for DE molecules |
| $h(x)$ | $\frac{x}{x+100}$ | $\mu M/s$ | Transformation from BE to DE |

**Computation of** $\int_\Gamma (k_u u R - l_s s)\mathrm{d}\sigma_y$ **and** $\int_\Gamma (k_v v R - l_w w)\mathrm{d}\sigma_y$

The functions $s$, $w$ and $R$ need to solve the partial differential equation

$$\begin{aligned}
\partial_t s - D_s \Delta_\Gamma s - k_u u R + l_s s &= -h(s) & \text{on } \Omega \times \Gamma \\
\partial_t w - D_w \Delta_\Gamma w - k_v v R + l_w w &= h(s) & \text{on } \Omega \times \Gamma \\
\partial_t R - R|k_u u + k_v v + k_s s + k_w w| &= \overline{R}|k_s s + k_w w| & \text{on } \Omega \times \Gamma,
\end{aligned}$$

for every discretization point for $u$, $v$ on $\Omega$. Since $s_I$, $w_I$, $R_I$ and $u$, $v$ are independent of the variable $y \in \Gamma$, and the circle $\Gamma$ is radially symmetric, the functions $s$, $w$ and $R$ also remain $y$-independent. Hence, the spatial derivatives $\Delta_\Gamma s$ and $\Delta_\Gamma w$ are equal to zero. This simplifies the equations above to the ordinary differential equations

$$\begin{aligned}
\partial_t s - k_u u R + l_s s &= -h(s) & \text{on } \Omega \times \Gamma \\
\partial_t w - k_v v R + l_w w &= h(s) & \text{on } \Omega \times \Gamma \\
\partial_t R - R|k_u u + k_v v + k_s s + k_w w| &= \overline{R}|k_s s + k_w w| & \text{on } \Omega \times \Gamma.
\end{aligned}$$

For $P^u$, $P^v$, $|Y^*|$ and $|\Gamma|$ we use the same values as in the linear case in section 2.8,

$$P^u = D_u \begin{pmatrix} 0.19649 & 0 \\ 0 & 0.19649 \end{pmatrix}, \quad P^v = D_v \begin{pmatrix} 0.19649 & 0 \\ 0 & 0.19649 \end{pmatrix},$$

and

$$|Y^*| = 1 - \pi r^2 = 1 - \pi(0.45)^2 = 0.363827$$

$$|\Gamma| = 2\pi r = 2\pi \cdot 0.45 = 2.82743.$$

The time interval is chosen as $[0, T] = [0s, 300s] = [0\text{min}, 5\text{min}]$. We use semi-discretization in space and the finite element method with linear Lagrange elemtents to solve the boundary value problem. The spatial one-dimensional interval $\Omega$ is given by $[1.5, 5]$ and the spatial mesh is fixed with mesh size 0.07.

In the time variable the MATLAB ode solver *ode15s* is used, which uses Gear's method to solve the ordinary differential equations. For every adaptive time step and for every spatial discretization point in $\Omega = [1.5, 5]$, at first the integrals $\int_\Gamma k_u u R - l_s s d\sigma_y$ and $\int_\Gamma k_v v R - l_w w d\sigma_y$ are calculated applying the MATLAB ode solver *ode45*, which uses the Runke-Kutta method. Only then, calculations for $u$ and $v$ in $\Omega$ are continued for the next time step.

**Results and Simulations**

We see simulations made with MATLAB. On the $x$ axis we see the radius of the cell in the interval $[1.5\mu m, 5\mu m]$, on the $y$ axis we see the time in the interval $[0s, 300s]$ and on the $z$ axis we see the concentration of the molecules in $\mu M$ or receptors in %.

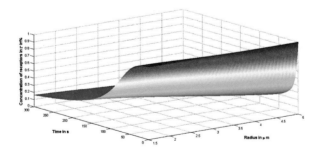

Figure 3.1: The solution $R$, the amount of free receptors at the membrane of the ER.

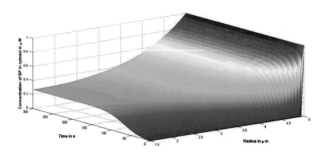

Figure 3.2: The solution $u$, the concentration of BP molecules in the cytosol.

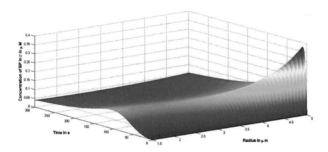

Figure 3.3: The solution $s$, the concentration of BP molecules bound to the membrane of the ER.

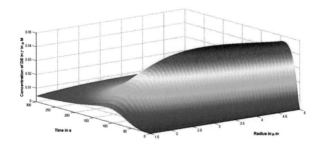

Figure 3.4: The solution $v$, the concentration of DE molecules in the cytosol.

144

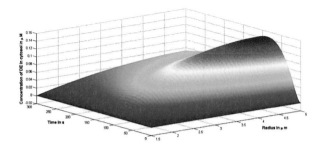

Figure 3.5: The solution $w$, the concentration of DE molecules bound to the membrane of the ER.

We are going to compare these numerical results with the ones from the linear model in section 2.8. For small times, the graph of the functions $u$, $v$, $s$ and $w$ is similar to the ones in the linear case. Then, in particular the solutions $s$, $w$ and $v$ behave differently. These changes originate in the nonlinear binding term to the surface of the ER,

$$\int_\Gamma (k_u u R - l_s s)\,d\sigma_y, \quad \text{and} \quad \int_\Gamma (k_v v R - l_w w)\,d\sigma_y.$$

The crucial point is the function $R$ denoting the amount of free receptors on the surface of the ER. For small values of $R$ BP or DE molecules $u$ and $v$ can hardly bind to the ER and transform to $s$ and $w$, respectively. This fact brings several consequences.

In the first figure 3.1 we see the solution $R$, the proportion of free receptors on the surface of the ER. Initially, the function is 1, because all receptors are free. With increasing time the amount decreases steadily. This means that more and more BP and DE molecules are bound to the ER.

In the second figure 3.2 we see the solution $u$ denoting the concentration of BP molecules in cytosol. The shape of the graph is very similar to the one in the linear case. Despite cleaning mechanisms the solution $u$ has slightly larger values as in the linear case. The reason is the function $R$ which decreases and prevents molecules in cytosol $u$ and $v$ from binding to the surface of the ER.

145

Next we see in figure 3.3 the concentration of ER-bound BP molecules $s$. In the very beginning, for time $0 \leq t \leq 10s$, the shape of the graph is similar to the linear case. Then, the curve reaches a global maximum of $\approx 0.36\mu M$ at the plasma membrane. After this peak the function is decreasing and for $t = 300s$ it diminished to $\approx 0.05\mu M$. This decline is due to the transformation of BP molecules $s$ into DE molecules $w$ by the transformation function $h(s)$ and due to the fact that no BP molecules from cytosol follow.

Figure 3.4 illustrates the concentration of DE molecules $w$ bound to the ER. The shape of this graph is also similar to the linear one for time $0 \leq t \leq 10s$. Then we see the impact of the modified function $s$ and here we have also a delayed peak with $\approx 0.045\mu M$ at time $t \approx 40s$. DE molecules on the surface of the ER arise from BP molecules through the transformation function $h(s)$. But because less BP molecules follow, and DE molecules unbind from ER, the function $w$ decreases. At time $t = 300s$ there remains a concentration of $\approx 0.007\mu M$.

In the last figure 3.5 we see the solution $v$ denoting the concentration of DE molecules in cytosol. The shape of the graph is similar to the graph of solution $w$, but we clearly see the Dirichlet boundary condition $v = 0$ at the nuclear membrane. The function reaches a global maximum of $\approx 0.15\mu M$ at the plasma membrane and time $t \approx 80s$. Then, it decreases because no DE molecules from ER follow, the cleaning mechanisms diminish the DE concentration and also DE molecules flow over the nuclear membrane into the nucleus.
For $t = 300s$, the DE concentration in cytosol at the plasma membrane is $\approx 0.05\mu M$.

Through the natural constraint of receptors, the appearance of DE molecules in cytosol, which can reach the nucleas, declines. Additionally, the already emerged DE molecules are diminished by cleaning mechanisms. In summary it can be stated, that incorporating the natural bounds of a cell promotes the health of the cell and discourages the carcinogenesis.

# FOUR

# CALCIUM-STIM1 MODEL

## Introduction

In section 4.1 we deduce a way to homogenize a system with global diffusion on a surface using a technique, which we call two-step convergence. We consider a partial differential equation on a smooth surface. We change this model by blowing up the surface to a volume with small width $\delta$. Then we perform homogenization and subsequently take the limit for $\delta$ tending to zero. In this way we find a homogenized partial differential equation with global diffusion on a surface. A direct way to find global diffusion on a surface is deduce in theorem 3.27 in section 3.2, see remark 4.3 for more details.

In section 4.2 a new result about the limit behavior in homogenization of a Neumann or Robin boundary condition is derived. We prove three theorems that describe how to handle those boundary terms. In the following application we apply the new results to several Robin boundary terms.

Then, it follows an application to the two-step convergence. Patrick Fletcher and Yue Xian Li developed a new model that describes signaling in lymphocytes as it is explained in section 4.3, see [23]. As in the previous chapters, direct numerical studies of this model are not feasible because of the fine structure of the cell. In this model molecules that live in the cytosol, lumen of the endoplasmic reticulum or surface of the endoplasmic reticulum are involved and build a highly nonlinear system of equations.

In section 4.4 well-posedness of the model developed in section 4.3 is proven. The main steps are to show that the functions of the model are bounded in suitable spaces. For the nonlinear terms the functions need to converge strongly and the functionals must be continuous.

In section 4.5 we show that a solution of the system of equations exists.

In section 4.6 we use homogenization to calculate the $\varepsilon$-limit of the model.

Once, the $\varepsilon$-limit is derived we deduce the $\delta$-limit of the equations in section 4.7.

In the end, in section 4.8 uniqueness of the limit equation is proven.

After these steps, the limit equation is suitable for numerical studies and we show some results by using the matlab code implemented by Patrick Fletcher.

# 4.1   Two-step convergence

The two-step convergence is a mathematical tool to determine macroscopic diffusion on hypersurfaces in the context of periodic homogenization. The idea is to regard the hypersurface as a thin domain with thickness $\delta > 0$. Using homogenization on this blown up domains we are able to apply well known homogenization results valid on subsets of $\mathbb{R}^n$ with positive volume. After homogenization the limit equation is defined on the homogeneous domain $\Omega$ and the unit cell $Y$, which is a characteristical part of the still blown up hypersurface.

To get the initial shape of the domain back, we let $\delta$ tend to zero in the unit cell $Y$ of the homogenized system. A direct way to find global diffusion on a surface is deduce in theorem 3.27 in section 3.2, see remark 4.3 for more details.

**Example in $\mathbb{R}^3$**

Let $\delta > 0$ and $\Gamma \subset \mathbb{R}^3$ be a smooth compact hypersurface with outer normal $n_p$ for every $p \in \Gamma$. Then $Y^\delta = \{p + d \cdot n_p \mid p \in \Gamma, d \in (-\delta, \delta)\} \subset \mathbb{R}^3$ is the blown up domain for a small $\delta$. We consider the domain $\Omega_\varepsilon = \bigcup_{k \in \mathbb{Z}^3} \varepsilon(Y^\delta + k)$.

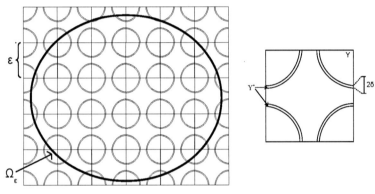

After homogenization, $\varepsilon \to 0$,

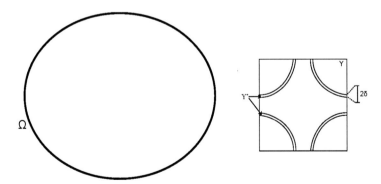

we let $\delta$ tend to 0. This means for the domain $Y^\delta$ that $Y^\delta = \{p + d \cdot n_p \mid p \in \Gamma, d \in (-\delta, \delta)\} \xrightarrow{\delta \to 0} \Gamma$.

149

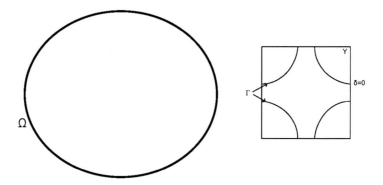

Other ways to let the unit cell $Y$ or coefficients of the given partial differential equation depend on another small parameter $\delta$ one finds in the articles [9, 8]. There also first the system is homogenized and afterwards the limit of $\delta$ tending to zero is calculated.

This method of the two-step convergence is a possible way to derive the two-scale limit of the gradient $\nabla u_\varepsilon$ of a bounded sequence $u_\varepsilon$ in $H^1(\Gamma_\varepsilon)$, where $\Gamma_\varepsilon := \bigcup_{k \in \mathbb{Z}^3} \varepsilon(\Gamma + k)$ is a smooth and connected manifold. Here, we want to use proposition 2.5 (i) on the blown up domain. Homogenization works as usual. A helpful paper for the convergence of the thin domain to a hypersurface is [37]. We are going to use some ideas from this article.

We define $\Omega_\varepsilon := \bigcup_{k \in \mathbb{Z}^3} \varepsilon(Y^\delta + k) \cap \Omega$ as usual. Let $u_\varepsilon^\delta$ be a bounded sequence in $H^1(\Omega_\varepsilon)$. Then, see proposition 2.5, $u_\varepsilon^\delta$ two-scale converges to $u^\delta \in H^1(\Omega)$, and there exists a function $u_1^\delta \in L^2(\Omega, H_\#^1(Y^\delta))$ such that, up to a subsequence, $\nabla u_\varepsilon^\delta$ two-scale converges to $\nabla_x u^\delta + \nabla_y u_1^\delta$.

Now we look what happens to the solution $u^\delta$ and the gradient $\nabla_x u^\delta + \nabla_y u_1^\delta$ for $Y^\delta \xrightarrow{\delta \to 0} \Gamma$. The weak formulation of the diffusion term in the standard case is given by

$$\int_\Omega \int_{Y^\delta} [\nabla_x u^\delta + \nabla_y u_1^\delta][\nabla_x \psi + \nabla_y \psi_1] dy dx = 0 \qquad (4.1)$$

150

for all $(\psi, \psi_1) \in H^1(\Omega) \times L^2(\Omega, H^1_\#(Y^\delta))$. This is now an easy example, a homogeneous Laplace equation, to focus on the details happening in the gradient for $\delta$ tending to zero. Later in theorem 4.1 we consider a more general equation on more general domains.

Let us consider the cell problem by setting $\psi = 0$.

$$\int_\Omega \int_{Y^\delta} [\nabla_x u^\delta + \nabla_y u^\delta_1] \nabla_y \psi_1 \mathrm{d}y\mathrm{d}x = 0$$

for all $\psi_1 \in L^2(\Omega, H^1_\#(Y^\delta))$. Before we work with more complicated manifolds, we consider a warm-up problem with $\Gamma = [0,1]^2$ and $Y^\delta = [0,1]^2 \times (-\delta, \delta)$.

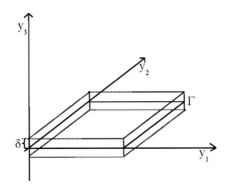

We set $u^\delta_1 = \sum_{j=1}^3 \mu^\delta_j(y)\partial_{x_j} u^\delta$ and $\nabla_y u^\delta_1 = \sum_{j=1}^3 \nabla_y \mu^\delta_j(y)\partial_{x_j} u^\delta$, where $\mu_j : H^1_\#(Y^\delta) \to \mathbb{R}$ for $j = 1,2,3$. This substitution in (4.1) leads to

$$\int_\Omega \int_{Y^\delta} \left[ \nabla_x u^\delta + \sum_{j=1}^3 \nabla_y \mu^\delta_j(y)\partial_{x_j} u^\delta \right] \nabla_y \psi_1 \mathrm{d}y\mathrm{d}x$$

$$= \int_\Omega \int_{Y^\delta} \left[ \sum_{j=1}^3 \partial_{x_j} u^\delta e_j + \sum_{j=1}^3 \nabla_y \mu^\delta_j(y)\partial_{x_j} u^\delta \right] \nabla_y \psi_1 \mathrm{d}y\mathrm{d}x$$

$$= \int_\Omega \partial_{x_j} u^\delta \int_{Y^\delta} \sum_{j=1}^3 [e_j + \nabla_y \mu^\delta_j(y)] \nabla_y \psi_1 \mathrm{d}y\mathrm{d}x = 0$$

151

for $j = 1, 2, 3$.

**The substitution**

With $Y^\delta = \Gamma \times (-\delta, \delta)$ and the variables $(y_1, y_2, y_3) \in [0,1] \times [0,1] \times (-\delta, \delta)$ we find

$$\int_\Omega \int_{-\delta}^{\delta} \int_\Gamma \left( e_j + \begin{pmatrix} \partial_{y_1} \mu_j^\delta(y) \\ \partial_{y_2} \mu_j^\delta(y) \\ \partial_{y_3} \mu_j^\delta(y) \end{pmatrix} \right) \begin{pmatrix} \partial_{y_1} \psi_1(y) \\ \partial_{y_2} \psi_1(y) \\ \partial_{y_3} \psi_1(y) \end{pmatrix} dy_1 dy_2 dy_3 dx = 0,$$

for $j = 1, 2, 3$. With the substitution $(y_1, y_2, y_3/\delta) = z = (z_1, z_2, z_3)$ we continue with

$$\int_\Omega \int_{-1}^{1} \int_\Gamma \left( e_j + \begin{pmatrix} \partial_{z_1} \mu_j^\delta(z_1, z_2, \delta z_3) \\ \partial_{z_2} \mu_j^\delta(z_1, z_2, \delta z_3) \\ \frac{1}{\delta} \partial_{z_3} \mu_j^\delta(z_1, z_2, \delta z_3) \end{pmatrix} \right)$$

$$\begin{pmatrix} \partial_{z_1} \psi_1(z_1, z_2, \delta z_3) \\ \partial_{z_2} \psi_1(z_1, z_2, \delta z_3) \\ \frac{1}{\delta} \partial_{z_3} \psi_1(z_1, z_2, \delta z_3) \end{pmatrix} dz_1 dz_2 \delta dz_3 dx = 0.$$

**Boundedness and limit formation**

Before we consider the limit for $\delta$ tending to zero we need to see that $\nabla \mu_j$ remains bounded independently of $\delta$. Therefore, we test the equation with $\psi_1 = \mu_j^\delta$ and divide by $\delta$,

$$\left\| \begin{pmatrix} \partial_{z_1} \mu_j^\delta \\ \partial_{z_2} \mu_j^\delta \\ \frac{1}{\delta} \partial_{z_3} \mu_j^\delta \end{pmatrix} \right\|_{\Omega \times \Gamma \times [-1,1]}^2 = -\int_\Omega \int_{-1}^{1} \int_\Gamma e_j \begin{pmatrix} \partial_{z_1} \mu_j^\delta \\ \partial_{z_2} \mu_j^\delta \\ \frac{1}{\delta} \partial_{z_3} \mu_j^\delta \end{pmatrix} dz dx$$

$$\leq \| e_j \|_{\Gamma \times [-1,1]} \left\| \begin{pmatrix} \partial_{z_1} \mu_j^\delta \\ \partial_{z_2} \mu_j^\delta \\ \frac{1}{\delta} \partial_{z_3} \mu_j^\delta \end{pmatrix} \right\|_{\Omega \times \Gamma \times [-1,1]}$$

which yields

$$\left\| \begin{pmatrix} \partial_{z_1}\mu_j^\delta \\ \partial_{z_2}\mu_j^\delta \\ \frac{1}{\delta}\partial_{z_3}\mu_j^\delta \end{pmatrix} \right\|_{\Omega\times\Gamma\times[-1,1]} \leq \|e_j\|_{\Omega\times\Gamma\times[-1,1]}.$$

Because the right-hand side is bounded, we find for $\delta$ tending to zero that

$$\partial_{z_3}\mu_j^\delta \to 0,$$

for all $j = 1,2,3$ and with $\mu_j^\delta$ bounded in $H_\#^1(Y^\delta)$, there exists a subsequence such that $\mu_j^\delta \overset{\delta\to 0}{\rightharpoonup} \mu_j$ and $\nabla_x\mu_j^\delta \overset{\delta\to 0}{\rightharpoonup} \nabla_\Gamma\mu_j$, where $\nabla_\Gamma$ is the gradient on the tangent space $T_y\Gamma$.

Now we consider again equation (4.1) and test with the functions $\psi = u^\delta$ and $\psi_1 = u_1^\delta$, which yields,

$$\|\nabla_x u^\delta + \nabla_y u_1^\delta\|_{\Omega\times Y^\delta}^2 = 0. \tag{4.2}$$

Hence $\nabla_x u^\delta + \nabla_y u_1^\delta$ is bounded independently of $\delta$. Again, we perform the substitution $(y_1, y_2, y_3/\delta) = z = (z_1, z_2, z_3)$ in the cell problem, test now with functions $\psi_1 \in C^\infty(\Omega, C_\#^\infty(Y))$, and arrive at

$$\int_\Omega \int_{-1}^1 \int_\Gamma \left( \nabla_x u^\delta + \begin{pmatrix} \partial_{z_1}u_1^\delta(z_1, z_2, \delta z_3) \\ \partial_{z_2}u_1^\delta(z_1, z_2, \delta z_3) \\ \frac{1}{\delta}\partial_{z_3}u_1^\delta(z_1, z_2, \delta z_3) \end{pmatrix} \right)$$
$$\begin{pmatrix} \partial_{z_1}\psi_1(z_1, z_2, \delta z_3) \\ \partial_{z_2}\psi_1(z_1, z_2, \delta z_3) \\ \frac{1}{\delta}\partial_{z_3}\psi_1(z_1, z_2, \delta z_3) \end{pmatrix} \, \mathrm{d}z\mathrm{d}x = 0.$$

This leads to

$$\int_\Omega \int_{-1}^1 \int_\Gamma \begin{pmatrix} \partial_{x_1}u^\delta + \partial_{z_1}u_1^\delta(z_1, z_2, \delta z_3) \\ \partial_{x_2}u^\delta + \partial_{z_2}u_1^\delta(z_1, z_2, \delta z_3) \\ \frac{1}{\delta}(\partial_{x_3}u^\delta + \frac{1}{\delta}\partial_{z_3}u_1^\delta(z_1, z_2, \delta z_3)) \end{pmatrix}$$
$$\begin{pmatrix} \partial_{z_1}\psi_1(z_1, z_2, \delta z_3) \\ \partial_{z_2}\psi_1(z_1, z_2, \delta z_3) \\ \partial_{z_3}\psi_1(z_1, z_2, \delta z_3) \end{pmatrix} \, \mathrm{d}z\mathrm{d}x = 0$$

for all $\psi_1 \in C^\infty(\Omega, C^\infty_\#(Y))$. With estimation (4.2) we deduce with $u^\delta_1 = \sum_{i=1}^{3} \partial_{x_i} u^\delta \mu^\delta_i$ that

$$\frac{1}{\delta}\left(\partial_{x_3} u^\delta + \frac{1}{\delta}\partial_{z_3} u^\delta_1\right) \leq C$$

$$\partial_{x_3} u^\delta + \sum_{i=1}^{3} \partial_{x_i} u^\delta \frac{1}{\delta} \underbrace{\partial_{z_3} \mu^\delta_i}_{\overset{\delta\to 0}{\to} 0} \leq \delta C \to 0.$$

This means, that $\partial_{x_3} u^\delta \overset{\delta\to 0}{\longrightarrow} 0$ and with $u^\delta$ bounded in $H^1(\Omega)$ there exists a subsequence such that $u^\delta \overset{\delta\to 0}{\rightharpoonup} u$ and $\nabla_x u^\delta \overset{\delta\to 0}{\rightharpoonup} \nabla_x u = P_\Gamma \nabla_x u$, where $P_\Gamma$ is the projection onto the tangent space $T_y\Gamma$.

**The limit cell problem**
We deduce with $\sum_{i=1}^{3}\partial_{x_i}u\nabla_\Gamma\mu_i(z_1,z_2,0) = \sum_{i=1}^{2}\partial_{x_i}u\nabla_\Gamma\mu_i(z_1,z_2,0) = \nabla_\Gamma u_1(z_1,z_2,0)$ the limit equation for $\psi = 0$, which is given by

$$0 = \int_{-1}^{1} dz_3 \int_\Omega \int_\Gamma (P_\Gamma\nabla_x u + \nabla_\Gamma u_1)\nabla_\Gamma\psi_1 dz_1 dz_2 dx$$

$$= 2\int_\Omega \sum_{i=1}^{3}\partial_{x_i}u \int_\Gamma (P_\Gamma e_i + \nabla_\Gamma\mu_i)\nabla_\Gamma\psi_1 d\sigma_z dx$$

for all $\psi_1 \in C^\infty(\Omega, C^\infty_\#(\Gamma))$. This yields the limit cell problem for $i = 1,2,3$

$$\int_\Omega \int_\Gamma (P_\Gamma e_i + \nabla_\Gamma\mu_j)\nabla_\Gamma\psi_1 d\sigma_z dx = 0.$$

The strong formulation of the limit cell problem for $i = 1,2,3$ is given by

$$\nabla_\Gamma \cdot (P_\Gamma e_i + \nabla_\Gamma\mu_i) = 0 \qquad \text{in } \Gamma$$
$$(P_\Gamma e_i + \nabla_\Gamma\mu_i) \cdot n = 0 \qquad \text{on } \partial\Gamma,$$

and $\mu_i$ being $Y$-periodic.

**The limit diffusion tensor**

Now we set $\psi_1 = 0$ to find the diffusion tensor $P$.

$$\int_\Omega \int_\Gamma \sum_j \partial_{x_j} u P_\Gamma (e_j + \nabla_\Gamma \mu_j) \nabla_x \psi \, d\sigma_y dx$$

$$= \int_\Omega \int_\Gamma \sum_{i,j} \partial_{x_j} u (P_\Gamma (e_j + \nabla_\Gamma \mu_j))_i \partial_{x_i} \psi \, d\sigma_y dx$$

$$= \int_\Omega \sum_{i,j} \partial_{x_j} u \underbrace{\int_\Gamma (P_\Gamma (e_j + \nabla_\Gamma \mu_j))_i d\sigma_y}_{=P_{ij}} \partial_{x_i} \psi \, dx.$$

$$= \int_\Omega P \nabla_x u \nabla_x \psi \, dx.$$

The diffusion tensor $P$ is given by

$$P_{ij} = \int_\Gamma (P_\Gamma (e_j + \nabla_\Gamma \mu_j))_i d\sigma_y.$$

Now we transfer this concept to general manifolds.

**A generic problem.**

Let $\Omega \subset \mathbb{R}^n$ and $\Gamma \subset Y = [0,1]^n$ be a smooth, compact and periodic hypersurface, such that $\Gamma_\varepsilon = \bigcup_{k \in \mathbb{Z}^n} \varepsilon(k + \Gamma)$ is smooth and connected. We define $Y^\delta = \{y + d \cdot n_y | \, y \in \Gamma, d \in (-\delta, \delta)\} \subset Y$ for a small $\delta > 0$, which means that the manifold gets a volume through an additional component pointing in the normal $n_y$-direction in every point $y \in \Gamma$.

Let $f \in C(\Omega, C_\#(Y))$ with $f_\varepsilon(x) := f(x, x/\varepsilon)$ and $h \in C(\partial\Omega, C_\#(\partial Y))$ with $h_\varepsilon(x) := h(x, x/\varepsilon)$. Furthermore, let $\Omega_\varepsilon = \bigcup_{k \in \mathbb{Z}^n} \varepsilon(k + Y^\delta) \cap \Omega \subset \mathbb{R}^n$ and $u_\varepsilon$ be the solution of the partial differential equation

$$\begin{aligned}
\partial_t u_\varepsilon - D \Delta u_\varepsilon + u_\varepsilon &= f_\varepsilon && \text{in } \Omega_\varepsilon \\
-D \nabla u_\varepsilon \cdot n &= a(u_\varepsilon - h_\varepsilon) && \text{on } \partial\Omega_\varepsilon \cap \partial\Omega
\end{aligned} \tag{4.3}$$

with constant diffusion coefficient $D > 0$ and constant $a > 0$. With standard estimations and two-scale convergence in [2] and theorem 4.9 we find the weak limit equation for $\varepsilon \to 0$ with $u \in$

$L^2([0,T], H^1(\Omega))$ and $u_1 \in L^2([0,T] \times \Omega, H^1_\#(Y^\delta))$ satisfying

$$|Y^\delta| \int_\Omega \partial_t u^\delta \psi dx + D \int_\Omega \int_{Y^\delta} [\nabla_x u^\delta + \nabla_y u_1^\delta][\nabla_x \psi + \nabla_y \psi_1] dy dx$$

$$+ |Y^\delta| \int_\Omega u^\delta \psi dx + \int_{\partial\Omega} \int_{\partial Y^\delta} a(u^\delta - h)\psi d\sigma_y d\sigma_x = \int_\Omega \int_{Y^\delta} f\psi dy dx \tag{4.4}$$

for all $\psi \in H^1(\Omega)$ and $\psi_1 \in L^2(\Omega, H^1_\#(Y^\delta))$, where $u^\delta$ and $u_1^\delta$ denotes the dependence on $\delta$.

**Theorem 4.1** *Let there be given a generic problem as (4.3), which two-scale converges to (4.4),*

$$|Y^\delta| \int_\Omega \partial_t u^\delta \psi dx + D \int_\Omega \int_{Y^\delta} [\nabla_x u^\delta + \nabla_y u_1^\delta][\nabla_x \psi + \nabla_y \psi_1] dy dx$$

$$+ |Y^\delta| \int_\Omega u^\delta \psi dx + \int_{\partial\Omega} \int_{\partial Y^\delta} a(u^\delta - h)\psi d\sigma_y d\sigma_x = \int_\Omega \int_{Y^\delta} f\psi dy dx$$

*for all $\psi \in H^1(\Omega)$ and $\psi_1 \in L^2(\Omega, H^1_\#(Y^\delta))$.*
*Then, for $\delta \to 0$, the solution $u$ must satisfy the weak limit equation*

$$|\Gamma| \int_\Omega \partial_t u \psi dx + \int_\Omega P\nabla_x u \nabla_x \psi dx + |\Gamma| \int_\Omega u\psi dx + |\partial\Gamma| \int_{\partial\Omega} au\psi d\sigma_x$$

$$= \int_\Omega \int_\Gamma f\psi d\sigma_y dx + \int_{\partial\Omega} \int_{\partial\Gamma} ah\psi d\sigma_y d\sigma_x \tag{4.5}$$

*with*

$$P_{ij} = D \int_\Gamma \left(P_\Gamma(e_j + \nabla_\Gamma \mu_j)\right)_i d\sigma_y$$

*for $i,j = 1,\ldots,n$ and $\mu_j$ satisfying*

$$\nabla_\Gamma \cdot \left(P_\Gamma(e_j + \nabla_\Gamma \mu_j)\right) = 0 \qquad \text{in } \Gamma,$$

$$P_\Gamma(e_j + \nabla_\Gamma \mu_j) \cdot n = 0 \qquad \text{on } \partial\Gamma,$$

*with $\mu_j$ being $Y$-periodic. Here, $P_\Gamma$ is the orthogonal projection from $Y$ to $\Gamma$.*

Before we start the proof, we briefly illustrate the setting.

Let $\{(\tilde{U}_\lambda, \tilde{\alpha}_\lambda)\}$ be an atlas of the manifold $\Gamma$ such that $\bigcup_\lambda \tilde{U}_\lambda = \Gamma$ and $\tilde{\alpha}_\lambda : \tilde{U}_\lambda \to V_\lambda \subset \mathbb{R}^{n-1}$. With $n_y$ being the normal vector of $\Gamma$ in the point $y \in \Gamma$ we define another atlas $\{(U_\lambda, \alpha_\lambda)\}$ of the blown up domain $Y^\delta$ with $\bigcup_\lambda U_\lambda = Y^\delta$ by

$$\alpha_\lambda^{-1} : V_\lambda \times (-\delta, \delta) \to U_\lambda$$

$$\alpha_\lambda^{-1}(\xi_1, \ldots, \xi_{n-1}, \xi_n) = \underbrace{\tilde{\alpha}_\lambda^{-1}(\xi_1, \ldots, \xi_{n-1})}_{\in \tilde{U}_\lambda} + \xi_n n_y.$$

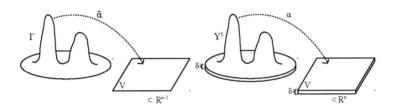

Note that the last component of the local coordinates affects just the normal direction of $\Gamma$, which yields for the Riemannian metric tensor $g_{ij}$, $i, j = 1, \ldots, n$ that

$$g_{in} = g^{in} = g_{ni} = g^{ni} = 0, \qquad i = 1, \ldots, n-1,$$

where $g^{ij}$, $i, j = 1, \ldots n$ are the components of the inverse of the matrix $(g_{ij})_{i,j=1,\ldots,n}$. The $n$th basis vector is given by

$$\frac{\mathrm{d}}{\mathrm{d}\xi^n} = \frac{\mathrm{d}}{\mathrm{d}t}\Big|_{t=0} \alpha_\lambda^{-1}(\xi + t e_n)$$

$$= \frac{\mathrm{d}}{\mathrm{d}t}\Big|_{t=0} \left(\tilde{\alpha}_\lambda^{-1}(\xi_1, \ldots, \xi_{n-1}) + (\xi_n + t)n_y\right) = n_y$$

and consequently, $g_{nn} = n_y^T n_y = 1$. We choose the Riemannian metric tensor $g_{ij}$, such that $g_{ij} = g^{ij} = 0$ also holds for any $i \neq j$ on the manifold $\Gamma$, which means that the basic vectors $\frac{\mathrm{d}}{\mathrm{d}\xi^i}$, $i = 1, \ldots, n$ are orthogonal. Note that $g_{ii} = \left\langle \frac{\mathrm{d}}{\mathrm{d}\xi^i}, \frac{\mathrm{d}}{\mathrm{d}\xi^i} \right\rangle > 0$. The gradient of a

157

function $\mu : Y^\delta \to \mathbb{R}$ in $U_\lambda$ in new coordinates is given by

$$\nabla \mu = \sum_{i=1}^{n} g^{ii} \frac{\partial(\mu \circ \alpha_\lambda^{-1})}{\partial \xi_i} \frac{\mathrm{d}}{\mathrm{d}\xi^i}.$$

The divergence on $Y^\delta$ of a function $\mu : Y^\delta \to \mathbb{R}^n$ in new coordinates is given by

$$\nabla \circ \mu = \frac{1}{\sqrt{\det g}} \sum_{i=1}^{n} \frac{\mathrm{d}}{\mathrm{d}\xi^i} \left( \mu^i \sqrt{\det g} \right).$$

Here $\mu^i$ is the $i$-th component of $\mu$ in the basis vectors $\frac{\mathrm{d}}{\mathrm{d}\xi^i}$.

**Proof of theorem 4.1**
The proof is composed of three steps. First we show that the terms, where the limit formation takes place, are bounded. In the second step we let $\delta$ tend to zero and consider the consequences in the various terms. In the last step we deduce the cell problem and the macroscopic limit equation.
For the charts we use the abbreviating notation $\alpha(x) = \alpha_\lambda(x)$ for $x \in U_\lambda \subset Y^\delta$. In the proof we indicate the $\delta$ dependence of the functions $u$ and $u_1$ by $u^\delta$ and $u_1^\delta$.

*Step 1. Boundedness of $\|\nabla \mu_k^\delta\|_{Y^\delta}$.*
In the third step, where we perform the limit formation for $\delta$ tending to zero, we need that the norm of the solution of the cell problem is bounded. We are going to ensure the boundedness now in the first step.
We consider the cell problem by setting $\psi = 0$ and use $\nabla_y u_1^\delta = \sum_{k=1}^{n} \nabla_y \mu_k^\delta \partial_{x_k} u^\delta$ for $\mu_k^\delta \in H^1_\#(Y^\delta)$ with $\int_{Y^\delta} \mu_k^\delta dy = 0$ in the equation (4.4),

$$D \int_\Omega \int_{Y^\delta} \left[ \nabla_x u^\delta + \sum_{k=1}^{n} \nabla_y \mu_k^\delta \partial_{x_k} u^\delta \right] \nabla_y \psi_1 dy dx = 0.$$

We rewrite this equation as

$$D \int_\Omega \sum_{k=1}^{n} \partial_{x_k} u^\delta \int_{Y^\delta} \left[ e_k + \nabla_y \mu_k^\delta \right] \nabla_y \psi_1 dy dx = 0.$$

Testing the equation with $\psi_1 = \mu_k^\delta$ for every $k = 1, \dots, n$, leads to

$$\int_\Omega \int_{Y^\delta} \nabla_y \mu_k^\delta \nabla_y \mu_k^\delta \mathrm{d}y\mathrm{d}x = - \int_\Omega \int_{Y^\delta} e_k \nabla_y \mu_k^\delta \mathrm{d}y\mathrm{d}x$$

$$\leq \|e_k\|_{\Omega \times Y^\delta} \|\nabla_y \mu_k^\delta\|_{\Omega \times Y^\delta}$$

$$= \sqrt{|\Omega| \cdot |Y^\delta|} \|\nabla_y \mu_k^\delta\|_{\Omega \times Y^\delta},$$

where we used the Cauchy-Schwarz inequality. It follows that

$$\frac{1}{|\Omega| \cdot |Y^\delta|} \|\nabla_y \mu_k^\delta\|_{\Omega \times Y^\delta}^2 \leq 1,$$

which means that the norm of $\nabla_y \mu_k^\delta$ remains bounded independently of the size of the domain $Y^\delta$.

*Boundedness of $\|\nabla_x u^\delta + \nabla_y u_1^\delta\|_{\Omega \times Y^\delta}$.*
To show boundedness also for the diffusion term in the macroscopic problem we consider equation (4.4),

$$|Y^\delta| \int_\Omega \partial_t u^\delta \psi \mathrm{d}x + D \int_{\Omega \times Y^\delta} [\nabla_x u^\delta + \nabla_y u_1^\delta][\nabla_x \psi + \nabla_y \psi_1] \mathrm{d}y\mathrm{d}x$$

$$+ |Y^\delta| \int_\Omega u^\delta \psi \mathrm{d}x + a|\partial Y^\delta| \int_{\partial\Omega} u^\delta \psi \mathrm{d}\sigma_x$$

$$= \int_\Omega \int_{Y^\delta} f\psi \mathrm{d}y\mathrm{d}x + \int_{\partial\Omega} \int_{\partial Y^\delta} h\psi \mathrm{d}\sigma_y \mathrm{d}\sigma_x.$$

We perform a substitution by using charts $\alpha : V \times (-\delta, \delta) \to Y^\delta$. Thereby, the terms $|Y^\delta|$ and $|\partial Y^\delta|$ are equal to $2\delta|\Gamma|$ and $2\delta|\partial\Gamma|$ in their first approximation, respectively.

$$(2\delta|\Gamma| + \mathcal{O}(\delta^2)) \int_\Omega \partial_t u^\delta \psi \mathrm{d}x$$

$$+ D \int_{\Omega \times Y^\delta} [\nabla_x u^\delta + \nabla_y u_1^\delta][\nabla_x \psi + \nabla_y \psi_1] \mathrm{d}y\mathrm{d}x$$

$$+ (2\delta|\Gamma| + \mathcal{O}(\delta^2)) \int_\Omega u^\delta \psi \mathrm{d}x + a(2\delta|\partial\Gamma| + \mathcal{O}(\delta^2)) \int_{\partial\Omega} u^\delta \psi \mathrm{d}\sigma_x$$

$$= \int_\Omega \int_{V \times (-\delta, \delta)} f\psi \sqrt{\det g} \mathrm{d}\xi \mathrm{d}d\mathrm{d}x$$

$$+ \int_{\partial\Omega} \int_{\partial V \times (-\delta, \delta)} h\psi \sqrt{\det g} \mathrm{d}\sigma_\xi \mathrm{d}d\mathrm{d}\sigma_x$$

for all $(\psi, \psi_1) \in H^1(\Omega) \times L^2(\Omega, H^1_\#(Y^\delta))$. Now we test with the functions $\psi = u^\delta$ and $\psi_1 = u^\delta_1$,

$$(2\delta|\Gamma| + \mathcal{O}(\delta^2)) \int_\Omega \partial_t u^\delta u^\delta \mathrm{d}x + D\|\nabla_x u^\delta + \nabla_y u^\delta_1\|^2_{\Omega \times Y^\delta}$$
$$+ (2\delta|\Gamma| + \mathcal{O}(\delta^2))\|u^\delta\|^2_\Omega + a(2\delta|\partial\Gamma| + \mathcal{O}(\delta^2))\|u^\delta\|^2_{\partial\Omega}$$
$$= \int_\Omega u^\delta \int_{V \times (-\delta,\delta)} f\sqrt{\det g}\,\mathrm{d}\xi\mathrm{d}d\mathrm{d}x$$
$$+ \int_{\partial\Omega} u^\delta \int_{\partial V \times (-\delta,\delta)} h\sqrt{\det g}\,\mathrm{d}\sigma_\xi\mathrm{d}d\mathrm{d}\sigma_x$$

Furthermore, we substitute:

$$(\xi_1, \ldots, \xi_{n-1}, \xi_n/\delta) = z = (z_1, \ldots, z_{n-1}, z_n),$$
$$\mathrm{d}\xi = \mathrm{d}\xi_1 \ldots \mathrm{d}\xi_n = \mathrm{d}z_1 \ldots \mathrm{d}z_{n-1}\delta\mathrm{d}z_n,$$

and continue with

$$(2\delta|\Gamma| + \mathcal{O}(\delta^2)) \int_\Omega \partial_t u^\delta u^\delta \mathrm{d}x + D\|\nabla_x u^\delta + \nabla_y u^\delta_1\|^2_{\Omega \times Y^\delta}$$
$$+ (2\delta|\Gamma| + \mathcal{O}(\delta^2))\|u^\delta\|^2_\Omega + a(2\delta|\partial\Gamma| + \mathcal{O}(\delta^2))\|u^\delta\|^2_{\partial\Omega}$$
$$= \delta \int_\Omega u^\delta \int_{V \times (-1,1)} f\sqrt{\det g}\,\mathrm{d}z\mathrm{d}d\mathrm{d}x$$
$$+ \delta \int_{\partial\Omega} u^\delta \int_{\partial V \times (-1,1)} h\sqrt{\det g}\,\mathrm{d}\sigma_z\mathrm{d}d\mathrm{d}\sigma_x$$

and divide by $\delta$ to find

$$(2|\Gamma|+\mathcal{O}(\delta)) \int_\Omega \partial_t u^\delta u^\delta \mathrm{d}x + \frac{1}{\delta}D\|\nabla_x u^\delta + \nabla_y u^\delta_1\|^2_{\Omega \times Y^\delta}$$
$$+ (2|\Gamma| + \mathcal{O}(\delta))\|u^\delta\|^2_\Omega + a(2|\partial\Gamma| + \mathcal{O}(\delta))\|u^\delta\|^2_{\partial\Omega}$$
$$= \int_\Omega u^\delta \int_{V \times (-1,1)} f\sqrt{\det g}\,\mathrm{d}z\mathrm{d}d\mathrm{d}x$$
$$+ \int_{\partial\Omega} u^\delta \int_{\partial V \times (-1,1)} h\sqrt{\det g}\,\mathrm{d}\sigma_z\mathrm{d}d\mathrm{d}\sigma_x$$
$$\leq \|u^\delta\|_\Omega\|\int_{V \times (-1,1)} f\sqrt{\det g}\,\mathrm{d}z\mathrm{d}d\|_\Omega$$
$$+ \|u^\delta\|_{\partial\Omega}\|\int_{\partial V \times (-1,1)} h\sqrt{\det g}\,\mathrm{d}\sigma_z\mathrm{d}d\|_{\partial\Omega}$$

$$\leq \frac{1}{2}\|u^\delta\|_\Omega^2 + \underbrace{\frac{1}{2}\|\int_{V\times(-1,1)} f\sqrt{\det g}\,dzdd\|_\Omega^2}_{\text{bounded}}$$

$$+ \frac{1}{2\lambda}\|u^\delta\|_{\partial\Omega}^2 + \underbrace{\frac{\lambda}{2}\|\int_{\partial V\times(-1,1)} h\sqrt{\det g}\,d\sigma_z dd\|_{\partial\Omega}^2}_{\text{bounded}}$$

for any $\lambda > 0$. This yields, after integration with respect to time,

$$(2|\Gamma| + \mathcal{O}(\delta))\frac{1}{2}\|u^\delta\|_\Omega^2 + \frac{1}{\delta}D\|\nabla_x u^\delta + \nabla_y u_1^\delta\|_{\Omega\times Y^\delta,t}^2$$

$$+ (2|\Gamma| + \mathcal{O}(\delta))\|u^\delta\|_{\Omega,t}^2 + \left(a(2|\partial\Gamma| + \mathcal{O}(\delta)) - \frac{1}{2\lambda}\right)\|u^\delta\|_{\partial\Omega,t}^2$$

$$\leq c_1 + c_2\frac{\lambda}{2} + \frac{1}{2}\|u^\delta\|_{\Omega,t}^2 + \frac{1}{2}\|u(0)\|_\Omega^2,$$

where we choose $\lambda$ such that $\lambda > \frac{1}{4a|\partial\Gamma|}$, but finite. The constants $c_1, c_2$ and $\lambda$ are independent of $\delta$. We find with Gronwall's lemma that

$$(2|\Gamma| + \mathcal{O}(\delta))\frac{1}{2}\|u^\delta\|_\Omega^2 + \frac{1}{\delta}D\|\nabla_x u^\delta + \nabla_y u_1^\delta\|_{\Omega\times Y^\delta,t}^2$$

$$+ (2|\Gamma| + \mathcal{O}(\delta))\|u^\delta\|_{\Omega,t}^2 + \left(a(2|\partial\Gamma| + \mathcal{O}(\delta)) - \frac{1}{2\lambda}\right)\|u^\delta\|_{\partial\Omega,t}^2 < C$$

for a constant $C > 0$, independent of $\delta$.

*Step 2. Limit of the linear terms.*
Now, with $u^\delta$ bounded in $H^1(\Omega)$ we deduce the existence of a weak converging subsequence. The equation (4.4) is now tested with functions $(\psi, \psi_1) \in C^\infty(\Omega) \times C^\infty(\Omega, C_\#^\infty(Y))$.
First we consider the limit for $\delta \to 0$ of the first term,

$$(2|\Gamma| + \mathcal{O}(\delta))\int_\Omega \partial_t u^\delta \psi dx \to 2|\Gamma| \int_\Omega \partial_t u \psi dx.$$

It also easily follow that

$$(2|\Gamma| + \mathcal{O}(\delta))\int_\Omega u^\delta \psi dx \xrightarrow{\delta\to 0} 2|\Gamma| \int_\Omega u\psi dx.$$

161

*Limit of the diffusion term.*

To perform the limit formation in the diffusion term we use the same substitutions, which we used in Step 1. We consider the diffusion term of equation (4.4) and use the gradient formula on manifolds.

$$\frac{1}{\delta}D\int_\Omega\int_{Y^\delta}[\nabla_x u^\delta + \nabla_y u_1^\delta][\nabla_x\psi + \nabla_y\psi_1]\,dy\,dx$$

$$= \frac{1}{\delta}D\int_\Omega\int_V\int_{(-\delta,\delta)}\left[\nabla_x u^\delta + \sum_{k=1}^n \partial_{x_k}u^\delta \sum_{i=1}^n g^{ii}\frac{\partial(\mu_k^\delta \circ \alpha^{-1})}{\partial\xi_i}\frac{\mathrm{d}}{\mathrm{d}\xi^i}\right]$$

$$\left[\nabla_x\psi + \sum_{j=1}^n g^{jj}\frac{\partial(\psi_1 \circ \alpha^{-1})}{\partial\xi_j}\frac{\mathrm{d}}{\mathrm{d}\xi^j}\right]\sqrt{\det g}\,d\xi\,dx$$

We substitute:

$$(\xi_1,\ldots,\xi_{n-1},\xi_n/\delta) = z = (z_1,\ldots,z_{n-1},z_n),$$

$$d\xi = d\xi_1\ldots d\xi_n = dz_1\ldots dz_{n-1}\delta dz_n.$$

We define functions $\bar\mu^\delta$ and $\bar\psi_1$ as

$$(\bar\mu^\delta \circ \alpha^{-1})(z) := (\mu^\delta \circ \alpha^{-1})(z_1, z_2, \delta z_3)$$

$$(\bar\psi_1 \circ \alpha^{-1})(z) := (\psi_1 \circ \alpha^{-1})(z_1, z_2, \delta z_3),$$

respectively, and continue with

$$\frac{\delta}{\delta}D\int_\Omega\int_V\int_{(-1,1)}\left[\nabla_x u^\delta + \sum_{k=1}^n \partial_{x_k}u^\delta \sum_{i=1}^n g^{ii}\frac{\partial(\bar\mu_k^\delta \circ \alpha^{-1})}{(1+\delta_{ni}(\delta-1))\partial z_i}\frac{\mathrm{d}}{\mathrm{d}z^i}\right]$$

$$\left[\nabla_x\psi + \sum_{j=1}^n g^{jj}\frac{\partial(\bar\psi_1 \circ \alpha^{-1})}{(1+\delta_{nj}(\delta-1))\cdot\partial z_j}\frac{\mathrm{d}}{\mathrm{d}z^j}\right]\sqrt{\det g}\,dz\,dx$$

with $\delta_{nj} = 0$, if $j = 1,\ldots,n-1$ and $\delta_{nj} = 1$, if $j = n$. From Step 1 we know that

$$\frac{1}{(2|\Gamma| + \mathcal{O}(\delta))}\int_{V\times(-1,1)}\left|\sum_{i=1}^n g^{ii}\frac{\partial(\bar\mu_k^\delta \circ \alpha^{-1})}{(1+\delta_{ni}(\delta-1))\cdot\partial z_i}\frac{\mathrm{d}}{\mathrm{d}z^i}\right|^2 d\xi\,dd$$

$$= \frac{1}{|Y^\delta|}\|\nabla_y\mu_k^\delta\|_{Y^\delta}^2 \le 1.$$

Taking a look at the $n$th summand we deduce:

$$\left|\frac{1}{\delta}\frac{\partial(\bar{\mu}_k^\delta \circ \alpha^{-1})}{\partial z_n}\right| \leq C \quad \text{yields} \quad \left|\frac{\partial(\bar{\mu}_k^\delta \circ \alpha^{-1})}{\partial z_n}\right| \leq C\delta \xrightarrow{\delta \to 0} 0.$$

This implies that $\frac{\partial(\bar{\mu}_k^\delta \circ \alpha^{-1})}{\partial z_n}$ converges strongly to zero and with $\mu_k^\delta \circ \alpha^{-1}$ bounded in $H_\#^1(V \times (-1,1))$ indepedently of $\delta$ there exists a weak converging subsequence $\mu_k^\delta \circ \alpha^{-1} \xrightarrow{\delta \to 0} \mu_k \circ \alpha^{-1}$ in $H_\#^1(V \times (-1,1))$ such that $\bar{\mu}_k \circ \alpha^{-1}$ is independent of $z_n$;

To deduce the limit of $\nabla_x u^\delta$ for $\delta$ tending to zero, we set $\psi = 0$ and arrive at

$$D \int_\Omega \int_{V \times (-1,1)} \left[ \nabla_x u^\delta + \sum_{k=1}^n \partial_{x_k} u^\delta \sum_{i=1}^n g^{ii} \frac{\partial(\bar{\mu}_k^\delta \circ \alpha^{-1})}{(1+\delta_{ni}(\delta-1))\partial z_i}\frac{\mathrm{d}}{\mathrm{d}z^i} \right]$$
$$\left[ \sum_{j=1}^n g^{jj} \frac{\partial(\bar{\psi}_1 \circ \alpha^{-1})}{(1+\delta_{nj}(\delta-1))\partial z_j}\frac{\mathrm{d}}{\mathrm{d}z^j} \right] \sqrt{\det g}\,\mathrm{d}z\mathrm{d}x.$$

Now $\nabla_x u^\delta$ is written as $\sum_i \langle \frac{\mathrm{d}}{\mathrm{d}z^i}, \nabla_x u^\delta \rangle \frac{\mathrm{d}}{\mathrm{d}z^i}$ and we use the definition of the scalar product on $\Gamma$, where here $\langle a, b \rangle = \sum_i g_{ii}a_i b_i$ for $a, b \in T_y Y^\delta$,

$$D \int_\Omega \int_{V \times (-1,1)} \sum_{i=1}^n g_{ii} \left[ \left\langle \frac{\mathrm{d}}{\mathrm{d}z^i}, \nabla_x u^\delta \right\rangle \right.$$
$$\left. + \sum_{k=1}^n \partial_{x_k} u^\delta g^{ii} \frac{\partial(\bar{\mu}_k^\delta \circ \alpha^{-1})}{(1+\delta_{ni}(\delta-1))\partial z_i} \right]$$
$$\left[ g^{ii} \frac{\partial(\bar{\psi}_1 \circ \alpha^{-1})}{(1+\delta_{ni}(\delta-1))\partial z_i} \right] \sqrt{\det g}\,\mathrm{d}z\mathrm{d}x$$

$$= D \int_{\Omega} \int_{V \times (-1,1)} \sum_{i=1}^{n-1} g_{ii} \left[ \left\langle \frac{d}{dz^i}, \nabla_x u^{\delta} \right\rangle + \sum_{k=1}^{n} \partial_{x_k} u^{\delta} g^{ii} \frac{\partial (\bar{\mu}_k^{\delta} \circ \alpha^{-1})}{\partial z_i} \right]$$

$$\left[ g^{ii} \frac{\partial (\bar{\psi}_1 \circ \alpha^{-1})}{\partial z_i} \right]$$

$$+ \left[ \frac{1}{\delta} \left\langle \frac{d}{dz^n}, \nabla_x u^{\delta} \right\rangle + \frac{1}{\delta} \sum_{k=1}^{n} \partial_{x_k} u^{\delta} g^{nn} \frac{\partial (\bar{\mu}_k^{\delta} \circ \alpha^{-1})}{\delta \partial z_n} \right]$$

$$g^{nn} \frac{\partial (\bar{\psi}_1 \circ \alpha^{-1})}{\partial z_n} \sqrt{\det g} dz dx$$

for any $\psi_1 \in C^{\infty}(\Omega, C_{\#}^{\infty}(Y))$. Because $\|\nabla_x u^{\delta} + \nabla_y u_1^{\delta}\|_{\Omega \times V \times (-1,1)}^2$ is bounded (see Step 1) and $\frac{\partial (\bar{\mu}_k^{\delta} \circ \alpha^{-1})}{\partial z_n} \xrightarrow{\delta \to 0} 0$ for $k = 1, \dots, n$, we deduce by considering the $n$th summand that

$$\underbrace{\left\langle \frac{d}{dz^n}, \nabla_x u^{\delta} \right\rangle}_{\to 0 !!} + \sum_{k=1}^{n} \underbrace{\partial_{x_k} u^{\delta} g^{nn}}_{\text{bounded}} \underbrace{\frac{\partial (\bar{\mu}_k^{\delta} \circ \alpha^{-1})}{\delta \partial z_n}}_{\to 0} \leq C\delta \to 0$$

and conclude that

$$\nabla_x u^{\delta} = \sum_{i=1}^{n} \left\langle \frac{d}{dz^i}, \nabla_x u^{\delta} \right\rangle \xrightarrow{\delta \to 0} \sum_{i=1}^{n-1} \left\langle \frac{d}{dz^i}, \nabla_x u \right\rangle = P_{\Gamma} \nabla_x u$$

where $P_{\Gamma}$ is the projection onto the tangent space $T_y \Gamma$.

Since we know that $g^{nn} = g_{nn} = 1$, we use $\sqrt{\det g} = \sqrt{\det(g_{ij})_{i,j=1,\dots,n-1}}$. Because $\bar{\mu}_k$ and $\mu_k$ just differ in the last component, but also are independent of this component, we rewrite the integral using $\mu_k, k = 1, \dots, n$ and $\int_{-1}^{1} dz_n = 2$ for $\delta = 0$.

$$2D \int_{\Omega} \int_{V} \left[ P_{\Gamma} \nabla_x u + \sum_{k=1}^{n} \partial_{x_k} u \sum_{i=1}^{n-1} g^{ii} \frac{\partial (\mu_k \circ \alpha^{-1})}{\partial z_i} \frac{d}{dz^i} \right]$$

$$\sum_{j=1}^{n-1} g^{jj} \frac{\partial (\psi_1 \circ \alpha^{-1})}{\partial z_j} \frac{d}{dz^j} \sqrt{\det g} dz_1 \dots dz_{n-1} dx$$

$$= 2D \int_{\Omega} \int_{\Gamma} \left[ P_{\Gamma} \nabla_x u + \sum_{k=1}^{n} \partial_{x_k} u \nabla_{\Gamma} \mu_k \right] \nabla_{\Gamma} \psi_1 d\sigma_y dx,$$

for all $\psi_1 \in C^\infty(\Omega, C^\infty_\#(\Gamma))$, where $\nabla_\Gamma$ is the gradient respective to the tangent space.

The limit diffusion term for $\psi \neq 0$ is given by

$$2D \int_{\Omega \times [0,t]} \int_\Gamma \sum_{k=1}^n \partial_{x_k} u \left[ P_\Gamma e_k + \nabla_\Gamma \mu_k \right] \left[ \nabla_x \psi + \nabla_\Gamma \psi_1 \right] \mathrm{d}\sigma_y \mathrm{d}x \mathrm{d}t.$$

*Limit of the right-hand side.*

For $\delta$ tending to zero, the right-hand side has the following behavior. With $f$ continuous, it easily holds that

$$\int_\Omega \int_{V \times (-1,1)} f(x; z_1, ., z_{n-1}, \delta z_n; t) \psi(x; z_1, ., z_{n-1}, \delta z_n) \sqrt{\det g} \mathrm{d}z \mathrm{d}x$$

$$\overset{\delta \to 0}{\to} 2 \int_\Omega \int_V f(x; z_1, ., z_{n-1}, 0) \psi(x; z_1, ., z_{n-1}, 0) \sqrt{\det g} \mathrm{d}z \mathrm{d}x$$

$$= 2 \int_\Omega \int_\Gamma f(x, y) \psi(x, y) \mathrm{d}\sigma_y \mathrm{d}x.$$

Moreover, we find because of $h$ continuous

$$\int_{\partial\Omega} \int_{\partial V \times (-1,1)} h(x; z_1, ., z_{n-1}, \delta z_n; t) \psi(x; z_1, ., z_{n-1}, \delta z_n) \sqrt{\det g} \mathrm{d}\sigma_z \mathrm{d}\sigma_x$$

$$\overset{\delta \to 0}{\to} 2 \int_{\partial\Omega} \int_{\partial V} h(x; z_1, ., z_{n-1}, 0) \psi(x; z_1, ., z_{n-1}, 0) \sqrt{\det g} \mathrm{d}\sigma_z \mathrm{d}\sigma_x$$

$$= 2 \int_{\partial\Omega} \int_{\partial\Gamma} h(x, y) \psi(x, y) \mathrm{d}\sigma_y \mathrm{d}\sigma_x.$$

Hence, for $\delta = 0$ we arrive at the equation

$$2|\Gamma| \int_\Omega \partial_t u \psi \mathrm{d}x$$

$$+ 2D \int_\Omega \int_\Gamma \sum_{k=1}^n \partial_{x_k} u \left[ P_\Gamma e_k + \nabla_\Gamma \mu_k \right] \left[ \nabla_x \psi + \nabla_\Gamma \psi_1 \right] \mathrm{d}\sigma_y \mathrm{d}x$$

$$+ 2|\Gamma| \int_\Omega u \psi \mathrm{d}x + 2a|\partial\Gamma| \int_{\partial\Omega} u \psi \mathrm{d}\sigma_x$$

$$= 2 \int_\Omega \int_\Gamma f(x, y) \psi(x, y) \mathrm{d}\sigma_y \mathrm{d}x + 2a \int_{\partial\Omega} \int_{\partial\Gamma} h \psi \mathrm{d}\sigma_y \mathrm{d}\sigma_x$$

and may divide by 2.

*Step 3. The limit cell problem.*
It is left to find the cell problem and therefore we set again $\psi = 0$ and obtain for $k = 1, \ldots, n$

$$\int_\Gamma [P_\Gamma e_k + \nabla_\Gamma \mu_k] \nabla_\Gamma \psi_1 d\sigma_y = 0. \tag{4.6}$$

Then, the strong formulation of the cell problem is given by

$$\nabla_\Gamma \cdot (P_\Gamma (e_j + \nabla_\Gamma \mu_j)) = 0 \qquad \text{in } \Gamma,$$
$$P_\Gamma (e_j + \nabla_\Gamma \mu_j) \cdot n = 0 \qquad \text{on } \partial\Gamma,$$

and $\mu_j$ being $Y$-periodic for $k = 1, \ldots, n$.

*The limit diffusion tensor.*
By setting $\psi_1$ to zero we can find the diffusion tensor $P$ by considering the diffusion term

$$D \int_\Omega \int_\Gamma \sum_j \partial_{x_j} u P_\Gamma (e_j + \nabla_\Gamma \mu_j) \nabla_x \psi d\sigma_y dx$$

$$= D \int_\Omega \int_\Gamma \sum_{i,j} \partial_{x_j} u (P_\Gamma (e_j + \nabla_\Gamma \mu_j))_i \partial_{x_i} \psi d\sigma_y dx$$

$$= D \int_\Omega \sum_{i,j} \partial_{x_j} u \int_\Gamma (P_\Gamma (e_j + \nabla_\Gamma \mu_j))_i d\sigma_y \partial_{x_i} \psi dx$$

$$= \int_\Omega P \nabla_x u \nabla_x \psi dx$$

with the diffusion tensor $P = (P_{ij})_{ij}$ given by

$$P_{ij} = D \int_\Gamma (P_\Gamma (e_j + \nabla_\Gamma \mu_j))_i d\sigma_y. \tag{4.7}$$

This leads to the required result (4.5). $\qquad\qquad\qquad\square$

**Remark 4.2** If there are more linear or nonlinear terms, which are independent of $y$, i.e multiplied by a factor $|Y^\delta|$, the theorem 4.1 also holds and the factor $|Y^\delta|$ tends to $2|\Gamma|$ for $\delta \to 0$.

**Remark 4.3** To prove theorem 4.1 we detoured by assuming that the membrane $\Gamma_\varepsilon$, where we want to find global diffusion in the

$\varepsilon$-limit, has a thickness $\delta > 0$. Then, by letting the thickness tend to zero, the assertion follows.

In theorem 3.27 we obtained this result directly. In example 3.28 we start with equation (3.11) and derive the limit equation (3.14). Here, in theorem 4.1, we start with the same equation (4.3) but defined on the blown up domain $\Omega_\varepsilon$. In the limit we find the same macroscopic equation (4.5). The resulting effective diffusion coefficient (3.13) is the same as (4.7). Hence, there are two different ways to arrive at the same result.

Now we present two examples how to apply theorem 4.1.

**Example 4.4** As a first example we take a smooth surface $\Gamma$ embedded in $\mathbb{R}^3$ that is symmetric in $xy$-, $yz$-, and $xz$- plane (see figure below showing the unit cell $Y$).

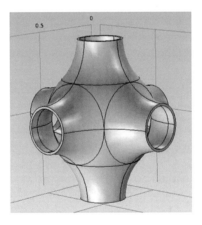

For a pde with diffusion term defined on $\Gamma_\varepsilon = \bigcup_{k \in \mathbb{Z}^3} \varepsilon(\Gamma + k)$, which is connected, we obtain after homogenization diffusion in $x$, $y$ and $z$ direction in the macroscopic equation.

In the following we give a numerical example.

**Example 4.5** In a domain $\Gamma_\varepsilon$ we consider the heat equation. On the right-hand side we see the unit cell $Y$ and on the left-hand side we see the domain $\Gamma_\varepsilon$, the sine-shaped lines. We assume a no-flux condition at the boundary of $\Gamma_\varepsilon$.

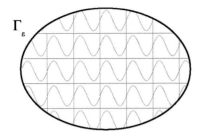
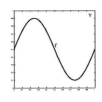

Usually, the domain $\Gamma_\varepsilon$ should be conntected, which is not the case in this example. But in two dimensions a smooth, periodic and connected manifold is hard to find, hence we still want to use this domain $\Gamma_\varepsilon$. With this example we want to give an easy idea and outline how to use theorem 4.1 and do not draw attention to theoretical subtleties.

The function $\alpha : [0,1] \to \Gamma$ is a parametrization of $\Gamma$:

$$\alpha(x) = \begin{pmatrix} x \\ \frac{1}{2} + \frac{2}{5}\sin(2\pi x) \end{pmatrix}, \qquad x \in [0,1].$$

The partial differential equation is given by

$$\partial_t u_\varepsilon - d\Delta u_\varepsilon = 1 \qquad \text{on } \Gamma_\varepsilon,$$
$$-d\nabla u_\varepsilon \cdot n = 0 \qquad \text{on } \partial\Gamma_\varepsilon.$$

with a constant diffusion coefficient $d \in \mathbb{R}$. The weak formulation is given by

$$\int_{\Gamma_\varepsilon} \partial_t u_\varepsilon \psi_\varepsilon \mathrm{d}x + d\int_{\Gamma_\varepsilon} \nabla u_\varepsilon \nabla \psi_\varepsilon \mathrm{d}x = \int_{\Gamma_\varepsilon} \psi_\varepsilon \mathrm{d}x$$

for every $\psi_\varepsilon \in C^\infty(\Omega, C_\#^\infty(\Gamma))$.

Homogenization works as usual and with theorem 4.1 we find the following limit equation

$$|\Gamma| \int_\Omega \partial_t u \psi dx + d \int_\Omega \int_\Gamma \sum_{i,j}^n d\left((P_\Gamma e_i) + \nabla_\Gamma \mu_i(y)\right)_j d\sigma_y \partial_{x_j} u \partial_{x_i} \psi dx$$

$$= |\Gamma| \int_\Omega \psi dx.$$

Now we solve the cell problem

$$\nabla_\Gamma \cdot (\nabla_\Gamma \mu_i + P_\Gamma e_i) = 0 \quad \text{in } \Gamma,$$
$$(\nabla_\Gamma \mu_i + P_\Gamma e_i) \cdot n = 0 \quad \text{in } \partial\Gamma,$$

where $\mu_i$ is $Y$-periodic for $i, j = 1, 2$.

The domain $\Omega_\epsilon$ is constructed such that there is no connection in the $x_2$-direction. Hence, there cannot be a diffusion in the $x_2$-direction and also the $y_2$-direction. The function $\mu_2(y)$ does not need to be calculated, and we already know that

$$\int_\Gamma d((P_\Gamma e_2 + \nabla_\Gamma \mu_2(y))_{1,2} d\sigma_y = 0.$$

The function $\mu_1(y)$ must satisfy

$$\nabla_\Gamma \cdot (\nabla_\Gamma \mu_1 + P_\Gamma e_1) = 0 \quad \text{on } \Gamma,$$

and $\mu_1$ must be $Y$-periodic.

To find the Riemannian metric we differentiate the parametrization $\alpha$

$$\frac{\partial \alpha}{\partial x} = \begin{pmatrix} 1 \\ \frac{4}{5}\pi \cos(2\pi x) \end{pmatrix},$$

$$g_{11}(x) = \left\langle \frac{\partial \alpha}{\partial x}, \frac{\partial \alpha}{\partial x} \right\rangle = 1 + \frac{16}{25}\pi^2 \cos^2(2\pi x),$$

$$g^{11}(x) = \frac{1}{g_{11}(x)} = \frac{1}{1 + \frac{16}{25}\pi^2 \cos^2(2\pi x)},$$

$$\sqrt{\det g} = \sqrt{1 + \frac{16}{25}\pi^2 \cos^2(2\pi x)}.$$

Now we find the projection of $e_1$ on $\Gamma$:

$$P_\Gamma e_1 = P_\Gamma \begin{pmatrix} 1 \\ 0 \end{pmatrix} = \left( \frac{\mathrm{d}}{\mathrm{d}y_1} \circ \begin{pmatrix} 1 \\ 0 \end{pmatrix} \right) \frac{\mathrm{d}}{\mathrm{d}y_1}$$

$$= \underbrace{\frac{1}{\sqrt{1 + \frac{16}{25}\pi^2 \cos^2(2\pi x)}} \begin{pmatrix} 1 \\ \frac{4}{5}\pi \cos(2\pi x) \end{pmatrix}}_{\text{tangent vector on } \Gamma = \frac{\mathrm{d}}{\mathrm{d}y_1}} \circ \begin{pmatrix} 1 \\ 0 \end{pmatrix} \frac{\mathrm{d}}{\mathrm{d}y_1}$$

$$= \frac{1}{\sqrt{1 + \frac{16}{25}\pi^2 \cos^2(2\pi x)}} \frac{\mathrm{d}}{\mathrm{d}y_1}.$$

In the local coordinates the gradient takes the form

$$\nabla_\Gamma \mu_1 = g^{11} \frac{\partial(\mu_1 \circ \alpha)}{\partial x} \frac{\mathrm{d}}{\mathrm{d}y_1} = g^{11} \tilde{\mu}_1' \frac{\mathrm{d}}{\mathrm{d}y_1},$$

where we define $\tilde{\mu}_1 = \mu_1 \circ \alpha$. With the divergence the cell problem in local coordinates takes the form

$$\nabla_\Gamma \cdot (\nabla_\Gamma \mu_1 + P_\Gamma e_1) = \nabla_\Gamma \cdot \left( g^{11} \frac{\partial(\mu_1 \circ \alpha)}{\partial x} \frac{\mathrm{d}}{\mathrm{d}y_1} + P_\Gamma e_1 \right)$$

$$= \nabla_\Gamma \cdot \left( g^{11} \tilde{\mu}_1' \frac{\mathrm{d}}{\mathrm{d}y_1} + \frac{1}{\sqrt{1 + \frac{16}{25}\pi^2 \cos^2(2\pi x)}} \frac{\mathrm{d}}{\mathrm{d}y_1} \right)$$

$$= \frac{1}{\sqrt{\det g}} \frac{\mathrm{d}}{\mathrm{d}x} \left( g^{11} \tilde{\mu}_1' \sqrt{\det g} + \frac{\sqrt{\det g}}{\sqrt{1 + \frac{16}{25}\pi^2 \cos^2(2\pi x)}} \right)$$

$$= 0.$$

Multiplying by $\sqrt{\det g}$ and considering the derivative $\frac{\mathrm{d}}{\mathrm{d}x}$ yields

$$g^{11} \tilde{\mu}_1' \sqrt{\det g} + \frac{\sqrt{\det g}}{\sqrt{1 + \frac{16}{25}\pi^2 \cos^2(2\pi x)}} = \text{constant} =: a$$

and

$$\frac{1}{\sqrt{1 + \frac{16}{25}\pi^2 \cos^2(2\pi x)}} \tilde{\mu}_1' + \frac{\sqrt{1 + \frac{16}{25}\pi^2 \cos^2(2\pi x)}}{\sqrt{1 + \frac{16}{25}\pi^2 \cos^2(2\pi x)}} = a.$$

Further we find

$$\tilde{\mu}_1' = (a-1)\sqrt{1 + \frac{16}{25}\pi^2\cos^2(2\pi x)}$$

for $x \in [0,1]$.

To satisfy periodicity it must hold that $\tilde{\mu}_1(0) = \tilde{\mu}_1(1)$ and $\tilde{\mu}_1'(0) = \tilde{\mu}_1'(1)$. This yields that $a = 1$ and

$$\tilde{\mu}_1 = 0 = \mu_1.$$

Hence, it remains to calculate

$$\int_\Gamma (\nabla_\Gamma \mu_1 + P_\Gamma e_1)_1 \mathrm{d}\sigma_y = \int_\Gamma (P_\Gamma e_1)_1 \mathrm{d}\sigma_y$$

$$= \int_0^1 \frac{1}{1 + \frac{16}{25}\pi^2\cos^2(2\pi x)}\sqrt{\det g}\,\mathrm{d}x$$

$$= \int_0^1 \frac{1}{\sqrt{1 + \frac{16}{25}\pi^2\cos^2(2\pi x)}}\mathrm{d}x \approx 0.572405,$$

$$\int_\Gamma (\nabla_\Gamma \mu_1 + P_\Gamma e_1)_2 \mathrm{d}\sigma_y = \int_\Gamma (P_\Gamma e_1)_2 \mathrm{d}\sigma_y$$

$$= \int_0^1 \frac{\frac{4}{5}\pi\cos(2\pi x)}{1 + \frac{16}{25}\pi^2\cos^2(2\pi x)}\sqrt{\det g}\,\mathrm{d}x$$

$$= \int_0^1 \frac{\frac{4}{5}\pi\cos(2\pi x)}{\sqrt{1 + \frac{16}{25}\pi^2\cos^2(2\pi x)}}\mathrm{d}x = 0.$$

Therefore, the diffusion tensor is given by

$$P \approx \begin{pmatrix} 0.572405 & 0 \\ 0 & 0 \end{pmatrix}.$$

We obtain for the Lebesgue measure of $\Gamma$ that $|\Gamma| = \int_0^1 1\sqrt{\det g}\,\mathrm{d}x \approx 1.9519$. The resulting macroscopic equation is

$$1.9519\int_\Omega \partial_t u \psi \mathrm{d}x + 0.572405 \cdot d \int_\Omega \partial_{x_1} u \partial_{x_1} \psi \mathrm{d}x = 1.9519\int_\Omega \psi \mathrm{d}x.$$

(4.8)

The strong formulation of this equation is given by

$$\partial_t u - d\nabla \cdot \left[ \begin{pmatrix} 0.2933 & 0 \\ 0 & 0 \end{pmatrix} \nabla u \right] = 1 \qquad \text{in } \Omega,$$

$$d \left[ \begin{pmatrix} 0.2933 & 0 \\ 0 & 0 \end{pmatrix} \nabla u \right] \cdot n = 0 \qquad \text{on } \partial\Omega.$$

To be able to compare the numerical and analytical results of the homogenized system of equations, we also solve the cell problem numerically.

In most numerical solvers it is not possible to define a partial differential equation on a hypersurface. Therefore we use the same trick as we used in the analytical method ealier in this section and we consider a thin tube around the sine-curve. Here we see the unit cell $Y$ with decreasing thickness.

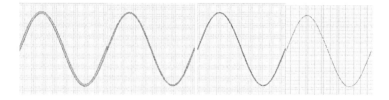

The question is now, which equation should be solved in the blown up domain. In the manifold $\Gamma$ we want to solve the pde with diffusion tensor $D$ for the cell problem

$$\nabla_\Gamma \cdot (P_\Gamma(e_j + \nabla_\Gamma \mu_j)) = 0 \qquad \text{in } \Gamma,$$
$$P_\Gamma(e_j + \nabla_\Gamma \mu_j) = 0 \qquad \text{on } \partial\Gamma,$$

and $\mu_j$ should be $Y$-periodic. For the blown up domain - let us call it $V$ - we denote the two new boundaries, which have distance $\delta$, by $\partial V_\delta$. From the calculations in the proof of theorem 4.1 we know that the derivative in normal direction of $\Gamma$ must be zero. Hence, the boundary condition on $\partial V_\delta$ is

$$\nabla \mu_j \cdot n = 0 \qquad \text{on } \partial V_\delta.$$

In the interior of the domain $V$ and the other boundaries - let us call them $\partial V_\Gamma$ - we adopt the conditions

$$\nabla \cdot (e_j + \nabla \mu_j) = 0 \qquad \text{in } V,$$
$$(e_j + \nabla \mu_j) \cdot n = 0 \qquad \text{on } \partial V_\Gamma,$$

from the original differential equation and as always, $\mu_j$ must be $Y$-periodic for $j = 1, 2$. We solve this equation for the sine-shaped domain and obtain the constant zero solution. We calculate that $\int_\Gamma (\nabla \mu_1 + P_\Gamma e_1)_1 d\sigma_y \approx 0.57241$ and $|\Gamma| = 1.9519$. Then, the homogenized system of equations is

$$1.9519 \int_\Omega \partial_t u \psi dx + 0.572405 \cdot d \int_\Omega \partial_{x_1} u \partial_{x_1} \psi dx = 1.9519 \int_\Omega \psi dx.$$

This is exactly the same one we obtained from the analytical calculations (4.8).

## 4.2 Limit behavior on Neumann and Robin boundaries

One important question is what happens on the boundaries with Neumann and Robin boundary condition in a domain $\Omega \subset \mathbb{R}^n$ by performing homogenization. If we consider the outer boundary as a periodic domain - the shape of the unit cell is the shape of the outer boundaries of the unit cell $Y$ - and if the functions living on that boundary are elements of $L^2(\partial\Omega)$, then we could use two-scale convergence in dimension $n - 1$. Therefore, the boundary $\partial\Omega_\varepsilon$ must not depend on $\varepsilon$ and also must be $\varepsilon$-periodic. Hence, the domain $\Omega_\varepsilon$ is a union of squares. For example, the shape of a circle or ellipsoid is not possible.

We define $\partial\Omega_\varepsilon$ as the outer boundary of $\Omega_\varepsilon$ and $\partial Y$ as one side of the outer boundary of the unit cell $Y$.

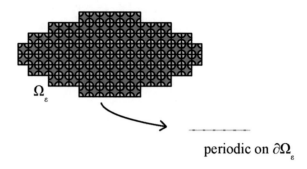

periodic on $\partial\Omega_\varepsilon$

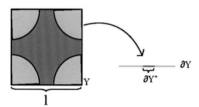

Then we have a periodic structure on $\partial\Omega_\varepsilon$ with unit cell $\partial Y$. We find the following theorem describing the two-scale convergence on $\partial\Omega_\varepsilon$.

**Theorem 4.6** *Let $\Omega_\varepsilon \subset \mathbb{R}^n$ be a domain as described above and let $g \in C(\partial\Omega, C_\#(\partial Y))$ with $g_\varepsilon(x) = g(x, x/\varepsilon)$ be $\partial Y$-periodic in its second argument. Then*

$$\int_{\partial\Omega_\varepsilon} g_\varepsilon(x)\varphi\left(x, \frac{x}{\varepsilon}\right) d\sigma_x \xrightarrow{\text{2-scale}} \int_{\partial\Omega} \int_{\partial Y} g(x,y)\varphi(x,y)d\sigma_y d\sigma_x$$

*for all $\varphi \in C^\infty(\Omega, C_\#^\infty(Y))$.*

**Proof** In the given setting, the domain $\partial\Omega_\varepsilon$ is $\varepsilon$-periodic with period $\partial Y$. Because the test functions $\varphi \in C^\infty(\Omega, C_\#^\infty(Y))$ are smooth, they also work as test functions on $\partial\Omega_\varepsilon$. Then, with classical homogenization, see section 2.1, the claim follows. $\quad\square$

**Remark 4.7** Theorem 4.6 can be used for homogenization of partial differential equations with Neumann boundary condition with right-hand side $g_\varepsilon$ at the outer boundary.

The situation is more complicated for Robin boundary conditions, because we need to identify the function in the boundary term at the outer boundary with the solution of the partial differential equation in the domain $\Omega_\varepsilon$. The following theorem secures two-scale convergence of the function $u_\varepsilon$ on $\partial\Omega_\varepsilon$, if $u_\varepsilon$ satisfies special conditions.

**Theorem 4.8**    *a) Let $u_\varepsilon$ be bounded in $H^1(\Omega_\varepsilon)$ independently of $\varepsilon$, and $\nabla u_\varepsilon \overset{\varepsilon\to 0}{\longrightarrow} \nabla_x u_0 + \nabla_y u_1$ for $u_0 \in H^1(\Omega)$ and $u_1 \in L^2(\Omega, H^1_\#(Y))$.*
*Let $u_1$ be of the form $u_1 = \sum_{i=1}^{n} \partial_{x_i} u_0(x,t)\mu_i(y)$ for $\mu_i \in H^1_\#(Y)$. Then,*

$$\left[ u_\varepsilon(x) - u_0(x) - \varepsilon u_1\left(x, \frac{x}{\varepsilon}\right) \right] \to 0 \qquad \text{strongly in } H^1(\Omega).$$

*b) If further $u_0 \in H^{3/2}(\Omega)$, then*

$$\gamma(u_\varepsilon) \to \gamma(u_0) \qquad \text{strongly in } L^2(\partial\Omega)$$

*and it holds that*

$$\int_{\partial\Omega_\varepsilon} \gamma(u_\varepsilon)\varphi\left(x, \frac{x}{\varepsilon}\right) d\sigma_x \overset{\text{2-scale}}{\longrightarrow} \int_{\partial\Omega}\int_{\partial Y} \gamma(u_0)\varphi(x,y)d\sigma_y d\sigma_x$$

*for all $\varphi \in C^\infty(\partial\Omega, C^\infty_\#(\partial Y))$. Here $\gamma$ is the trace operator.*

**Proof**
*a)* All the tools used in the proof of part a) are found in [2] by G. Allaire. We perform a combination of his results.
To prove this theorem we use proposition 2.5 and theorem 2.6 from article [2]. With proposition 2.5 we find that $\nabla u_\varepsilon$ two-scale converges to $\nabla_x u_0 + \nabla_y u_1$ with $u_0 \in H^1(\Omega)$ and $u_1 \in L^2(\Omega, H^1_\#(Y))$.
The assertion of theorem 4.8 holds once we showed that

$$\lim_{\varepsilon\to 0} \int_\Omega \left| \varphi\left(x, \frac{x}{\varepsilon}\right) \right|^2 dx = \int_\Omega \int_Y |\varphi(x,y)|^2 \, dy dx$$

for $\varphi = u_1$, $\nabla_x u_1$ and $\nabla_y u_1$, see theorem 2.6.

175

- For the following equalities we use parts 2,3,4,5 of lemma 3.9.

$$\int_\Omega u_1\left(x, \frac{x}{\varepsilon}\right)^2 dx = \int_\Omega \left(\sum_{i=1}^n \partial_{x_i} u_0(x,t)\mu_i\left(\frac{x}{\varepsilon}\right)\right)^2 dx$$

$$= \int_{\Omega \times Y}\left(\sum_{i=1}^n \mathcal{T}_\varepsilon(\partial_{x_i} u_0)(x,t)\mathcal{T}_\varepsilon(\mu_i)(y)\right)^2 dy dx$$

$$= \int_{\Omega \times Y}\left(\sum_{i=1}^n \partial_{x_i} u_0(x,t)\mu_i(y)\right)^2 dy dx$$

$$= \int_{\Omega \times Y} u_1(x,y)^2 dy dx.$$

- For the following equalities we use parts 2,3,4,5,8 of lemma 3.9 and $\varepsilon \nabla_x \mu_i(x/\varepsilon) = \nabla_y \mu_i(x/\varepsilon)$.

$$\int_\Omega |\nabla_y u_1\left(x, \frac{x}{\varepsilon}\right)|^2 dx = \int_\Omega \left|\sum_{i=1}^n \partial_{x_i} u_0(x,t)\varepsilon \nabla_x \mu_i\left(\frac{x}{\varepsilon}\right)\right|^2 dx$$

$$= \int_{\Omega \times Y}\left|\sum_{i=1}^n \mathcal{T}_\varepsilon(\partial_{x_i} u_0)(x,t)\varepsilon \mathcal{T}_\varepsilon\left(\nabla_x \mu_i\right)(y)\right|^2 dy dx$$

$$= \int_{\Omega \times Y}\left|\sum_{i=1}^n \mathcal{T}_\varepsilon(\partial_{x_i} u_0)(x,t)\nabla_y \mathcal{T}_\varepsilon\left(\mu_i\right)(y)\right|^2 dy dx$$

$$= \int_{\Omega \times Y}\left|\sum_{i=1}^n \partial_{x_i} u_0(x,t)\nabla_y \mu_i(y)\right|^2 dy dx$$

$$= \int_{\Omega \times Y} |\nabla_y u_1(x,y)|^2 dy dx.$$

- Finally we use again parts 2,3,4,5 of lemma 3.9 and find

$$\int_\Omega |\nabla_x u_1\left(x, \frac{x}{\varepsilon}\right)|^2 dx = \int_\Omega \left|\sum_{i=1}^n \nabla_x \partial_{x_i} u_0(x,t)\mu_i\left(\frac{x}{\varepsilon}\right)\right|^2 dx$$

$$= \int_{\Omega \times Y}\left|\sum_{i=1}^n \mathcal{T}_\varepsilon(\nabla_x \partial_{x_i} u_0)(x,t)\mathcal{T}_\varepsilon\left(\mu_i\right)(y)\right|^2 dy dx$$

176

$$= \int_{\Omega \times Y} \left| \sum_{i=1}^{n} \nabla_x \partial_{x_i} u_0(x,t) \mu_i(y) \right|^2 \mathrm{d}y \mathrm{d}x$$

$$= \int_{\Omega \times Y} |\nabla_x u_1(x,y)|^2 \mathrm{d}y \mathrm{d}x.$$

*b)* With $u_\varepsilon - \varepsilon u_1 \to u_0$ strongly in $H^1(\Omega)$, see a), we find, applying the trace operator $\gamma : H^1(\Omega) \to L^2(\partial\Omega)$, that

$$\gamma(u_\varepsilon - \varepsilon u_1) \overset{\varepsilon \to 0}{\longrightarrow} \gamma(u_0) \qquad \text{strongly in } L^2(\partial\Omega).$$

Now we use the trace operator $\gamma : H^s(\Omega) \to H^{s-1/2}(\partial\Omega)$ which is linear and bounded for $1/2 < s \leq 3/2$. Since $u_0 \in H^{3/2}(\Omega)$ it follows that $\gamma(u_0) \in H^1(\partial\Omega)$. With $\mu_i \in H^1_\#(Y)$ and hence $\gamma(\mu_i) \in H^{1/2}_\#(\partial Y)$, we deduce $\gamma(u_1) = \sum_{i=1}^{n} \gamma(\partial_{x_i} u_0(x,t) \mu_i(y)) \in L^2(\partial\Omega, L^2_\#(\partial Y))$. The periodicity in $H^{1/2}_\#(\partial Y)$ is ment in one dimension less. Because of $\gamma(u_1) \in L^2(\partial\Omega)$ it holds that

$$\varepsilon \int_{\partial\Omega_\varepsilon} \gamma(u_1)^2 \mathrm{d}\sigma_x \to 0 \quad \text{for} \quad \varepsilon \to 0$$

and we deduce that

$$\gamma(u_\varepsilon) \overset{\varepsilon \to 0}{\longrightarrow} \gamma(u_0) \quad \text{strongly in } L^2(\partial\Omega).$$

Hence it follows for two-scale convergence

$$\int_{\partial\Omega_\varepsilon} \gamma(u_\varepsilon) \varphi\left(x, \frac{x}{\varepsilon}\right) \mathrm{d}\sigma_x \overset{\text{2-scale}}{\longrightarrow} \int_{\partial\Omega} \int_{\partial Y} \gamma(u_0) \varphi(x,y) \mathrm{d}\sigma_y \mathrm{d}\sigma_x$$

for all $\varphi \in C^\infty(\partial\Omega, C^\infty_\#(\partial Y))$. $\qquad\qquad\qquad\qquad\qquad\square$

There is another way to interpret the situation. Let us introduce the following theorem.

**Theorem 4.9**     *a) Let $u_\varepsilon \in H^1(\Omega)$ be a sequence of functions such that $\|u_\varepsilon\|_\Omega + \|\nabla u_\varepsilon\|_\Omega < C$ for a constant $C > 0$ independent of $\varepsilon$. Let $u_\varepsilon$ weakly converge to a limit function $u_0 \in H^1(\Omega)$ and let $\partial Y^* \subset \partial Y$.*

*Then, up to a subsequence,*

$$\chi_\varepsilon \gamma(u_\varepsilon) \rightharpoonup |\partial Y^*| \gamma(u_0) \qquad \text{weakly in } L^2(\partial\Omega),$$

*where $\chi_\varepsilon$ is the characteristic function on $\bigcup_{k\in\mathbb{Z}^n} \varepsilon(k + \partial Y^*) \cap \partial\Omega$ and $\gamma : H^1(\Omega) \to H^{1/2}(\partial\Omega)$ is the trace operator.*

177

b) Let $u_\varepsilon \in L^2([0,T],H^1(\Omega)) \cap H^1([0,T],H^{-1}(\Omega))$, then there exists a subsequence of $u_\varepsilon$, also denoted by $u_\varepsilon$, such that

$$\gamma(u_\varepsilon) \to \gamma(u_0) \quad \text{strongly in } L^2([0,T],L^2(\partial\Omega))$$

and

$$\chi_\varepsilon f(\gamma(u_\varepsilon)) \rightharpoonup |\partial Y^*| f(u_0) \quad \text{weakly in } L^2([0,T],L^2(\partial\Omega))$$

for any bounded and continuous function $f : \mathbb{R} \to \mathbb{R}$.

**Proof**

a) We know that the function $u_\varepsilon \in H^1(\Omega)$ has a weak limit $u_0$ in $H^1(\Omega)$ such that up to a subsequence,

$$(u_\varepsilon - u_0, \varphi)_{H^1(\Omega) \times H^1(\Omega)'} \xrightarrow{\varepsilon \to 0} 0$$

for all $\varphi \in H^1(\Omega)'$ using classical weak convergence.

With the trace operator $\gamma : H^1(\Omega) \to H^{1/2}(\partial\Omega)$, which is linear and bounded, we obtain

$$\langle \gamma(u_\varepsilon) - \gamma(u_0), \gamma(\varphi) \rangle_{H^{1/2}(\partial\Omega) \times H^{1/2}(\partial\Omega)'} \xrightarrow{\varepsilon \to 0} 0$$

for all $\varphi \in C^\infty(\Omega)$. Moreover, $H^{1/2}(\partial\Omega)$ is compactly embedded in $L^2(\partial\Omega)$ and hence, there exists a strongly converging subsequence $\gamma(u_\varepsilon) \to \gamma(u_0)$ in $L^2(\partial\Omega)$. Using the characteristic function $\chi_\varepsilon$ on $\bigcup_{k \in \mathbb{Z}^n} \varepsilon(k + \partial Y^*) \cap \partial\Omega$ we ensure that the domain, where the convergence holds, does not change with $\varepsilon$. Now we conclude with lemma 6 in [34] that

$$\langle \chi_\varepsilon \gamma(u_\varepsilon), \varphi \rangle_{L^2(\partial\Omega)} \xrightarrow{\varepsilon \to 0} |\partial Y^*| \langle \gamma(u_0), \varphi \rangle_{L^2(\partial\Omega)}$$

for all $\varphi \in C^\infty(\Omega)$.

b) With the condition $u_\varepsilon \in L^2([0,T],H^1(\Omega)) \cap H^1([0,T],H^{-1}(\Omega))$ we find with the trace operator that $\gamma(u_\varepsilon) \in L^2([0,T],H^{1/2}(\partial\Omega)) \cap H^1([0,T],H^{-3/2}(\partial\Omega))$. The embedding theorems in Sobolev spaces and theorem

3.5 yield a strongly converging subsequence of $\gamma(u_\varepsilon)$ in $L^2([0,T], L^2(\partial\Omega))$.

With theorem 3.6, we also find that $f(\gamma(u_\varepsilon))$ strongly converges to $f(\gamma(u_0))$ in $L^2([0,T], L^2(\partial\Omega))$. Because $\chi_\varepsilon$ is only a weak converging sequence, we deduce that

$$\langle \chi_\varepsilon f(\gamma(u_\varepsilon)), \varphi \rangle_{L^2(\partial\Omega)} \to |\partial Y^*| \langle f(\gamma(u_0)), \varphi \rangle_{L^2(\partial\Omega)}$$

for all $\varphi \in C^\infty(\Omega)$.

$\square$

With theorems 4.8 and 4.9 we found two ways to find strong convergence on the outer boundary of $\Omega_\varepsilon$. We use the second one to deduce the limit of a Robin boundary term in the following application in the context of homogenization.

## 4.3 Signaling in Lymphocytes: Stim1 and Orai1

Our immune system is a very complex machinery which is orchestrated by different kinds of cells and organs. Still many functions and procedures are not completely or just partially understood. A leading part of the immune system are the T cells (or thymus lymphocytes). Their purpose is to pour out messengers if they find alien substances in the body (helper T cell) or to kill the intruder directly (cytotoxic T cell).

To accomplish their tasks, complex signaling cascades take place inside these cells. One important step is the store-operated calcium entry through CRAC (Calcium Release-Activated Calcium) channels. If this step is defective, it can come to immunodeficiency syndromes in human patients.

To understand the function of the CRAC channels we briefly need to explain the situation in T cells. In a non-activated T cell the calcium concentration in the cytosol is $[Ca^{2+}]_i \approx 50 - 100nM$, the calcium concentration in the intercellular space is $[Ca^{2+}]_e \approx 1mM$, and in the lumen of the endoplasmic reticulum it is $[Ca^{2+}]_{ER} \approx 500\mu M$, see [32]. This means that the concentration in the cytosol is at least 5000 times lower than in the neighboring domains. To sustain that strong gradient there are several pumps working to permanently pump calcium out of the cell (PMCA, NCX) or into the lumen of the endoplasmic reticulum (SERCA). The pump PMCA pumps calcium with the aid of ATP, the pump NCX exchanges calcium with sodium.

On the surface of the endoplasmic reticulum (ER) the molecule Stim1 (Stromal interaction molecule 1) exists. Usually it binds to two calcium molecules $Ca^{2+}$ which are in the lumen of the ER. Furthermore, on the plasma membrane of the cell there are molecules Orai1 (calcium release-activated calcium channel protein 1) to which Stim1 can also bind to.

To get the procedure of the activation of the T cell started, the lumen of the ER must be induced to release its calcium. That can happen through molecules named IP3 directly, or a molecule TG closes the SERCA pumps and calcium is not pumped back into the lumen of the ER. But in general IP3 is the trigger.
After depletion of the ER there is no calcium left for the Stim1 molecules to bind to. But on the surface of the ER, that is near to the plasma membrane, unbound Stim1 bind to Orai1. There two Stim1 molecules can bind to one Orai1 molecule. Stim1 molecules diffuse on the surface of the ER and, in this way, reach the plasma membrane.

**Question** A question we want to answer with the following mathematical model is if diffusion of Stim1 molecules is sufficient or if an additional driving force or mechanism moves Stim1 molecules towards the plasma membrane. This means, for a time given from biological experiments, we consider, if for our mathematical model using just diffusion of Stim1 molecules the T cell activates after expiration of this time. If the T cell in the mathematical model takes longer to activate, then only diffusion is likely not sufficient for Stim1 molecules.

Once four Stim1 are connected to two Orai1, they build a CRAC channel, which lets calcium diffuse from the intercellular space into the cytosol. This state holds on as long as IP3 is present in the T cell.

When IP3 is gone, calcium moves back into in the lumen of the ER and can bind to Stim1 again. A Stim1 molecule, that binds to Orai1 and $Ca^{2+}$, quickly breaks away from Orai1 and the CRAC channel closes. The calcium pumps restore the original state soon.

This information and more details on the biological background can be found in [32, 31, 33, 39, 55].
Now we build a mathematical model from this knowledge. We will use the following notions.

| | |
|---|---|
| Cytosol | $\Omega^1$, |
| Lumen of the ER | $\Omega^2$, |
| Surface of the ER | $\Gamma^{ER}$, |
| Plasma membrane | $\Gamma^1$, |
| Nuclear membrane | $\Gamma^2$, |

| | |
|---|---|
| Calcium in the cytosol | $C$, |
| Calcium in the lumen of the ER | $C_e$, |
| Two unbound molecules of Stim1 on the surface of the ER | $S$, |
| Two molecules of Stim1 bound to 4 calcium on the surface of the ER | $S_C$, |
| Two molecules of Stim1 bound to Orai1 on the plasma membrane | $S_O$, |
| Two molecules of Stim1 bound to Orai1 and 4 calcium on the plasma membrane | $S_{CO}$. |

The following effective model is due to Patrick Fletcher and Yue-Xian Li [23]. It was derived phenomenologically without taking into account the cell microstructure. Here we want to regard the microstructure of the cell.

At first we take a look at the calcium dynamics. In the cytosol $\Omega^1$ calcium $C$ satisfies the diffusion equation with diffusion coefficient $D_C$

$$\partial_t C - D_C \Delta C = 0 \quad \text{in } \Omega^1.$$

At $\Gamma^{ER}$, the SERCA pump is working and there is a natural osmosis rate $L_0$ through the membrane and - when the cell is in contact with IP3 - an additional osmotic flux $L_{IP3}$

$$-D_C \nabla C \cdot n = \underbrace{(L_0 + L_{IP3})(C - C_e)}_{-j_{rel}} + \underbrace{v_{SERCA} \frac{C^2}{K_{SERCA}^2 + C^2}}_{j_{fill}} \quad \text{on } \Gamma_{ER}$$

for constants $v_{SERCA}$, $K_{SERCA} > 0$.

At the plasma membrane two pumps (PMCA, NCX) pump calcium out of the cell. Also the CRAC channels are working at the plasma membrane. The current $I_{CRAC}$ is negative, so $\alpha I_{CRAC}$ is an influx,

$$- D_C \nabla C \cdot n$$

$$= \underbrace{\alpha I_{CRAC}}_{j_{in}} + \underbrace{v_P \frac{C^2}{C^2 + K_P^2} + v_{NCX} \frac{C^4}{C^4 + K_{NCX}^4}}_{j_{out}} \quad \text{on } \Gamma^1 \cap \partial\Omega^1$$

for constants $\alpha$, $v_P$, $K_P$, $v_{NCX}$, $K_{NCX} > 0$. Again, the parameter values of the pumps can be found experimentally, for example by curve fitting. At the nuclear membrane we have a no-flux condition for the calcium molecules

$$-D_C \nabla C \cdot n = 0 \quad \text{on } \Gamma^2 \cap \partial\Omega^1.$$

In the lumen of the endoplasmic reticulum $\Omega^2$, calcium $C_e$ satisfies the diffusion equation with diffusion coefficient $D_{ER}$

$$\partial_t C_e - D_{ER} \Delta C_e = 0 \quad \text{in } \Omega^2.$$

At the surface of the endoplasmic reticulum $\Gamma_{ER}$ we need to have $-D_{ER} \nabla C_e \cdot n = D_C \nabla C \cdot n$,

$$- D_{ER} \nabla C_e \cdot n = j_{rel} - j_{fill}$$

$$= (L_0 + L_{IP3})(C_e - C) - v_{SERCA} \frac{C^2}{K_{SERCA}^2 + C^2} \quad \text{on } \Gamma^{ER}.$$

At the plasma membrane and the nuclear membrane we have a no-flux condition for the calcium molecules

$$-D_{\mathrm{ER}}\nabla C_e \cdot n = 0 \quad \text{on } \Gamma^1 \cap \partial\Omega^2,$$

$$-D_{\mathrm{ER}}\nabla C_e \cdot n = 0 \quad \text{on } \Gamma^2 \cap \partial\Omega^2.$$

For convenience we always consider two Stim1 molecules as one unit, because two Stim1 molecules must bind to one Orai1 molecule to open a CRAC channel. The unbound Stim1 molecules $S$ diffuse on the membrane of the endoplasmic reticulum with diffusion coefficient $D_S$, they can bind to 4 calcium molecules, which live in the lumen of the ER, with rate $k_C^+$ and transform to $S_C$. We assume that the transformation follows the law of mass action. Therefore, we find that $S$ needs to be multiplied by $C_e^4$. Because the law of mass action only makes sense for a finite set of particles, we use a function $f_e$, that reaches a threshold $v_e$ for too great a concentration of $C_e$ molecules,

$$f_e(x) = v_e \frac{x^4}{x^4 + K_e^4}$$

for constants $K_e, v_e > 0$. If we choose $K_e$ and $v_e$ large, the function $f_e(C_e)$ resembles $C_e^4$ for $C_e \ll \infty$. If $S_C$ breaks away from calcium, which happens with rate $k_C^-$, we get again unbound Stim1 molecules $S$.

$$\partial_t S = D_S \Delta_\Gamma S - k_C^+ f_e(C_e)S + k_C^- S_C \quad \text{on } \Gamma^{\mathrm{ER}},$$

$$\partial_t S_C = D_{SC} \Delta_\Gamma S_C + k_C^+ f_e(C_e)S - k_C^- S_C \quad \text{on } \Gamma^{\mathrm{ER}}.$$

We do not have corresponding terms considering the binding of Stim1 molecules to calcium in the equation for $C_e$, because the concentration of calcium $C_e$ in the lumen of the ER is so much bigger than the concentration of Stim1 molecules that these terms are negligible.

At the plasma membrane $S$ and $S_C$ can bind to Orai1. We assume that there are many Orai1 molecules on the plasma membrane compared to $S$. Hence, Stim1 appears to bind to the plasma membrane with constant rate $k_O^+$ and unbind with constant rate $k_O^-$. Once bound to the plasma membrane we denote $S$ with $S_O$.

We assume to have an analogous behavior for $S_C$, which are the Stim1 molecules bound to calcium, with constant rates $k_{CO}^+$ and $k_{CO}^-$.

$$-D_S \nabla S \cdot n = k_O^+ S - k_O^- S_O \quad \text{on } \Gamma^1 \cap \Gamma^{ER},$$
$$-D_{SC} \nabla S_C \cdot n = k_{CO}^+ S_C - k_{CO}^- S_{CO} \quad \text{on } \Gamma^1 \cap \Gamma^{ER}.$$

At the nuclear boundary we have a no-flux condition,

$$-D_S \nabla S \cdot n = 0 \quad \text{on } \Gamma^2 \cap \Gamma^{ER},$$
$$-D_{SC} \nabla S_C \cdot n = 0 \quad \text{on } \Gamma^2 \cap \Gamma^{ER}.$$

Finally, we consider the dynamics for $S_O$ and $S_{CO}$ at the plasma membrane. Once Stim1 is bound to Orai1 it is "at puncta" and spatially fixed, so we do not have a diffusion term in the equation for $S_O$ and $S_{CO}$. By binding to 4 calcium, which live in the lumen of the ER, $S_O$ changes to $S_{CO}$ with rate $k_C^+$ and $S_{CO}$ changes to $S_O$ through unbinding from calcium with rate $k_C^-$,

$$\partial_t S_O = k_O^+ S - k_O^- S_O - k_C^+ f_e(C_e) S_O + k_C^+ S_{CO} \quad \text{on } \Gamma^1 \cap \Gamma^{ER},$$
$$\partial_t S_{CO} = k_{CO}^+ S_C - k_{CO}^- S_{CO} + k_C^+ f_e(C_e) S_O - k_C^+ S_{CO} \quad \text{on } \Gamma^1 \cap \Gamma^{ER}.$$

**CRAC channel**

Now we specify the flux $I_{CRAC}$ using $S_O$ in the equation for $C$. This calcium influx is due to the opening CRAC channels. It is important to know that the flux through the channels at the plasma membrane always depends on the potential gradient $V$. In resting state the membrane potential is about $\sim -70mV$, the inside of the cell is negatively charged.

Before we can build the $I_{CRAC}$, we must extend the function $S_O$ from $\Gamma^1 \cap \partial \Gamma^{ER}$ to $\Gamma^1 \cap \partial \Omega^1$, because we need to pick up $S_O$ values on $\Gamma^1 \cap \partial \Omega^1$.

The $I_{CRAC}$ on $\Gamma^1 \cap \partial \Omega^1$ is given by

$$I_{CRAC} = g_{CRAC} S_O^2 (V - V_{Ca}),$$

where $V_{Ca}$ is the resting potential for calcium and $1/g_{CRAC}$ stands for the resistance of the channel. Later we define the extension for $S_O$ more precisely in (4.11).

Ionic channels are mainly responsible for the potential fluxes, amongst others for example the CRAC channel with flux $I_{CRAC}$. But also the CAN channel with flux/current $I_{CAN}$, the K channel with flux/current $I_K$ and the K(Ca) channel with flux/current $I_{K(Ca)}$ are important.

The plasma membrane acts as a capacitor with capacity $C_m$. Since the relation between the current $I$ and the potential $V$ at a capacitor is $dV/dt = -I/C_m$, we have the following formula for the membrane potential

$$\frac{dV}{dt} = -\frac{I_K + I_{K(Ca)} + I_{CRAC} + I_{CAN}}{C_m}.$$

All the fluxes through the channels depend on the membrane potential. We already described the opening mechanism of the CRAC channel. Now we take a look at the opening mechanisms of the other channels.

### K channel
The K channel is a potassium channel at the plasma membrane. The resistance of the K channel is $1/g_K$ and $V_K$ is the resting potential for potassium. The current $I_K$ is given by

$$I_K := g_K \frac{1}{1 + e^{\frac{-(V-V_n)}{K_n}}} (V - V_K),$$

where $V_n$ and $K_n$ are constants. Depending on $V$, the channel opens more or less.

### K(Ca) channel
This is another potassium channel at the plasma membrane, but the degree of opening of the channel depends on the calcium concentration on the plasma membrane. The resistance of the K(Ca) channel is $1/g_{K(Ca)}$.

$$I_{K(Ca)} = g_{K(Ca)} \frac{C^4}{C^4 + K^4_{K(Ca)}} (V - V_K)$$

185

for a constant $K_{K(Ca)} > 0$. The higher the calcium concentration $C$ is on the plasma membrane the more the channel opens.

### CAN channel

The CAN channel is a sodium channel at the plasma membrane and the resting potential of sodium is $V_{Na}$. The resistance of the CAN channel is $1/g_{CAN}$. The degree of opening of this channel also depends on calcium $C$.

$$I_{CAN} = g_{CAN}a(V - V_{Na}),$$

with

$$\frac{da}{dt} = \frac{C^4}{\tau_{CAN}(C^4 + K_{CAN}^4)} - \frac{a}{\tau_{CAN}},$$

where $\tau_{CAN}$ and $K_{CAN}$ are constants.

## Model for Homogenization

To handle the finestructure of the endoplasmic reticulum, we use periodic homogenization and let the domains $\Omega^1$, $\Omega^2$ and $\Gamma^{ER}$ depend on a small parameter $\varepsilon \ll 1$. Consequently, all the solutions $C, C_e, S, S_C, S_O, S_{CO}, V$ and $a$ are dependent on $\varepsilon$.

Additionally we modify the manifold $\Gamma_\varepsilon^{ER}$ to a thin layer $\Omega_\varepsilon^{ER}$ with width $\delta > 0$ as described in section 4.1. The new domain $\Omega_\varepsilon^{ER}$ extends into the domain $\Omega_\varepsilon^2$ and overlaps with it, i.e $\Omega_\varepsilon^{ER} \subset \Omega_\varepsilon^2$. The manifold $\Gamma_\varepsilon^{ER}$ is now the boundary between $\Omega_\varepsilon^1$ and $\Omega_\varepsilon^{ER}$ and also the boundary between $\Omega_\varepsilon^1$ and $\Omega_\varepsilon^2$. We label the new inner boundary of $\Omega_\varepsilon^{ER}$ with $\Gamma_{e,\varepsilon}^{ER}$.

Also the subsets $Y^1$, $Y^2$, and $Y^{ER}$ are constructed such that the new domain $Y^{ER}$ extends into and overlaps with $Y^2$. The manifold $\Gamma^{ER}$ is the boundary between $Y^1$ and $Y^2$ and the boundary between $Y^1$ and $Y^{ER}$. See the following figures, where the new domains are sketched.

We want to perform homogenization first, and then let $\delta$ tend to 0 afterwards to recover the manifold $\Gamma^{ER}$ in the end. We constructed the subsets $Y^1$ and $Y^2$ of the unit cell $Y$ such that their shape is independent of $\delta$, only $Y^{ER}$ depends on $\delta$.

We assume that the equations for Stim1 molecules living in the domain $\Omega_\varepsilon^{ER}$ satisfy a no-flux condition at the boundaries $\Gamma_\varepsilon^{ER}$ and $\Gamma_{e,\varepsilon}^{ER}$. The calcium molecules $C_{e,\varepsilon}$ live in $\Omega_\varepsilon^2$, hence also in $\Omega_\varepsilon^{ER}$, which is important for the Stim1 molecules $S_\varepsilon$ to bind to $C_{e,\varepsilon}$.

On the left figure, we see the unit cell $Y = [0,1]^n$ with the thin layer of the endoplasmic reticulum. The right figure shows a cross section (horizontal and vertical) of the left picture.

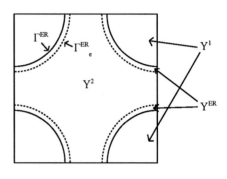

The domains for the human cell $\Omega \subset \mathbb{R}^n$ are defined as the following

$$\Omega_\varepsilon^1 := \bigcup_{k \in \mathbb{Z}^n} \varepsilon(Y^1 + k) \cap \Omega, \qquad \Omega_\varepsilon^2 := \bigcup_{k \in \mathbb{Z}^n} \varepsilon(Y^2 + k) \cap \Omega,$$

$$\Omega_\varepsilon^{ER} := \bigcup_{k \in \mathbb{Z}^n} \varepsilon(Y^{ER} + k) \cap \Omega.$$

The manifolds are defined as

$$\Gamma_\varepsilon^{ER} := \bigcup_{k \in \mathbb{Z}^n} \varepsilon(\Gamma^{ER} + k) \cap \Omega, \qquad \Gamma_{e,\varepsilon}^{ER} := \bigcup_{k \in \mathbb{Z}^n} \varepsilon(\Gamma_e^{ER} + k) \cap \Omega.$$

We summarize that the domains $\Omega_\varepsilon^1$ and $Y^1$ stand for the cytosol of the cell, the domains $\Omega_\varepsilon^2$ and $Y^2$ stand for the lumen of the ER, and the domains $\Omega_\varepsilon^{ER}$ and $Y^{ER}$ for the membrane of the ER; as subsets of the macroscopic domain $\Omega$ and the unit cell $Y$, respectively. A macroscopic picture of the cell is illustrated below.

187

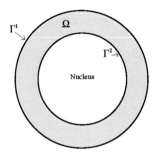

For convenience we introduce several abbreviations:

$$f_{\text{SERCA}}(C_\varepsilon) = v_{\text{SERCA}} \frac{C_\varepsilon^2}{C_\varepsilon^2 + K_{\text{SERCA}}^2},$$

$$f_{\text{P}}(C_\varepsilon) = v_{\text{P}} \frac{C_\varepsilon^2}{C_\varepsilon^2 + K_{\text{P}}^2},$$

$$f_{\text{NCX}}(C_\varepsilon) = v_{\text{NCX}} \frac{C_\varepsilon^4}{C_\varepsilon^4 + K_{\text{NCX}}^4}.$$

All theses functions are nonnegative, smooth and bounded by $v_{\text{SERCA}}$, $v_{\text{P}}$, $v_{\text{NCX}} > 0$, respectively. The $\varepsilon$-dependent version of the system described in the current section is the following.

**Dynamics for calcium $C_\varepsilon$**

$$\partial_t C_\varepsilon - D_C \Delta C_\varepsilon = 0 \qquad \qquad \text{in } \Omega_\varepsilon^1,$$

$$-D_C \nabla C_\varepsilon \cdot n = \varepsilon(L_0 + L_{\text{IP3}})(C_\varepsilon - C_{e,\varepsilon}) + \varepsilon f_{\text{SERCA}}(C_\varepsilon) \qquad \text{on } \Gamma_\varepsilon^{\text{ER}},$$

$$-D_C \nabla C_\varepsilon \cdot n = \alpha I_{\text{CRAC}} + f_{\text{P}}(C_\varepsilon) + f_{\text{NCX}}(C_\varepsilon) \qquad \text{on } \Gamma^1 \cap \partial\Omega_\varepsilon^1,$$

$$-D_C \nabla C_\varepsilon \cdot n = 0 \qquad \qquad \text{on } \Gamma^2 \cap \partial\Omega_\varepsilon^1.$$

**Dynamics for calcium $C_{e,\varepsilon}$**

$$\partial_t C_{e,\varepsilon} - D_{\text{ER}} \Delta C_{e,\varepsilon} = 0 \qquad \qquad \text{in } \Omega_\varepsilon^2,$$

$$-D_{\text{ER}} \nabla C_{e,\varepsilon} \cdot n = \varepsilon(L_0 + L_{\text{IP3}})(C_{e,\varepsilon} - C_\varepsilon) - \varepsilon f_{\text{SERCA}}(C_\varepsilon) \text{ on } \Gamma_\varepsilon^{\text{ER}},$$

$$-D_{\text{ER}} \nabla C_{e,\varepsilon} \cdot n = 0 \qquad \qquad \text{on } \Gamma^1 \cap \partial\Omega_\varepsilon^2,$$

$$-D_{\text{ER}} \nabla C_{e,\varepsilon} \cdot n = 0 \qquad \qquad \text{on } \Gamma^2 \cap \partial\Omega_\varepsilon^2.$$

**Dynamics for Stim1 $S_\varepsilon$**

$$\partial_t S_\varepsilon - D_S \Delta S_\varepsilon = -k_C^+ f_e(C_{e,\varepsilon}) S_\varepsilon + k_C^- S_{C,\varepsilon} \qquad \text{on } \Omega_\varepsilon^{ER},$$

$$-D_S \nabla S_\varepsilon \cdot n = k_O^+ S_\varepsilon - k_O^- S_{O,\varepsilon} \qquad \text{on } \Gamma^1 \cap \partial\Omega_\varepsilon^{ER},$$

$$-D_S \nabla S_\varepsilon \cdot n = 0 \qquad \text{on } \Gamma^2 \cap \partial\Omega_\varepsilon^{ER},$$

$$-D_S \nabla S_\varepsilon \cdot n = 0 \qquad \text{on } \Gamma_\varepsilon^{ER},$$

$$-D_S \nabla S_\varepsilon \cdot n = 0 \qquad \text{on } \Gamma_{e,\varepsilon}^{ER}.$$

**Dynamics for Stim1 $S_{C,\varepsilon}$**

$$\partial_t S_{C,\varepsilon} - D_{SC} \Delta S_{C,\varepsilon} = k_C^+ f_e(C_{e,\varepsilon}) S_\varepsilon - k_C^- S_{C,\varepsilon} \qquad \text{on } \Omega_\varepsilon^{ER},$$

$$-D_{SC} \nabla S_{C,\varepsilon} \cdot n = k_{CO}^+ S_{C,\varepsilon} - k_{CO}^- S_{CO,\varepsilon} \qquad \text{on } \Gamma^1 \cap \partial\Omega_\varepsilon^{ER},$$

$$-D_{SC} \nabla S_{C,\varepsilon} \cdot n = 0 \qquad \text{on } \Gamma^2 \cap \partial\Omega_\varepsilon^{ER},$$

$$-D_{SC} \nabla S_{C,\varepsilon} \cdot n = 0 \qquad \text{on } \Gamma_\varepsilon^{ER},$$

$$-D_{SC} \nabla S_{C,\varepsilon} \cdot n = 0 \qquad \text{on } \Gamma_{e,\varepsilon}^{ER}.$$

**Dynamics for Stim1 $S_{O,\varepsilon}$**

$$\partial_t S_{O,\varepsilon}$$
$$= k_O^+ S_\varepsilon - k_O^- S_{O,\varepsilon} - k_C^+ f_e(C_{e,\varepsilon}) S_{O,\varepsilon} + k_C^- S_{CO,\varepsilon} \text{ on } \Gamma^1 \cap \partial\Omega_\varepsilon^{ER}.$$

**Dynamics for Stim1 $S_{CO,\varepsilon}$**

$$\partial_t S_{CO,\varepsilon}$$
$$= k_{CO}^+ S_{C,\varepsilon} - k_{CO}^- S_{CO,\varepsilon} + k_C^+ f_e(C_{e,\varepsilon}) S_{O,\varepsilon} - k_C^- S_{CO,\varepsilon} \text{ on } \Gamma^1 \cap \partial\Omega_\varepsilon^{ER}.$$

**Dynamics for the channels on $\Gamma^1 \cap \partial\Omega_\varepsilon^1$**

$$
\begin{aligned}
\partial_t V_\varepsilon &= -\frac{I_K + I_{K(Ca)} + I_{CRAC} + I_{CAN}}{C_m}, \\
I_K &= g_K \frac{1}{1 + e^{\frac{-(V_\varepsilon - V_n)}{K_n}}} (V_\varepsilon - V_K), \\
I_{K(Ca)} &= g_{K(Ca)} \frac{C_\varepsilon^4}{C_\varepsilon^4 + K_{K(Ca)}^4} (V_\varepsilon - V_K), \\
I_{CRAC} &= g_{CRAC} (\text{ext}_\varepsilon S_{O,\varepsilon})^2 (V_\varepsilon - V_{Ca}), \\
I_{CAN} &= g_{CAN} a_\varepsilon (V_\varepsilon - V_{Na}), \\
\partial_t a_\varepsilon &= \frac{C_\varepsilon^4}{\tau_{CAN}(C_\varepsilon^4 + K_{CAN}^4)} - \frac{a_\varepsilon}{\tau_{CAN}}.
\end{aligned}
\tag{4.9}
$$

with positive, bounded and smooth initial conditions $C_\varepsilon(0), C_{e,\varepsilon}(0), S_\varepsilon(0), S_{C,\varepsilon}(0), S_{O,\varepsilon}(0), S_{CO,\varepsilon}(0), a_\varepsilon(0)$ and bounded and smooth initial condition $V_\varepsilon(0)$.

Before we continue to consider this system of equations, we take a look at the extension for $S_{O,\varepsilon}$.

## Extension from $S_{O,\varepsilon}$ on $\Gamma^1 \cap \partial\Omega_\varepsilon^{ER}$ to *ext* $S_{O,\varepsilon}$ on $\Gamma^1 \cap \partial\Omega_\varepsilon^1$

We consider the boundary $\Gamma^1$ of $\Omega_\varepsilon^{ER}$ and take a look at the boundary of one cell in $\varepsilon$-size. The function $S_{O,\varepsilon}$ lives in $\Gamma^1 \cap \partial\Omega_\varepsilon^{ER}$. But the influx $I_{CRAC}$ for $C_\varepsilon$ picks up $S_{O,\varepsilon}$ values in $\Gamma^1 \cap \partial\Omega_\varepsilon^1$. Therefore, we construct an extension from $\Gamma^1 \cap \partial\Omega_\varepsilon^{ER}$ to $\Gamma^1 \cap \partial\Omega_\varepsilon^1$.

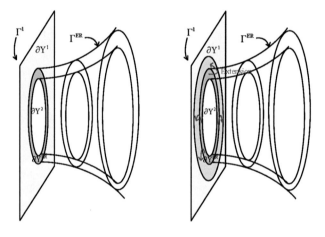

Figure 4.1: Extension from $\Gamma^1 \cap \partial\Omega_\varepsilon^{ER}$ to $\Gamma^1 \cap \partial\Omega_\varepsilon^1$

This also makes sense from the biological point of view, since the CRAC channels appear where the plasma membrane $\Gamma^1$ and the surface of the endoplasmic reticulum $\Gamma^{ER}$ overlap. As we can see on the figure below, there are parts of the plasma membrane (no null sets) that overlap with the endoplasmic reticulum, see [32].

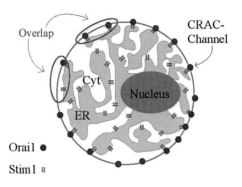

From the mathematical point of view, these overlapping parts must be part of the boundary of $\partial\Omega^1_\varepsilon$, since the calcium passing the CRAC channels flows into the cytosol.

We follow the proof of theorem 4.1 in chapter 4.2 in [30]. Here, an exemplary extension is built for $\Omega \subset \mathbb{R}^3$ and a specific shape of the unit cell $Y = [0,1]^3$. We consider one face of the unit cell $Y$ and take the following domains for $\partial Y^1$, $\partial Y^2$ and $\partial Y^{ER}$, see figure 4.1,

$$\partial Y^2 := B_{(1/2,1/2)}(\rho), \qquad \partial Y^{ER} := \partial Y^2 \backslash B_{(1/2,1/2)}(\rho - \delta),$$

$$\partial Y^1 := [0,1]^2 \backslash \partial Y^2, \quad \text{and} \quad \partial\Gamma^{ER} := \{x \in [0,1]^2 | \, |x - \left(\frac{1}{2},\frac{1}{2}\right)| = \rho\}$$

for some $0 < \rho < 1/2$.

As subsets of the domain $\Omega_\varepsilon$, the unit cells $Y$ are scaled with the factor $\varepsilon$.

Every $\varepsilon$-sized square of the boundary can be shifted by a translation and rotation to the cell $[0,\varepsilon]^2 \subset \mathbb{R}^2$. So we assume, without loss of generality, that we now work on $[0,\varepsilon]^2 \subset \mathbb{R}^2$.

The extension on $[0,\varepsilon]^2$ is denoted by $\text{ext}_\varepsilon$ and is going to be constructed such that values of a function defined on the ring-shaped domain $\varepsilon\partial Y^{ER}$ are folded outwards to a larger ring-shaped domain $\varepsilon B_{(1/2,1/2)}(R + \rho)\backslash(\varepsilon\partial Y^2) \subset \varepsilon\partial Y^1$. Here, $R$ is chosen such that $R > 0$ and $R + \rho < 1/2$.

We denote this support of the extension by $\partial Y^{CRAC}$. On the remaining subset of $\partial Y^1$ we extend the extension by zero.

To construct the extension, we define sets $K_j$, $j = 1, 2, 3, 4$, given by

$$K_{j,\varepsilon} := \{\varepsilon(r\cos\varphi, r\sin\varphi)|\ r \in [\rho - \delta, \rho + R],$$
$$\varphi \in [\pi/2(j-1) - \pi/4, \pi/2(j-1) + \pi/4)\},$$

such that

$$\varepsilon\partial Y_\varepsilon^{\mathrm{ER}} \subset \bigcup_{j=1}^4 K_{j,\varepsilon} \quad \text{and} \quad \varepsilon\Gamma^{\mathrm{ER}} \subset \bigcup_{j=1}^4 K_{j,\varepsilon} \quad \text{and} \quad \varepsilon\partial Y_\varepsilon^{\mathrm{CRAC}} \subset \bigcup_{j=1}^4 K_{j,\varepsilon}.$$

Further, we define functions

$$\varphi_j(x) := \chi_{K_{j,\varepsilon}}(x), \qquad j = 1, 2, 3, 4,$$

where $\chi_{K_{j,\varepsilon}}$ is the characteristic function on $K_{j,\varepsilon}$.

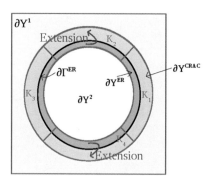

We construct functions $\psi^{(j)} : K_{j,\varepsilon} \to \mathbb{R}^2$ such that $\psi^{(j)}(K_{j,\varepsilon} \cap \varepsilon\partial Y^{\mathrm{ER}}) \subset \mathbb{R} \times \mathbb{R}^+$ and $\psi^{(j)}(K_{j,\varepsilon} \cap \varepsilon\partial\Gamma^{\mathrm{ER}}) \subset \mathbb{R} \times \{0\}$.

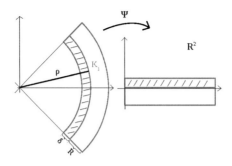

We define

$$\psi(x,y) := \left( \arctan\left(y/x\right), \varepsilon\rho - \sqrt{x^2 + y^2} \right) = (\varphi, r)$$

for $(x,y) \in K_{1,\varepsilon}$ and

$$\psi^{-1}(\varphi, r) = ((\varepsilon\rho - r)\cos(\varphi), (\varepsilon\rho - r)\sin(\varphi)) = (x,y)$$

for $(\varphi, r) \in [-\pi/4, \pi/4) \times [-\varepsilon R, \varepsilon\delta]$. Further, we define using translation functions and rotation matrices the functions $Q_2$, $Q_3$ and $Q_4$ that translate a vector to the origin $(0,0)$ and rotate it by the angle $-\frac{\pi}{2}$, $\pi$ and $\frac{\pi}{2}$, respectively.

$$Q_2(x,y) = \begin{pmatrix} 0 & 1 \\ -1 & 0 \end{pmatrix} \begin{pmatrix} x - \varepsilon/2 \\ y - \varepsilon/2 \end{pmatrix},$$

$$Q_2^{-1}(x,y) = \begin{pmatrix} 0 & -1 \\ 1 & 0 \end{pmatrix} \begin{pmatrix} x \\ y \end{pmatrix} + \begin{pmatrix} \varepsilon/2 \\ \varepsilon/2 \end{pmatrix},$$

$$Q_3(x,y) = \begin{pmatrix} -1 & 0 \\ 0 & -1 \end{pmatrix} \begin{pmatrix} x - \varepsilon/2 \\ y - \varepsilon/2 \end{pmatrix},$$

$$Q_3^{-1}(x,y) = \begin{pmatrix} -1 & 0 \\ 0 & -1 \end{pmatrix} \begin{pmatrix} x \\ y \end{pmatrix} + \begin{pmatrix} \varepsilon/2 \\ \varepsilon/2 \end{pmatrix},$$

$$Q_4(x,y) = \begin{pmatrix} 0 & -1 \\ 1 & 0 \end{pmatrix} \begin{pmatrix} x - \varepsilon/2 \\ y - \varepsilon/2 \end{pmatrix},$$

193

$$Q_4^{-1}(x,y) = \begin{pmatrix} 0 & 1 \\ -1 & 0 \end{pmatrix} \begin{pmatrix} x \\ y \end{pmatrix} + \begin{pmatrix} \varepsilon/2 \\ \varepsilon/2 \end{pmatrix}.$$

Then,

$$\psi^{(1)} = \psi, \qquad\qquad \psi^{(2)} = \psi \circ Q_2,$$
$$(\psi^{(1)})^{-1} = \psi^{-1}, \qquad (\psi^{(2)})^{-1} = Q_2^{-1} \circ \psi^{-1},$$

$$\psi^{(3)} = \psi \circ Q_3, \qquad\qquad \psi^{(4)} = \psi \circ Q_4$$
$$(\psi^{(3)})^{-1} = Q_3^{-1} \circ \psi^{-1}, \qquad (\psi^{(4)})^{-1} = Q_4^{-1} \circ \psi^{-1}.$$

Then we define

$$g_j : [-\pi/4, \pi/4] \times [-\varepsilon R, \varepsilon \delta] \to \mathbb{R}$$
$$g_j(y) := \left( \varphi_j \cdot S_{O,\varepsilon} \circ \left( \psi^{(j)} \right)^{-1} \right)(y)$$

with its extension

$$\text{ext}_\varepsilon : L^2(\varepsilon \partial Y^{ER}) \hookrightarrow L^2(\varepsilon \partial Y^{ER} \cup \varepsilon \partial Y^1),$$

$$\text{ext}_\varepsilon\, g_j(y) := \begin{cases} g_j(y), & y_2 \geq 0 \\ g_j\left(y_1, -\frac{\delta}{R}y_2\right), & y_2 < 0 \end{cases} \tag{4.10}$$

with $\text{supp}(\text{ext}_\varepsilon\, g_j) \subset \psi^{(j)}(K_{j,\varepsilon})$. The extension mirrors a function at $y_2 = 0$; see for example the following figure.

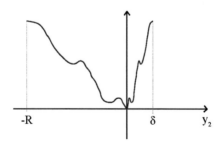

We translate $g_j$ back to $K_{j,\varepsilon}$ with

$$h_j(x) = (\mathrm{ext}_\varepsilon\, g_j) \circ \psi^{(j)}(x)$$

with $\mathrm{supp}(h_j) \subset K_{j,\varepsilon}$ and

$$h_j|_{\varepsilon\partial Y^{\mathrm{ER}} \cap K_{j,\varepsilon}} = \varphi_j S_{O,\varepsilon}.$$

We extend $h_j$ by $h_j(x) := 0$ for $x \in \mathbb{R}^2 \backslash K_{j,\varepsilon}$. Then

$$\mathrm{ext}_\varepsilon\, S_{O,\varepsilon} = \sum_{j=1}^{4} h_j. \tag{4.11}$$

We define $\partial Y^{\mathrm{CRAC}} := \bigcup_j \frac{1}{\varepsilon} K_j \cap \partial Y^1 \subset \partial Y$ and $\partial\Omega_\varepsilon^{\mathrm{CRAC}} := \bigcup_{k \in \mathbb{Z}^3} \varepsilon(Y^{\mathrm{CRAC}} + k) \cap \partial\Omega$.

**Continuity of the extension in the $L^p$-spaces**
Later we need the extension $\mathrm{ext}_\varepsilon$ to be continuous. To show continuity of the linear extension $\mathrm{ext}_\varepsilon$, we prove that $\|\mathrm{ext}_\varepsilon S_{O,\varepsilon}\|_{L^p(\varepsilon\partial Y^1)}^p$ is bounded. Let $p \geq 1$ and $\delta > 0$. In the following equalities we use four times substitution.

$$\|\mathrm{ext}_\varepsilon S_{O,\varepsilon}\|_{L^p(\varepsilon\partial Y^1)}^p = \sum_{j=1}^{4} \int_{K_{j,\varepsilon} \cap \varepsilon\partial Y^1} h_j(x)^p \mathrm{d}x$$

$$= \sum_{j=1}^{4} \int_{K_{j,\varepsilon} \cap \varepsilon\partial Y^1} \mathrm{ext}_\varepsilon\, g_j(\psi^{(j)}(x))^p \mathrm{d}x$$

$$= \sum_{j=1}^{4} \int_{\psi^{(j)}(K_{j,\varepsilon} \cap \varepsilon\partial Y^1)} \mathrm{ext}_\varepsilon\, g_j(y)^p |\det D(\psi^{(j)})^{-1}(y)| \mathrm{d}y$$

$$= \sum_{j=1}^{4} \int_{[-\frac{\pi}{4},\frac{\pi}{4}) \times [-\varepsilon R, 0]} g_j\, (y_1, -\delta/R \cdot y_2)^p\, |\det D(\psi^{(j)})^{-1}(y)| \mathrm{d}y$$

$$= \sum_{j=1}^{4} \int_{[-\frac{\pi}{4},\frac{\pi}{4}) \times [0,\varepsilon\delta]} g_j(y)^p |\det D(\psi^{(j)})^{-1}\, (y_1, -R/\delta \cdot y_2)|\frac{R}{\delta} \mathrm{d}y$$

195

$$= \sum_{j=1}^{4} \int_{(\psi^{(j)})^{-1}([-\frac{\pi}{4},\frac{\pi}{4}]\times[0,\varepsilon\delta])} S_{O,\varepsilon}(x)^p$$

$$\left| \det D(\psi^{(j)})^{-1} \left( \begin{pmatrix} 1 & 0 \\ 0 & -R/\delta \end{pmatrix} \psi^{(j)}(x) \right) \right| \left| \det D\psi^{(j)}(x) \right| \frac{R}{\delta} dx$$

$$= \sum_{j=1}^{4} \int_{K_{j,\varepsilon} \cap \partial Y^{\mathrm{ER}}} S_{O,\varepsilon}(x)^p \left| \det D(\psi^{(j)})^{-1} \left( \begin{pmatrix} 1 & 0 \\ 0 & -R/\delta \end{pmatrix} \psi^{(j)}(x) \right) \right|$$

$$\left| \det D\psi^{(j)}(x) \right| \frac{R}{\delta} dx.$$

We calculate the values of the determinants. With

$$\begin{pmatrix} 1 & 0 \\ 0 & -R/\delta \end{pmatrix} \psi^{(j)}(x,y) = \begin{pmatrix} \arctan y/x \\ -R/\delta(\varepsilon\rho - \sqrt{x^2+y^2}) \end{pmatrix},$$

and

$$\left| \det D(\psi^{(j)})^{-1}(r,\varphi) \right|$$

$$= \underbrace{\left| \det Q_j^{-1} \right|}_{=1} \left| \det \begin{pmatrix} -(\varepsilon\rho-r)\sin\varphi & -\cos\varphi \\ (\varepsilon\rho-r)\cos\varphi & -\sin\varphi \end{pmatrix} \right| = |\varepsilon\rho - r|$$

we find

$$\left| \det D(\psi^{(j)})^{-1} \left( \begin{pmatrix} 1 & 0 \\ 0 & -R/\delta \end{pmatrix} \psi^{(j)}(x) \right) \right|$$

$$= |\varepsilon\rho + \underbrace{R/\delta(\varepsilon\rho - \sqrt{x^2+y^2})}_{\leq \varepsilon\delta \text{ in } \Gamma^1 \cap \partial\Omega_\varepsilon^{\mathrm{ER}}} | \leq \varepsilon|\rho + R|.$$

Further,

$$\left| \det D\psi^{(j)} \right| = \left| \det \begin{pmatrix} -\frac{y}{x^2+y^2} & \frac{x}{x^2+y^2} \\ -\frac{x}{\sqrt{x^2+y^2}} & -\frac{y}{\sqrt{x^2+y^2}} \end{pmatrix} \right|$$

$$= \left| \frac{1}{\sqrt{x^2+y^2}} \right| \leq \frac{1}{\varepsilon|\rho - \delta|}.$$

This means that the product of the determinants is smaller than $\frac{|\rho+R|}{|\rho-\delta|}$, where the $\varepsilon$ cancels, and we obtain

$$\left(\int_{\varepsilon\partial Y^1}\text{ext}_\varepsilon\,S_{O,\varepsilon}(x)^p\mathrm{d}x\right)^{1/p}\leq\left(\frac{R(\rho+R)}{\delta(\rho-\delta)}\int_{\varepsilon\partial Y^{\text{ER}}}S_{O,\varepsilon}(x)^p\mathrm{d}x\right)^{1/p}.$$

For $p=2$ it follows that

$$\|\text{ext}_\varepsilon\,S_{O,\varepsilon}\|_{L^2(\Gamma^1\cap\partial\Omega^1_\varepsilon)}\leq\sqrt{\frac{R(\rho+R)}{\delta(\rho-\delta)}}\|S_{O,\varepsilon}\|_{L^2(\Gamma^1\cap\partial\Omega^{\text{ER}}_\varepsilon)}.$$

For $\delta>0$, we immediately see that the extension is continuous. Because $\delta$ is very small, $\delta\ll1$, and $0<\rho<1/2$ is a constant, we never have $\rho=\delta$.
For $p=\infty$ we find for the $L^\infty$ norm that

$$\|\text{ext}_\varepsilon\,S_{O,\varepsilon}\|_{L^\infty(\Gamma^1\cap\partial\Omega^1_\varepsilon)}=\|S_{O,\varepsilon}\|_{L^\infty(\Gamma^1\cap\partial\Omega^{\text{ER}}_\varepsilon)}.\qquad(4.12)$$

**Remark 4.11** With this extension (4.11) we obtain the function $\text{ext}_\varepsilon S_{O,\varepsilon}$ in $L^2(\Gamma^1\cap\partial\Omega^1_\varepsilon)$. One can even show that the function $S_{O,\varepsilon}$ is an element of $H^{1/2}(\Gamma^1\cap\partial\Omega^{\text{ER}}_\varepsilon)$. With some more effort we could construct an extension

$$\text{ext}_\varepsilon:H^{1/2}(\varepsilon\partial Y^{\text{ER}})\hookrightarrow H^{1/2}(\varepsilon(\partial Y^1\cup\partial Y^{\text{ER}})).$$

We would need to work more for the functions $\text{ext}_\varepsilon g_j$, $j=1,2,3,4$. To be in $H^{1/2}(\varepsilon(\partial Y^1\cup\partial Y^{\text{ER}}))$, $\text{ext}_\varepsilon g_j$ needs to satisfy more conditions at $y_2=0$, where the domain $\Gamma^1\cap\partial\Omega^1_\varepsilon$ touches the domain $\Gamma^1\cap\partial\Omega^{\text{ER}}_\varepsilon$. We would have

$$\text{ext}_\varepsilon g_j(y)=a_1 g_j(y_1,\lambda_1 y_2)+a_2 g_j(y_1,\lambda_2 y_2),$$

and choose $\lambda_{1,2}$ and $a_{1,2}$ such that $\text{ext}_\varepsilon g_j\in H^{1/2}$.

We decided to avoid these tedious calculations for the following reasons. Firstly, the $L^2$-extension makes more sense from the biological point of view. Because the function $S_{O,\varepsilon}$ is reflected from the originally domain at the boundary to the new domain, and the function values on $\varepsilon\partial Y^1$ are the same ones as on $\varepsilon\partial Y^{\text{ER}}$ with

197

the $L^2$-extension, we ensure the right values for the CRAC influx. With the $H^{1/2}$-extension the values would change.

Secondly, we do not need a $H^{1/2}$-extension, because for the influx in $\Omega^1_\varepsilon$, the function

$$I_{\text{CRAC}} = g_{\text{CRAC}} \, (\text{ext}_\varepsilon S_{O,\varepsilon})^2 (V - V_{Ca}) \qquad (4.13)$$

just needs to be an $L^2$-function to be well-posed.

## Weak formulation

We define the following function spaces

$$
\begin{aligned}
\mathcal{V}(\Omega^1_\varepsilon) &:= L^2([0,T], H^1(\Omega^1_\varepsilon)) \\
\mathcal{V}(\Omega^2_\varepsilon) &:= L^2([0,T], H^1(\Omega^2_\varepsilon)), \\
\mathcal{V}(\Omega^{ER}_\varepsilon) &:= L^2([0,T], H^1(\Omega^{ER}_\varepsilon)) \\
\mathcal{V}(\Gamma^1 \cap \partial\Omega^{ER}_\varepsilon) &:= L^2([0,T] \times (\Gamma^1 \cap \partial\Omega^{ER}_\varepsilon)).
\end{aligned}
$$

For the test functions we define the function spaces

$$
\begin{aligned}
V(\Omega^1_\varepsilon) &:= H^1(\Omega^1_\varepsilon) \\
V(\Omega^2_\varepsilon) &:= H^1(\Omega^2_\varepsilon), \\
V(\Omega^{ER}_\varepsilon) &:= H^1(\Omega^{ER}_\varepsilon) \\
V(\Gamma^1 \cap \partial\Omega^{ER}_\varepsilon) &:= L^2(\Gamma^1 \cap \partial\Omega^{ER}_\varepsilon).
\end{aligned}
$$

The weak formulation is given by finding $(C_\varepsilon, C_{e,\varepsilon}, S_\varepsilon, S_{C,\varepsilon}, S_{O,\varepsilon}, S_{CO,\varepsilon}) \in \mathcal{V}(\Omega^1_\varepsilon) \times \mathcal{V}(\Omega^2_\varepsilon) \times \mathcal{V}(\Omega^{ER}_\varepsilon) \times$

$\mathcal{V}(\Omega_\varepsilon^{ER}) \times \mathcal{V}(\Gamma^1 \cap \partial\Omega_\varepsilon^{ER}) \times \mathcal{V}(\Gamma^1 \cap \partial\Omega_\varepsilon^{ER})$ such that

$$(\partial_t C_\varepsilon, \psi_\varepsilon^1)_{\Omega_\varepsilon^1} + D_C(\nabla C_\varepsilon, \nabla\psi_\varepsilon^1)_{\Omega_\varepsilon^1}$$
$$+ \varepsilon\langle (L_0 + L_{IP3})(C_\varepsilon - C_{e,\varepsilon}) + f_{SERCA}, \psi_\varepsilon^1\rangle_{\Gamma_\varepsilon^{ER}}$$
$$+ \langle I_{CRAC} + f_P + f_{NCX}, \psi_\varepsilon^1\rangle_{\Gamma^1 \cap \partial\Omega_\varepsilon^1} = 0$$

$$(\partial_t C_{e,\varepsilon}, \psi_\varepsilon^2)_{\Omega_\varepsilon^2} + D_{ER}(\nabla C_{e,\varepsilon}, \nabla\psi_\varepsilon^2)_{\Omega_\varepsilon^2}$$
$$+ \varepsilon\langle (L_0 + L_{IP3})(C_{e,\varepsilon} - C_\varepsilon) - f_{SERCA}, \psi_\varepsilon^2\rangle_{\Gamma_\varepsilon^{ER}} = 0$$

$$(\partial_t S_\varepsilon, \psi_\varepsilon^3)_{\Omega_\varepsilon^{ER}} + D_S(\nabla S_\varepsilon, \nabla\psi_\varepsilon^3)_{\Omega_\varepsilon^{ER}}$$
$$+ (k_C^+ f_e(C_{e,\varepsilon}) S_\varepsilon - k_C^- S_{C,\varepsilon}, \psi_\varepsilon^3)_{\Omega_\varepsilon^{ER}}$$
$$+ \langle k_O^+ S_\varepsilon - k_O^- S_{O,\varepsilon}, \psi_\varepsilon^3\rangle_{\Gamma^1 \cap \partial\Omega_\varepsilon^{ER}} = 0$$

$$(\partial_t S_{C,\varepsilon}, \psi_\varepsilon^3)_{\Omega_\varepsilon^{ER}} + D_{SC}(\nabla S_{C,\varepsilon}, \nabla\psi_\varepsilon^3)_{\Omega_\varepsilon^{ER}}$$
$$+ (k_C^- S_{C,\varepsilon} - k_C^+ f_e(C_{e,\varepsilon}) S_\varepsilon, \psi_\varepsilon^3)_{\Omega_\varepsilon^{ER}}$$
$$+ \langle k_{CO}^+ S_{C,\varepsilon} - k_{CO}^- S_{CO,\varepsilon}, \psi_\varepsilon^3\rangle_{\Gamma^1 \cap \partial\Omega_\varepsilon^{ER}} = 0$$

$$\langle \partial_t S_{O,\varepsilon}, \psi_\varepsilon^4\rangle_{\Gamma^1 \cap \partial\Omega_\varepsilon^{ER}} + \langle k_O^- S_{O,\varepsilon} - k_O^+ S_\varepsilon$$
$$- k_C^- S_{CO,\varepsilon} + k_C^+ f_e(C_{e,\varepsilon}) S_{O,\varepsilon}, \psi_\varepsilon^4\rangle_{\Gamma^1 \cap \partial\Omega_\varepsilon^{ER}} = 0$$

$$\langle \partial_t S_{CO,\varepsilon}, \psi_\varepsilon^4\rangle_{\Gamma^1 \cap \partial\Omega_\varepsilon^{ER}} + \langle k_{CO}^- S_{CO,\varepsilon} - k_{CO}^+ S_{C,\varepsilon}$$
$$- k_C^+ f_e(C_{e,\varepsilon}) S_{O,\varepsilon} + k_C^- S_{CO,\varepsilon}, \psi_\varepsilon^4\rangle_{\Gamma^1 \cap \partial\Omega_\varepsilon^{ER}} = 0$$

$$(4.14)$$

for all $\psi_\varepsilon^1 \in V(\Omega_\varepsilon^1)$, $\psi_\varepsilon^2 \in V(\Omega_\varepsilon^2)$, $\psi_\varepsilon^3 \in V(\Omega_\varepsilon^{ER})$ and $\psi_\varepsilon^4 \in V(\Gamma^1 \cap \partial\Omega_\varepsilon^{ER})$.

We omit the weak formulation for the dynamics of the channels, because it is a system of ordinary differential equations and appears only indirectly through the function $I_{CRAC}$.

**Remark 4.12** The mathematical model by Patrick Fletcher and Yue-Xian Li in [23] has a similar form to the one, which we find after homogenization, where the cytosol, the membrane and the lumen of the ER are merged to a homogeneous cytoplasmic domain. Here, in model (4.14), the process is decribed in more detail, because the living environments of the molecules are considered

more precisely.
By performing homogenization and finding a similar model to the one in article [23], the form of the system of partial differential equations is mathematically confirmed and the coefficients are improved.

In this section we explained the biological background of the process of the T cell signaling and derived a first mathematical model. Then we prepared this model to be able to use the technique of periodic homogenization (see section 2.1) and the two-step convergence (see section 4.1). Furthermore, we established an extension for the function $S_{O,\varepsilon}$ and showed continuity of this extension.
In the next section we show well-posedness of the system of equations (4.14).

## 4.4 Estimations for the Calcium-Stim1 model

In this section we show that the functions $C_\varepsilon$, $C_{e,\varepsilon}$, $S_\varepsilon$ and $S_{C,\varepsilon}$ are elements of $H^1$ and that the functions $S_{O,\varepsilon}$ and $S_{CO,\varepsilon}$ are elements of $L^2$. This is necessary to apply proposition 2.5 and theorem 2.3. Further, we show that there exist extensions of the functions $C_\varepsilon$, $C_{e,\varepsilon}$, $S_\varepsilon$, $S_{C,\varepsilon}$, $S_{O,\varepsilon}$ and $S_{CO,\varepsilon}$, which converge strongly in $L^2([0,T] \times \Omega)$ and $L^2([0,T] \times \Gamma^1)$, respectively . Therefore we apply lemmas 3.4, 3.5 and 3.6. Hence, we additionally prove that the functions are elements of $L^\infty$ and that the time-derivative of the functions are bounded in a Banach Space $B_1$ with $L^2 \hookrightarrow B_1$. We will choose $B_1 = H^{-1}(\Omega)$, respectively.

Before we start with the estimations, we prove that the inverse trace inequality, where the inverse trace operator maps from the outer boundary $\partial\Omega_\varepsilon \cap \partial\Omega$ of an $\varepsilon$-depending domain $\Omega_\varepsilon \subset \Omega \subset \mathbb{R}^n$ into $\Omega_\varepsilon$, does not depend on $\varepsilon$.

**Lemma 4.13** *Let $\Omega \subset \mathbb{R}^n$ and $\Omega_\varepsilon$ be an $\varepsilon$-periodic subset of $\Omega$. Then it holds for any function $f_\varepsilon \in H^1(\Omega_\varepsilon)$ that*

$$\|f_\varepsilon\|^2_{L^2(\partial\Omega_\varepsilon \cap \partial\Omega)} \leq c\|f_\varepsilon\|^2_{H^1(\Omega_\varepsilon)}$$

*with $c > 0$ independent of $\varepsilon$.*

**Proof** Lemma 3.4 says that there exists an extention $\tilde{f}_\varepsilon \in H^1(\Omega)$ with $f_\varepsilon = \tilde{f}_\varepsilon$ in $\Omega_\varepsilon$ such that

$$\|\tilde{f}_\varepsilon\|_{H^1(\Omega)} \leq \tilde{c}\|f_\varepsilon\|_{H^1(\Omega_\varepsilon)}$$

where $\tilde{c}$ is independent of $\varepsilon$.

The trace operator $\gamma_\varepsilon : H^1(\Omega_\varepsilon) \to L^2(\partial\Omega_\varepsilon \cap \partial\Omega)$ maps $f_\varepsilon \mapsto \gamma_\varepsilon(f_\varepsilon)$ and $\tilde{f}_\varepsilon \mapsto \gamma_\varepsilon(\tilde{f}_\varepsilon)$ with $\gamma_\varepsilon(f_\varepsilon) = \gamma_\varepsilon(\tilde{f}_\varepsilon)$ on $\partial\Omega_\varepsilon \cap \partial\Omega$, because $f_\varepsilon = \tilde{f}_\varepsilon$ in $\Omega_\varepsilon$. This means for the trace operator $\gamma : H^1(\Omega) \to L^2(\partial\Omega)$ that $\gamma_\varepsilon(\tilde{f}_\varepsilon) = \tilde{f}_\varepsilon|_{\partial\Omega_\varepsilon \cap \partial\Omega} = \gamma(\tilde{f}_\varepsilon)$ on $\partial\Omega_\varepsilon \cap \partial\Omega$.

We deduce the following estimation

$$\|\gamma_\varepsilon(f_\varepsilon)\|^2_{L^2(\partial\Omega_\varepsilon \cap \partial\Omega)} = \|\gamma_\varepsilon(\tilde{f}_\varepsilon)\|^2_{L^2(\partial\Omega_\varepsilon \cap \partial\Omega)}$$
$$\leq \|\gamma(\tilde{f}_\varepsilon)\|^2_{L^2(\partial\Omega)} \leq c_0\|\tilde{f}_\varepsilon\|^2_{H^1(\Omega)} \leq \tilde{c}c_0\|f_\varepsilon\|^2_{H^1(\Omega_\varepsilon)},$$

where $c_0$ is bounded, because $\gamma$ is linear and continuous, and $c_0$ is independent of $\varepsilon$, since $\Omega$ is independent of $\varepsilon$. $\qquad\square$

Now we start with the estimations for the system (4.14) and first prove that $I_{CRAC}$ is bounded. This statement is used in the subsequent lemmas.

### Lemma 4.14 (Boundedness of $I_{CRAC}$)
*Let $S_{O,\varepsilon}$ be bounded for almost every $x \in \Gamma^1 \cap \partial\Omega_\varepsilon^{ER}$ and $t \in [0, T]$. Then the influx $I_{CRAC}$, defined in (4.13), is bounded for almost every $t \in [0, T]$ and $x \in \Gamma^1 \cap \partial\Omega_\varepsilon^1$ by a constant $C > 0$.*

**Proof** The function $I_{CRAC}$ is bounded in $L^\infty$, if $\text{ext}_\varepsilon S_{O,\varepsilon}$ and $V_\varepsilon$ are bounded in $L^\infty$. We know with $S_{O,\varepsilon}$ bounded and equality (4.12) that also $\text{ext}_\varepsilon S_{O,\varepsilon}$ is bounded. To show that $V_\varepsilon \in L^\infty$, we need that $a_\varepsilon$ is bounded in $L^\infty$. Hence we start the proof with show that $a_\varepsilon$ has a lower and upper bound.

At first, we want to show that $a_\varepsilon$, given in (4.9), is a nonnegative function and test the equation of $a_\varepsilon$ with $a_{\varepsilon-}$. We find

$$\langle \partial_t a_\varepsilon, a_{\varepsilon-} \rangle_{\Gamma^1 \cap \partial \Omega^1_\varepsilon} + \left\langle \frac{1}{\tau_{\text{CAN}}} a_\varepsilon, a_{\varepsilon-} \right\rangle_{\Gamma^1 \cap \partial \Omega^1_\varepsilon}$$
$$= \left\langle \frac{1}{\tau_{\text{CAN}}} \frac{C^4_\varepsilon}{C^4_\varepsilon + K^4_{\text{CAN}}}, a_{\varepsilon-} \right\rangle_{\Gamma^1 \cap \partial \Omega^1_\varepsilon}.$$

This leads to

$$\langle \partial_t a_{\varepsilon-}, a_{\varepsilon-} \rangle_{\Gamma^1 \cap \partial \Omega^1_\varepsilon} + \frac{1}{\tau_{\text{CAN}}} \|a_{\varepsilon-}\|^2_{\Gamma^1 \cap \partial \Omega^1_\varepsilon}$$
$$= - \left\langle \frac{1}{\tau_{\text{CAN}}} \frac{C^4_\varepsilon}{C^4_\varepsilon + K^4_{\text{CAN}}}, a_{\varepsilon-} \right\rangle_{\Gamma^1 \cap \partial \Omega^1_\varepsilon} \leq 0.$$

Integration from 0 to $t$ gives

$$\frac{1}{2} \|a_{\varepsilon-}\|^2_{\Gamma^1 \cap \partial \Omega^1_\varepsilon} + \frac{1}{\tau_{\text{CAN}}} \|a_{\varepsilon-}\|^2_{\Gamma^1 \cap \partial \Omega^1_\varepsilon, t} \leq 0.$$

Hence $a_{\varepsilon-}$ is zero and $a_\varepsilon$ is positive almost everywhere in $[0, T] \times \Gamma^1 \cap \partial \Omega^1_\varepsilon$.

Further, we want to show that $a_\varepsilon$ has an upper bound and, to that end, test the weak formulation of $a_\varepsilon$ with $(a_\varepsilon - \bar{a})_+$, where $\bar{a} := \|a_\varepsilon(t = 0)\|_{L^\infty(\Gamma^1 \cap \partial \Omega^1_\varepsilon)} e^{kt}$ for a $k > 0$.

$$\langle \partial_t a_\varepsilon, (a_\varepsilon - \bar{a})_+ \rangle_{\Gamma^1 \cap \partial \Omega^1_\varepsilon} + \underbrace{\left\langle \frac{1}{\tau_{\text{CAN}}} a_\varepsilon, (a_\varepsilon - \bar{a})_+ \right\rangle_{\Gamma^1 \cap \partial \Omega^1_\varepsilon}}_{\geq 0}$$
$$= \left\langle \frac{1}{\tau_{\text{CAN}}} \frac{C^4_\varepsilon}{C^4_\varepsilon + K^4_{\text{CAN}}}, (a_\varepsilon - \bar{a})_+ \right\rangle_{\Gamma^1 \cap \partial \Omega^1_\varepsilon}.$$

Considering the time derivative leads to

$$\langle \partial_t (a_\varepsilon - \bar{a}), (a_\varepsilon - \bar{a})_+ \rangle_{\Gamma^1 \cap \partial \Omega^1_\varepsilon}$$
$$\leq \left\langle \frac{1}{\tau_{\text{CAN}}} \overbrace{\frac{C^4_\varepsilon}{C^4_\varepsilon + K^4_{\text{CAN}}}}^{\in [0,1]} - k\bar{a}, (a_\varepsilon - \bar{a})_+ \right\rangle_{\Gamma^1 \cap \partial \Omega^1_\varepsilon}.$$

We choose $k$ large enough such that $k\overline{a}$ is greater than $\frac{1}{\tau_{CAN}}$ and integrate from 0 to $t$ to find

$$\frac{1}{2}\|(a_\varepsilon - \overline{a})_+\|^2_{\Gamma^1 \cap \partial\Omega^1_\varepsilon} \leq 0.$$

We find that $a_\varepsilon$ is bounded by $\overline{a}$ for almost every $x \in \Gamma^1 \cap \partial\Omega^1_\varepsilon$ and $t \in [0, T]$.

Further, we show that the potential $V_\varepsilon$ is bounded. We define $\overline{V} := \|V_\varepsilon(t=0)\|_{L^\infty(\Gamma^1\cap\partial\Omega^1_\varepsilon)}e^{kt}$ for another $k > 0$ and test the weak equation for $V_\varepsilon$ with $(V_\varepsilon - \overline{V})_+$.

$$\langle \partial_t V_\varepsilon, (V_\varepsilon - \overline{V})_+ \rangle_{\Gamma^1\cap\partial\Omega^1_\varepsilon}$$
$$= -\left\langle \frac{g_k}{C_m}\frac{1}{1+e^{-\frac{V_\varepsilon - V_n}{K_n}}}(V_\varepsilon - V_K), (V_\varepsilon - \overline{V})_+ \right\rangle_{\Gamma^1\cap\partial\Omega^1_\varepsilon}$$
$$-\left\langle \frac{g_{K(Ca)}}{C_m}\frac{C_\varepsilon^4}{C_\varepsilon^4 + K_{K(Ca)}^4}(V_\varepsilon - V_K), (V_\varepsilon - \overline{V})_+ \right\rangle_{\Gamma^1\cap\partial\Omega^1_\varepsilon}$$
$$-\left\langle \frac{g_{CRAC}}{C_m}(\text{ext}_\varepsilon S_{O,\varepsilon})^2(V_\varepsilon - V_{Ca}), (V_\varepsilon - \overline{V})_+ \right\rangle_{\Gamma^1\cap\partial\Omega^1_\varepsilon}$$
$$-\left\langle \frac{g_{CAN}}{C_m}a_\varepsilon(V_\varepsilon - V_{Na}), (V_\varepsilon - \overline{V})_+ \right\rangle_{\Gamma^1\cap\partial\Omega^1_\varepsilon}.$$

Considering the time derivative leads to

$$\langle \partial_t(V_\varepsilon - \overline{V})_+, (V_\varepsilon - \overline{V})_+ \rangle_{\Gamma^1\cap\partial\Omega^1_\varepsilon}$$
$$= \left\langle \frac{g_k}{C_m}\underbrace{\frac{1}{1+e^{-\frac{V_\varepsilon - V_n}{K_n}}}}_{\in[0,1]}(V_K - V_\varepsilon), (V_\varepsilon - \overline{V})_+ \right\rangle_{\Gamma^1\cap\partial\Omega^1_\varepsilon}$$
$$+ \left\langle \frac{g_{K(Ca)}}{C_m}\underbrace{\frac{C_\varepsilon^4}{C_\varepsilon^4 + K_{K(Ca)}^4}}_{\in[0,1]}(V_K - V_\varepsilon), (V_\varepsilon - \overline{V})_+ \right\rangle_{\Gamma^1\cap\partial\Omega^1_\varepsilon}$$
$$+ \left\langle \frac{g_{CRAC}}{C_m}\underbrace{(\text{ext}_\varepsilon S_{O,\varepsilon})^2}_{\text{bounded}, \geq 0}(V_{Ca} - V_\varepsilon), (V_\varepsilon - \overline{V})_+ \right\rangle_{\Gamma^1\cap\partial\Omega^1_\varepsilon}$$

$$+ \left\langle \underbrace{\frac{g_{\mathrm{CAN}}}{C_{\mathrm{m}}} \quad a_{\varepsilon}}_{\text{bounded, } \geq 0} (V_{\mathrm{Na}} - V_{\varepsilon}), (V_{\varepsilon} - \overline{V})_{+} \right\rangle_{\Gamma^1 \cap \partial\Omega^1_{\varepsilon}}$$

$$- \left\langle k\overline{V}, (V_{\varepsilon} - \overline{V})_{+} \right\rangle_{\Gamma^1 \cap \partial\Omega^1_{\varepsilon}}.$$

Merging the positive constants $g_{\mathrm{K}}, g_{\mathrm{K(Ca)}}, g_{\mathrm{CRAC}}, g_{\mathrm{CAN}}, C_{\mathrm{m}}$ and bounded functions to the constant $c_1$ and splitting $V_{\varepsilon}$ into $V_{\varepsilon+} - V_{\varepsilon-}$ gives

$$\left\langle \partial_t (V_{\varepsilon} - \overline{V})_{+}, (V_{\varepsilon} - \overline{V})_{+} \right\rangle_{\Gamma^1 \cap \partial\Omega^1_{\varepsilon}}$$

$$\leq \left\langle c_1 \left( |V_{\mathrm{K}}| + |V_{\mathrm{K}}| + |V_{Ca}| + |V_{\mathrm{Na}}| \right) + c_1 V_{\varepsilon-} - k\overline{V}, (V_{\varepsilon} - \overline{V})_{+} \right\rangle_{\Gamma^1 \cap \partial\Omega^1_{\varepsilon}}$$

$$- \left\langle V_{\varepsilon+} \left( \frac{1}{1 + e^{-\frac{V_{\varepsilon} - V_{\mathrm{n}}}{K_{\mathrm{n}}}}} + \frac{C_{\varepsilon}^4}{C_{\varepsilon}^4 + K_{\mathrm{K(Ca)}}^4} \right.\right.$$

$$\left.\left. + (\mathrm{ext}_{\varepsilon} S_{O,\varepsilon})^2 + a_{\varepsilon} \right), (V_{\varepsilon} - \overline{V})_{+} \right\rangle_{\Gamma^1 \cap \partial\Omega^1_{\varepsilon}}.$$

We drop the last term, since it is negative and choose $k$ large enough such that $|V_{\mathrm{K}}| + |V_{\mathrm{K}}| + |V_{Ca}| + |V_{\mathrm{Na}}| - 1/c_1 k\overline{V} < 0$.

$$\left\langle \partial_t (V_{\varepsilon} - \overline{V})_{+}, (V_{\varepsilon} - \overline{V})_{+} \right\rangle_{\Gamma^1 \cap \partial\Omega^1_{\varepsilon}}$$

$$\leq \left\langle \underbrace{c_1 \left( |V_{\mathrm{K}}| + |V_{\mathrm{K}}| + |V_{Ca}| + |V_{\mathrm{Na}}| \right) - k\overline{V}}_{\leq 0}, (V_{\varepsilon} - \overline{V})_{+} \right\rangle_{\Gamma^1 \cap \partial\Omega^1_{\varepsilon}}$$

$$+ \underbrace{\left\langle c_1 V_{\varepsilon-}, (V_{\varepsilon} - \overline{V})_{+} \right\rangle_{\Gamma^1 \cap \partial\Omega^1_{\varepsilon}}}_{=0},$$

where $V_{\varepsilon-}$ is zero, if $(V_{\varepsilon} - \overline{V})_{+} \neq 0$. Then, by integration from 0 to $t$, we get

$$\frac{1}{2} \| (V_{\varepsilon} - \overline{V})_{+} \|^2_{\Gamma^1 \cap \partial\Omega^1_{\varepsilon}} \leq 0.$$

This means that $V_{\varepsilon} \leq \overline{V}$ for all $t \in [0, T]$ and almost every $x \in \Gamma^1 \cap \partial\Omega^1_{\varepsilon}$.

Because $V_{\varepsilon}$ can also be negative (and mostly is), we also need to show that $-V_{\varepsilon} \leq \overline{V}$ has a lower bound. Hence, we test the equation of $V_\varepsilon$ with $(\overline{V} + V_\varepsilon)_-$,

$$\langle \partial_t (V_\varepsilon + \overline{V})_-, (V_\varepsilon + \overline{V})_- \rangle_{\Gamma^1 \cap \partial\Omega_\varepsilon^1}$$

$$= \left\langle \frac{g_k}{C_m} \frac{1}{1 + e^{-\frac{V_\varepsilon - V_n}{K_n}}} (V_K - V_\varepsilon), (V_\varepsilon + \overline{V})_- \right\rangle_{\Gamma^1 \cap \partial\Omega_\varepsilon^1}$$

$$+ \left\langle \frac{g_{K(Ca)}}{C_m} \frac{C_\varepsilon^4}{C_\varepsilon^4 + K_{K(Ca)}^4} (V_K - V_\varepsilon), (V_\varepsilon + \overline{V})_- \right\rangle_{\Gamma^1 \cap \partial\Omega_\varepsilon^1}$$

$$+ \left\langle \frac{g_{CRAC}}{C_m} (\text{ext}_\varepsilon S_{O,\varepsilon})^2 (V_{Ca} - V_\varepsilon), (V_\varepsilon + \overline{V})_- \right\rangle_{\Gamma^1 \cap \partial\Omega_\varepsilon^1}$$

$$+ \left\langle \frac{g_{CAN}}{C_m} a_\varepsilon (V_\varepsilon - V_{Na})(V_{Na} - V_\varepsilon), (V_\varepsilon + \overline{V})_- \right\rangle_{\Gamma^1 \cap \partial\Omega_\varepsilon^1}$$

$$- \langle k\overline{V}, (V_\varepsilon + \overline{V})_- \rangle_{\Gamma^1 \cap \partial\Omega_\varepsilon^1}.$$

As above we use the constant $c_1$ again and find

$$\langle \partial_t (V_\varepsilon + \overline{V})_-, (V_\varepsilon + \overline{V})_- \rangle_{\Gamma^1 \cap \partial\Omega_\varepsilon^1}$$

$$\leq \langle c_1(|V_K| + |V_K| + |V_{Ca}| + |V_{Na}|) - 4c_1 V_\varepsilon - k\overline{V}, (V_\varepsilon + \overline{V})_- \rangle_{\Gamma^1 \cap \partial\Omega_\varepsilon^1},$$

where we assume that $V_\varepsilon$ is negative; otherwise $(V_\varepsilon + \overline{V})_- = 0$. We take a constant $k_1 \in \mathbb{R}$ such that $k_1 \overline{V} \geq c_1(|V_K| + |V_K| + |V_{Ca}| + |V_{Na}|)$ and choose $k = k_1 + 4c_1$. Then we obtain

$$\langle \partial_t (V_\varepsilon + \overline{V})_-, (V_\varepsilon + \overline{V})_- \rangle_{\Gamma^1 \cap \partial\Omega_\varepsilon^1}$$

$$\leq \underbrace{\langle c_1(|V_K| + |V_K| + |V_{Ca}| + |V_{Na}|) - k_1 \overline{V}, (V_\varepsilon + \overline{V})_- \rangle_{\Gamma^1 \cap \partial\Omega_\varepsilon^1}}_{\leq 0}$$

$$- 4c_1 \langle V_\varepsilon + \overline{V}, (V_\varepsilon + \overline{V})_- \rangle_{\Gamma^1 \cap \partial\Omega_\varepsilon^1}.$$

Further, we integrate from 0 to $t$, and conclude

$$\frac{1}{2} \|(V_\varepsilon + \overline{V})_-\|_{\Gamma^1 \cap \partial\Omega_\varepsilon^1}^2 \leq 4c_1 \|(V_\varepsilon + \overline{V})_-\|_{\Gamma^1 \cap \partial\Omega_\varepsilon^1, t}^2.$$

Gronwall's lemma yields that $(V_\varepsilon + \overline{V})_- = 0$ almost everywhere and hence, $V_\varepsilon \in [-\overline{V}, \overline{V}]$ for almost every $x \in \Gamma^1 \cap \partial\Omega_\varepsilon^1$ and

$t \in [0, T]$.

Now, with $\text{ext}_\varepsilon S_{O,\varepsilon}$ and $V_\varepsilon$ being bounded in $L^\infty$, it follows that

$$|I_{\text{CRAC}}| = |g_{\text{CRAC}}(\text{ext}_\varepsilon S_{O,\varepsilon})^2 (V_\varepsilon - V_{Ca})| < \infty$$

almost everywhere. $\qquad\qquad\qquad\qquad\qquad\qquad\qquad\qquad$ □

We show that the number of Stim1 molecules is nonnegative. This means that the functions $S_\varepsilon$, $S_{C,\varepsilon}$, $S_{O,\varepsilon}$ and $S_{CO,\varepsilon}$ are equal to or greater than zero. The following lemma is necessary to find a lower bound for the functions. By obtaining an upper bound, too, in lemma 4.17 the functions $S_\varepsilon$, $S_{C,\varepsilon}$, $S_{O,\varepsilon}$ and $S_{CO,\varepsilon}$ are $L^\infty$-functions.

**Lemma 4.15 (Positivity of $S_\varepsilon$, $S_{C,\varepsilon}$, $S_{O,\varepsilon}$ and $S_{CO,\varepsilon}$)**

1. For almost every $x \in \Omega_\varepsilon^{ER}$ and $t \in [0, T]$ it holds that

$$S_\varepsilon(x,t) \geq 0, \qquad S_{C,\varepsilon}(x,t) \geq 0.$$

2. For almost every $x \in \Gamma^1 \cap \partial\Omega_\varepsilon^{ER}$ and $t \in [0, T]$ it holds that

$$S_{O,\varepsilon}(x,t) \geq 0, \qquad S_{CO,\varepsilon}(x,t) \geq 0.$$

**Proof** We test the weak formulations of $S_\varepsilon$, $S_{C,\varepsilon}$, $S_{O,\varepsilon}$, and $S_{CO,\varepsilon}$ with $k_O^+ S_{\varepsilon-}$, $k_{CO}^+ S_{C,\varepsilon-}$, $k_O^- S_{O,\varepsilon-}$, and $k_{CO}^- S_{CO,\varepsilon-}$, respectively. We start with $S_\varepsilon$ and $S_{O,\varepsilon}$, and then add the equations.

$$k_O^+ (\partial_t S_\varepsilon, S_{\varepsilon-})_{\Omega_\varepsilon^{ER}} + k_O^- \langle \partial_t S_{O,\varepsilon}, S_{O,\varepsilon-} \rangle_{\Gamma^1 \cap \partial\Omega_\varepsilon^{ER}} + D_S k_O^+ (\nabla S_\varepsilon, \nabla S_{\varepsilon-})_{\Omega_\varepsilon^{ER}}$$

$$+ k_C^+ k_O^+ (f_e(C_{e,\varepsilon}) S_\varepsilon, S_{\varepsilon-})_{\Omega_\varepsilon^{ER}} + k_C^+ k_O^- \langle f_e(C_{e,\varepsilon}) S_{O,\varepsilon}, S_{O,\varepsilon-} \rangle_{\Gamma^1 \cap \partial\Omega_\varepsilon^{ER}}$$

$$+ \langle k_O^+ S_\varepsilon - k_O^- S_{O,\varepsilon}, k_O^+ S_{\varepsilon-} - k_O^- S_{O,\varepsilon-} \rangle_{\Gamma^1 \cap \partial\Omega_\varepsilon^{ER}}$$

$$= k_C^- k_O^+ (S_{C,\varepsilon}, S_{\varepsilon-})_{\Omega_\varepsilon^{ER}} + k_C^- k_O^- \langle S_{CO,\varepsilon}, S_{O,\varepsilon-} \rangle_{\Gamma^1 \cap \partial\Omega_\varepsilon^{ER}}.$$

Multiplying both sides by $-1$ leads to

$$k_O^+ (\partial_t S_{\varepsilon-}, S_{\varepsilon-})_{\Omega_\varepsilon^{ER}} + k_O^- \langle \partial_t S_{O,\varepsilon-}, S_{O,\varepsilon-} \rangle_{\Gamma^1 \cap \partial \Omega_\varepsilon^{ER}} + D_S k_O^+ \| \nabla S_{\varepsilon-} \|^2_{\Omega_\varepsilon^{ER}}$$

$$+ k_C^+ k_O^+ \| \sqrt{f_e(C_{e,\varepsilon})} S_{\varepsilon-} \|^2_{\Omega_\varepsilon^{ER}} + k_C^+ k_O^- \| \sqrt{f_e(C_{e,\varepsilon})} S_{O,\varepsilon-} \|^2_{\Gamma^1 \cap \partial \Omega_\varepsilon^{ER}}$$

$$+ \| k_O^+ S_{\varepsilon-} - k_O^- S_{O,\varepsilon-} \|^2_{\Gamma^1 \cap \partial \Omega_\varepsilon^{ER}}$$

$$\leq -k_C^- k_O^+ (S_{C,\varepsilon}, S_{\varepsilon-})_{\Omega_\varepsilon^{ER}} - k_C^- k_O^- \langle S_{CO,\varepsilon}, S_{O,\varepsilon-} \rangle_{\Gamma^1 \cap \partial \Omega_\varepsilon^{ER}}$$

$$\leq k_C^- k_O^+ (S_{C,\varepsilon-}, S_{\varepsilon-})_{\Omega_\varepsilon^{ER}} + k_C^- k_O^- \langle S_{CO,\varepsilon-}, S_{O,\varepsilon-} \rangle_{\Gamma^1 \cap \partial \Omega_\varepsilon^{ER}}$$

$$\leq k_C^- k_O^+ \left( \| S_{C,\varepsilon-} \|^2_{\Omega_\varepsilon^{ER}} + \| S_{\varepsilon-} \|^2_{\Omega_\varepsilon^{ER}} \right)$$

$$+ k_C^- k_O^- \left( \| S_{CO,\varepsilon-} \|^2_{\Gamma^1 \cap \partial \Omega_\varepsilon^{ER}} + \| S_{O,\varepsilon-} \|^2_{\Gamma^1 \cap \partial \Omega_\varepsilon^{ER}} \right).$$

Integration from 0 to $t$, dropping some positive terms, and merging the constants yields

$$\| S_{\varepsilon-} \|^2_{\Omega_\varepsilon^{ER}} + \| S_{O,\varepsilon-} \|^2_{\Gamma^1 \cap \partial \Omega_\varepsilon^{ER}} \leq c_1 \left( \| S_{C,\varepsilon-} \|^2_{\Omega_\varepsilon^{ER}, t} + \| S_{\varepsilon-} \|^2_{\Omega_\varepsilon^{ER}, t} \right.$$

$$\left. + \| S_{CO,\varepsilon-} \|^2_{\Gamma^1 \cap \partial \Omega_\varepsilon^{ER}, t} + \| S_{O,\varepsilon-} \|^2_{\Gamma^1 \cap \partial \Omega_\varepsilon^{ER}, t} \right),$$

where we used that the initial conditions are nonnegative. We perform corresponding operations for the equations for $S_{C,\varepsilon}$ and $S_{CO,\varepsilon}$, and find

$$\| S_{C,\varepsilon-} \|^2_{\Omega_\varepsilon^{ER}} + \| S_{CO,\varepsilon-} \|^2_{\Gamma^1 \cap \partial \Omega_\varepsilon^{ER}}$$

$$\leq c_1 \left( \| S_{C,\varepsilon-} \|^2_{\Omega_\varepsilon^{ER}, t} + v_e \| S_{\varepsilon-} \|^2_{\Omega_\varepsilon^{ER}, t} \right.$$

$$\left. + \| S_{CO,\varepsilon-} \|^2_{\Gamma^1 \cap \partial \Omega_\varepsilon^{ER}, t} + v_e \| S_{O,\varepsilon-} \|^2_{\Gamma^1 \cap \partial \Omega_\varepsilon^{ER}, t} \right).$$

Then we add the two inequalities and find

$$\| S_{\varepsilon-} \|^2_{\Omega_\varepsilon^{ER}} + \| S_{O,\varepsilon-} \|^2_{\Gamma^1 \cap \partial \Omega_\varepsilon^{ER}} + \| S_{C,\varepsilon-} \|^2_{\Omega_\varepsilon^{ER}} + \| S_{CO,\varepsilon-} \|^2_{\Gamma^1 \cap \partial \Omega_\varepsilon^{ER}}$$

$$\leq c_1 \left( \| S_{C,\varepsilon-} \|^2_{\Omega_\varepsilon^{ER}, t} + \| S_{\varepsilon-} \|^2_{\Omega_\varepsilon^{ER}, t} \right.$$

$$\left. + \| S_{CO,\varepsilon-} \|^2_{\Gamma^1 \cap \partial \Omega_\varepsilon^{ER}, t} + \| S_{O,\varepsilon-} \|^2_{\Gamma^1 \cap \partial \Omega_\varepsilon^{ER}, t} \right).$$

Using the lemma of Gronwall 2.12 yields

$$\|S_{\varepsilon-}\|^2_{\Omega^{ER}_\varepsilon} + \|S_{O,\varepsilon-}\|^2_{\Gamma^1 \cap \partial\Omega^{ER}_\varepsilon} + \|S_{C,\varepsilon-}\|^2_{\Omega^{ER}_\varepsilon} + \|S_{CO,\varepsilon-}\|^2_{\Gamma^1 \cap \partial\Omega^{ER}_\varepsilon} \leq 0.$$

Hence, the functions $S_\varepsilon$, $S_{C,\varepsilon}$, $S_{O,\varepsilon}$ and $S_{CO,\varepsilon}$ are nonnegative.
$\square$

To be able to apply proposition 2.5 to the functions $S_\varepsilon$ and $S_{C,\varepsilon}$, we show that these functions are elements of $H^1$. To apply theorem 2.3 to the functions $S_{O,\varepsilon}$ and $S_{CO,\varepsilon}$, we prove that they are in $L^2$.

**Lemma 4.16 (Boundedness of $S_\varepsilon$, $S_{C,\varepsilon}$, $S_{O,\varepsilon}$ and $S_{CO,\varepsilon}$ in $H^1$ or $L^2$)**
*There exists a constant $C > 0$, independent of $\varepsilon$, such that*

$$\|S_\varepsilon\|^2_{\Omega^{ER}_\varepsilon} + \|S_{C,\varepsilon}\|^2_{\Omega^{ER}_\varepsilon} + \|\nabla S_\varepsilon\|^2_{\Omega^{ER}_\varepsilon, t}$$
$$+ \|\nabla S_{C,\varepsilon}\|^2_{\Omega^{ER}_\varepsilon, t} + \|S_{O,\varepsilon}\|^2_{\Gamma^1 \cap \partial\Omega^{ER}_\varepsilon} + \|S_{CO,\varepsilon}\|^2_{\Gamma^1 \cap \partial\Omega^{ER}_\varepsilon} \leq C.$$

**Proof** We test the weak formulations for $S_\varepsilon$ and $S_{C,\varepsilon}$ with the functions $S_\varepsilon$ and $S_{C,\varepsilon}$, respectively, and add the equations

$$(\partial_t S_\varepsilon, S_\varepsilon)_{\Omega^{ER}_\varepsilon} + (\partial_t S_{C,\varepsilon}, S_{C,\varepsilon})_{\Omega^{ER}_\varepsilon} + D_S(\nabla S_\varepsilon, \nabla S_\varepsilon)_{\Omega^{ER}_\varepsilon}$$
$$+ D_{SC}(\nabla S_{C,\varepsilon}, \nabla S_{C,\varepsilon})_{\Omega^{ER}_\varepsilon} + (k_C^+ f_e(C_{e,\varepsilon})S_\varepsilon - k_C^- S_{C,\varepsilon}, S_\varepsilon - S_{C,\varepsilon})_{\Omega^{ER}_\varepsilon}$$
$$+ \langle k_O^+ S_\varepsilon - k_O^- S_{O,\varepsilon}, S_\varepsilon \rangle_{\Gamma^1 \cap \partial\Omega^{ER}_\varepsilon} + \langle k_{CO}^+ S_{C,\varepsilon} - k_{CO}^- S_{CO,\varepsilon}, S_{C,\varepsilon} \rangle_{\Gamma^1 \cap \partial\Omega^{ER}_\varepsilon} = 0.$$

With integration from 0 to $t$ and the binomial theorem we get for any $\lambda > 0$

$$\frac{1}{2}\|S_\varepsilon\|^2_{\Omega^{ER}_\varepsilon} + \frac{1}{2}\|S_{C,\varepsilon}\|^2_{\Omega^{ER}_\varepsilon} + D_S\|\nabla S_\varepsilon\|^2_{\Omega^{ER}_\varepsilon, t} + D_{SC}\|\nabla S_{C,\varepsilon}\|^2_{\Omega^{ER}_\varepsilon, t}$$
$$+ k_O^+\|S_\varepsilon\|^2_{\Gamma^1 \cap \partial\Omega^{ER}_\varepsilon, t} + k_{CO}^+\|S_{C,\varepsilon}\|^2_{\Gamma^1 \cap \partial\Omega^{ER}_\varepsilon, t}$$
$$\leq \frac{1}{2}\|S_\varepsilon(0)\|^2_{\Omega^{ER}_\varepsilon} + \frac{1}{2}\|S_{C,\varepsilon}(0)\|^2_{\Omega^{ER}_\varepsilon} + c_1\|S_\varepsilon\|^2_{\Omega^{ER}_\varepsilon, t} + c_2\|S_{C,\varepsilon}\|^2_{\Omega^{ER}_\varepsilon, t}$$
$$+ c_3\frac{1}{\lambda}\|S_\varepsilon\|^2_{\Gamma^1 \cap \partial\Omega^{ER}_\varepsilon, t} + c_4\lambda\|S_{O,\varepsilon}\|^2_{\Gamma^1 \cap \partial\Omega^{ER}_\varepsilon, t}$$
$$+ c_5\frac{1}{\lambda}\|S_{C,\varepsilon}\|^2_{\Gamma^1 \cap \partial\Omega^{ER}_\varepsilon, t} + c_6\lambda\|S_{CO,\varepsilon}\|^2_{\Gamma^1 \cap \partial\Omega^{ER}_\varepsilon, t}.$$

Using the trace inequality with lemma 4.13 leads to

$$
\frac{1}{2}\|S_\varepsilon\|^2_{\Omega_\varepsilon^{ER}} + \frac{1}{2}\|S_{C,\varepsilon}\|^2_{\Omega_\varepsilon^{ER}} + D_S\|\nabla S_\varepsilon\|^2_{\Omega_\varepsilon^{ER},t} + D_{SC}\|\nabla S_{C,\varepsilon}\|^2_{\Omega_\varepsilon^{ER},t}
$$
$$
+ k_O^+\|S_\varepsilon\|^2_{\Gamma^1 \cap \partial\Omega_\varepsilon^{ER},t} + k_{CO}^+\|S_{C,\varepsilon}\|^2_{\Gamma^1 \cap \partial\Omega_\varepsilon^{ER},t}
$$
$$
\leq \frac{1}{2}\|S_\varepsilon(0)\|^2_{\Omega_\varepsilon^{ER}} + \frac{1}{2}\|S_{C,\varepsilon}(0)\|^2_{\Omega_\varepsilon^{ER}} + c_1\|S_\varepsilon\|^2_{\Omega_\varepsilon^{ER},t} + c_2\|S_{C,\varepsilon}\|^2_{\Omega_\varepsilon^{ER},t}
$$
$$
+ c_3 c_0 \frac{1}{\lambda}\|S_\varepsilon\|^2_{\Omega_\varepsilon^{ER},t} + c_3 c_0 \frac{1}{\lambda}\|\nabla S_\varepsilon\|^2_{\Omega_\varepsilon^{ER},t} + c_4\lambda\|S_{O,\varepsilon}\|^2_{\Gamma^1 \cap \partial\Omega_\varepsilon^{ER},t}
$$
$$
+ c_5 \frac{1}{\lambda}c_0\|S_{C,\varepsilon}\|^2_{\Omega_\varepsilon^{ER},t} + c_5 \frac{1}{\lambda}c_0\|\nabla S_{C,\varepsilon}\|^2_{\Omega_\varepsilon^{ER},t} + c_6\lambda\|S_{CO,\varepsilon}\|^2_{\Gamma^1 \cap \partial\Omega_\varepsilon^{ER},t}.
$$

Merging the constants and dropping some positive terms yields

$$
\|S_\varepsilon\|^2_{\Omega_\varepsilon^{ER}} + \|S_{C,\varepsilon}\|^2_{\Omega_\varepsilon^{ER}} + \left(D_S - c_1 c_0 \frac{1}{\lambda}\right)\|\nabla S_\varepsilon\|^2_{\Omega_\varepsilon^{ER},t}
$$
$$
+ \left(D_{SC} - c_2 c_0 \frac{1}{\lambda}\right)\|\nabla S_{C,\varepsilon}\|^2_{\Omega_\varepsilon^{ER},t} \leq c_3 + c_4\|S_\varepsilon\|^2_{\Omega_\varepsilon^{ER},t}
$$
$$
+ c_5\|S_{C,\varepsilon}\|^2_{\Omega_\varepsilon^{ER},t} + c_6\lambda\|S_{O,\varepsilon}\|^2_{\Gamma^1 \cap \partial\Omega_\varepsilon^{ER},t} + c_7\lambda\|S_{CO,\varepsilon}\|^2_{\Gamma^1 \cap \partial\Omega_\varepsilon^{ER},t}.
$$

Now we do similar estimates for the equations for $S_{O,\varepsilon}$ and $S_{CO,\varepsilon}$ and find for any $\lambda > 0$

$$
\frac{1}{2}\|S_{O,\varepsilon}\|^2_{\Gamma^1 \cap \partial\Omega_\varepsilon^{ER}} + \frac{1}{2}\|S_{CO,\varepsilon}\|^2_{\Gamma^1 \cap \partial\Omega_\varepsilon^{ER}}
$$
$$
\leq \frac{1}{2}\|S_{O,\varepsilon}(0)\|^2_{\Gamma^1 \cap \partial\Omega_\varepsilon^{ER}} + \frac{1}{2}\|S_{CO,\varepsilon}(0)\|^2_{\Gamma^1 \cap \partial\Omega_\varepsilon^{ER}} + c_1\lambda\|S_{O,\varepsilon}\|^2_{\Gamma^1 \cap \partial\Omega_\varepsilon^{ER},t}
$$
$$
+ c_2\lambda\|S_{CO,\varepsilon}\|^2_{\Gamma^1 \cap \partial\Omega_\varepsilon^{ER},t} + c_3 c_0 \frac{1}{\lambda}\|S_\varepsilon\|^2_{\Omega_\varepsilon^{ER},t} + c_3 c_0 \frac{1}{\lambda}\|\nabla S_\varepsilon\|^2_{\Omega_\varepsilon^{ER},t}
$$
$$
+ c_4 \frac{1}{\lambda}c_0\|S_{C,\varepsilon}\|^2_{\Omega_\varepsilon^{ER},t} + c_4 \frac{1}{\lambda}c_0\|\nabla S_{C,\varepsilon}\|^2_{\Omega_\varepsilon^{ER},t}.
$$

We sum up all inequalities to find

$$\|S_\varepsilon\|^2_{\Omega^{ER}_\varepsilon} + \|S_{C,\varepsilon}\|^2_{\Omega^{ER}_\varepsilon} + \left(D_S - c_0 c_1 \frac{1}{\lambda}\right) \|\nabla S_\varepsilon\|^2_{\Omega^{ER}_\varepsilon,t}$$
$$+ \left(D_{SC} - c_2 c_0 \frac{1}{\lambda}\right) \|\nabla S_{C,\varepsilon}\|^2_{\Omega^{ER}_\varepsilon,t} + \|S_{O,\varepsilon}\|^2_{\Gamma^1 \cap \partial\Omega^{ER}_\varepsilon} + \|S_{CO,\varepsilon}\|^2_{\Gamma^1 \cap \partial\Omega^{ER}_\varepsilon}$$
$$\leq c_3 + c_4 \|S_\varepsilon\|^2_{\Omega^{ER}_\varepsilon,t} + c_5 \|S_{C,\varepsilon}\|^2_{\Omega^{ER}_\varepsilon,t}$$
$$+ c_6 \lambda \|S_{O,\varepsilon}\|^2_{\Gamma^1 \cap \partial\Omega^{ER}_\varepsilon,t} + c_7 \lambda \|S_{CO,\varepsilon}\|^2_{\Gamma^1 \cap \partial\Omega^{ER}_\varepsilon,t}.$$

For $\lambda$ greater than $c_0 c_2 + c_0 c_1$, but finite, Gronwall's lemma yields

$$\|S_\varepsilon\|^2_{\Omega^{ER}_\varepsilon} + \|S_{C,\varepsilon}\|^2_{\Omega^{ER}_\varepsilon} + \|\nabla S_\varepsilon\|^2_{\Omega^{ER}_\varepsilon,t} + \|\nabla S_{C,\varepsilon}\|^2_{\Omega^{ER}_\varepsilon,t}$$
$$+ \|S_{O,\varepsilon}\|^2_{\Gamma^1 \cap \partial\Omega^{ER}_\varepsilon} + \|S_{CO,\varepsilon}\|^2_{\Gamma^1 \cap \partial\Omega^{ER}_\varepsilon} \leq C.$$

□

Next, we want to show that the functions that represent the concentration of Stim1 molecules are bounded, i.e. they are $L^\infty$ functions.

**Lemma 4.17 (Boundedness of $S_\varepsilon$, $S_{C,\varepsilon}$, $S_{O,\varepsilon}$ and $S_{CO,\varepsilon}$ in $L^\infty$)**
*There exists a constant $C > 0$, independent of $\varepsilon$, such that*

$$\|S_\varepsilon\|_{L^\infty(\Omega^{ER}_\varepsilon)} + \|S_{C,\varepsilon}\|_{L^\infty(\Omega^{ER}_\varepsilon)}$$
$$+ \|S_{O,\varepsilon}\|_{L^\infty(\Gamma^1 \cap \partial\Omega^{ER}_\varepsilon)} + \|S_{CO,\varepsilon}\|_{L^\infty(\Gamma^1 \cap \partial\Omega^{ER}_\varepsilon)} < C$$

*for almost every $t \in [0, T]$.*

**Proof** Let $k > 0$. We define the function

$$M(t) = \max\{\|S_\varepsilon(0)\|_{L^\infty}, \|S_{C,\varepsilon}(0)\|_{L^\infty}, \|S_{O,\varepsilon}(0)\|_{L^\infty}, \|S_{CO,\varepsilon}(0)\|_{L^\infty}\} e^{kt}$$

with $t \in [0, T]$. Note that the initial values are $L^\infty$ functions, so that $M$ is well-defined and finite for all $t \in [0, T]$.
We test the equations for $S_\varepsilon$ and $S_{C,\varepsilon}$ with $(k_O^+ S_\varepsilon - M)_+$ and

$(k_{CO}^+ S_{C,\varepsilon} - M)_+$, respectively, and add the results to obtain

$$(\partial_t S_\varepsilon, (k_O^+ S_\varepsilon - M)_+)_{\Omega_\varepsilon^{ER}} + (\partial_t S_{C,\varepsilon}, (k_{CO}^+ S_{C,\varepsilon} - M)_+)_{\Omega_\varepsilon^{ER}}$$

$$+ D_S(\nabla S_\varepsilon, \nabla(k_O^+ S_\varepsilon - M)_+)_{\Omega_\varepsilon^{ER}} + D_{SC}(\nabla S_{C,\varepsilon}, \nabla(k_{CO}^+ S_{C,\varepsilon} - M)_+)_{\Omega_\varepsilon^{ER}}$$

$$+ (k_C^+ f_e(C_{e,\varepsilon}) S_\varepsilon - k_C^- S_{C,\varepsilon}, (k_O^+ S_\varepsilon - M)_+ - (k_{CO}^+ S_{C,\varepsilon} - M)_+)_{\Omega_\varepsilon^{ER}}$$

$$+ \langle k_O^+ S_\varepsilon - k_O^- S_{O,\varepsilon}, (k_O^+ S_\varepsilon - M)_+ \rangle_{\Gamma^1 \cap \partial\Omega_\varepsilon^{ER}}$$

$$+ \langle k_{CO}^+ S_{C,\varepsilon} - k_{CO}^- S_{CO,\varepsilon}, (k_{CO}^+ S_{C,\varepsilon} - M)_+ \rangle_{\Gamma^1 \cap \partial\Omega_\varepsilon^{ER}} = 0.$$

Next, the equations for $S_{O,\varepsilon}$ and $S_{CO,\varepsilon}$ are tested with $(k_O^- S_{O,\varepsilon} - M)_+$ and $(k_{CO}^- S_{CO,\varepsilon} - M)_+$, respectively.

$$\langle \partial_t S_{O,\varepsilon}, (k_O^- S_{O,\varepsilon} - M)_+ \rangle_{\Gamma^1 \cap \partial\Omega_\varepsilon^{ER}} + \langle \partial_t S_{CO,\varepsilon}, (k_{CO}^- S_{CO,\varepsilon} - M)_+ \rangle_{\Gamma^1 \cap \partial\Omega_\varepsilon^{ER}}$$

$$+ \langle k_C^+ f_e(C_{e,\varepsilon}) S_{O,\varepsilon} - k_C^- S_{CO,\varepsilon}, (k_O^- S_{O,\varepsilon} - M)_+ - (k_{CO}^- S_{CO,\varepsilon} - M)_+ \rangle_{\Gamma^1 \cap \partial\Omega_\varepsilon^{ER}}$$

$$+ \langle k_O^- S_{O,\varepsilon} - k_O^+ S_\varepsilon, (k_O^- S_{O,\varepsilon} - M)_+ \rangle_{\Gamma^1 \cap \partial\Omega_\varepsilon^{ER}}$$

$$+ \langle k_{CO}^- S_{CO,\varepsilon} - k_{CO}^+ S_{C,\varepsilon}, (k_{CO}^- S_{CO,\varepsilon} - M)_+ \rangle_{\Gamma^1 \cap \partial\Omega_\varepsilon^{ER}} = 0$$

Adding the two reults above and considering the time and spatial derivatives of $M$ leads to

$$1/k_O^+ (\partial_t(k_O^+ S_\varepsilon - M)_+, (k_O^+ S_\varepsilon - M)_+)_{\Omega_\varepsilon^{ER}}$$

$$+ 1/k_{CO}^+ (\partial_t(k_{CO}^+ S_{C,\varepsilon} - M)_+, (k_{CO}^+ S_{C,\varepsilon} - M)_+)_{\Omega_\varepsilon^{ER}}$$

$$+ 1/k_O^+ D_S \|\nabla(k_O^+ S_\varepsilon - M)_+\|_{\Omega_\varepsilon^{ER}}^2 + 1/k_{CO}^+ D_{SC} \|\nabla(k_{CO}^+ S_{C,\varepsilon} - M)_+\|_{\Omega_\varepsilon^{ER}}^2$$

$$+ 1/k_O^- \langle \partial_t(k_O^- S_{O,\varepsilon} - M)_+, (k_O^- S_{O,\varepsilon} - M)_+ \rangle_{\Gamma^1 \cap \partial\Omega_\varepsilon^{ER}}$$

$$+ 1/k_{CO}^- \langle \partial_t(k_{CO}^- S_{CO,\varepsilon} - M)_+, (k_{CO}^- S_{CO,\varepsilon} - M)_+ \rangle_{\Gamma^1 \cap \partial\Omega_\varepsilon^{ER}}$$

$$+ (k_C^+ f_e(C_{e,\varepsilon}) S_\varepsilon - k_C^- S_{C,\varepsilon}, (k_O^+ S_\varepsilon - M)_+ - (k_{CO}^+ S_{C,\varepsilon} - M)_+)_{\Omega_\varepsilon^{ER}}$$

$$+ \langle k_C^+ f_e(C_{e,\varepsilon}) S_{O,\varepsilon} - k_C^- S_{CO,\varepsilon}, (k_O^- S_{O,\varepsilon} - M)_+ - (k_{CO}^- S_{CO,\varepsilon} - M)_+ \rangle_{\Gamma^1 \cap \partial\Omega_\varepsilon^{ER}}$$

$$+ \|(k_O^+ S_\varepsilon - M)_+ - (k_O^- S_{O,\varepsilon} - M)_+\|_{\Gamma^1 \cap \partial\Omega_\varepsilon^{ER}}^2$$

$$+ \|(k_{CO}^+ S_{C,\varepsilon} - M)_+ - (k_{CO}^- S_{CO,\varepsilon} - M)_+\|_{\Gamma^1 \cap \partial\Omega_\varepsilon^{ER}}^2$$

$$\leq -(1/k_O^+ kM, (k_O^+ S_\varepsilon - M)_+)_{\Omega_\varepsilon^{ER}} - (1/k_{CO}^+ kM, (k_{CO}^+ S_{C,\varepsilon} - M)_+)_{\Omega_\varepsilon^{ER}}$$

$$- \langle 1/k_O^- kM, (k_O^- S_{O,\varepsilon} - M)_+ \rangle_{\Gamma^1 \cap \partial\Omega_\varepsilon^{ER}} - \langle 1/k_{CO}^- kM, (k_{CO}^- S_{CO,\varepsilon} - M)_+ \rangle_{\Gamma^1 \cap \partial\Omega_\varepsilon^{ER}}.$$

We drop some positive terms, integrate from 0 to $t$ and use the binomial theorem. With $f_e(C_{e,\varepsilon})$ bounded and $\|(\varphi - M)_+\|_{\Omega_\varepsilon^{ER}} \leq \|\varphi\|_{\Omega_\varepsilon^{ER}}$ for $\varphi = S_\varepsilon, S_{C,\varepsilon}$ and $\|(\varphi - M)_+\|_{\Gamma^1 \cap \partial \Omega_\varepsilon^{ER}} \leq \|\varphi\|_{\Gamma^1 \cap \partial \Omega_\varepsilon^{ER}}$ for $\varphi = S_{O,\varepsilon}, S_{CO,\varepsilon}$ we find

$$
\begin{aligned}
& {}^{1\!}/{}_{2}k_O^+ \|(k_O^+ S_\varepsilon - M)_+\|^2_{\Omega_\varepsilon^{ER}} + {}^{1\!}/{}_{2}k_{CO}^+ \|(k_{CO}^+ S_{C,\varepsilon} - M)_+\|^2_{\Omega_\varepsilon^{ER}} \\
& + D_S {}^{1\!}/{}k_O^+ \|\nabla (k_O^+ S_\varepsilon - M)_+\|^2_{\Omega_\varepsilon^{ER},t} + D_{SC} {}^{1\!}/{}k_{CO}^+ \|\nabla (k_{CO}^+ S_{C,\varepsilon} - M)_+\|^2_{\Omega_\varepsilon^{ER},t} \\
& + {}^{1\!}/{}_{2}k_O^- \|(k_O^- S_{O,\varepsilon} - M)_+\|^2_{\Gamma^1 \cap \partial \Omega_\varepsilon^{ER}} + {}^{1\!}/{}_{2}k_{CO}^- \|(k_{CO}^- S_{CO,\varepsilon} - M)_+\|^2_{\Gamma^1 \cap \partial \Omega_\varepsilon^{ER}} \\
& \leq \underbrace{c_1 \|S_\varepsilon\|^2_{\Omega_\varepsilon^{ER},t} + c_2 \|S_{C,\varepsilon}\|^2_{\Omega_\varepsilon^{ER},t} + c_3 \|S_{O,\varepsilon}\|^2_{\Gamma^1 \cap \partial \Omega_\varepsilon^{ER},t} + c_4 \|S_{CO,\varepsilon}\|^2_{\Gamma^1 \cap \partial \Omega_\varepsilon^{ER},t}}_{\text{bounded, because of lemma 4.16}} \\
& - ({}^{1\!}/{}k_O^+ kM, (k_O^+ S_\varepsilon - M)_+)_{\Omega_\varepsilon^{ER},t} - ({}^{1\!}/{}k_{CO}^+ kM, (k_{CO}^+ S_{C,\varepsilon} - M)_+)_{\Omega_\varepsilon^{ER},t} \\
& \qquad - \langle {}^{1\!}/{}k_O^- kM, (k_O^- S_{O,\varepsilon} - M)_+ \rangle_{\Gamma^1 \cap \partial \Omega_\varepsilon^{ER},t} \\
& \qquad\qquad - \langle {}^{1\!}/{}k_{CO}^- kM, (k_{CO}^- S_{CO,\varepsilon} - M)_+ \rangle_{\Gamma^1 \cap \partial \Omega_\varepsilon^{ER},t}.
\end{aligned}
$$

Hence,

$$
\begin{aligned}
& {}^{1\!}/{}_{2}k_O^+ \|(k_O^+ S_\varepsilon - M)_+\|^2_{\Omega_\varepsilon^{ER}} + {}^{1\!}/{}_{2}k_{CO}^+ \|(k_{CO}^+ S_{C,\varepsilon} - M)_+\|^2_{\Omega_\varepsilon^{ER}} \\
& + {}^{1\!}/{}_{2}k_O^- \|(k_O^- S_{O,\varepsilon} - M)_+\|^2_{\Gamma^1 \cap \partial \Omega_\varepsilon^{ER}} + {}^{1\!}/{}_{2}k_{CO}^- \|(k_{CO}^- S_{CO,\varepsilon} - M)_+\|^2_{\Gamma^1 \cap \partial \Omega_\varepsilon^{ER}} \\
& \leq c_1 - ({}^{1\!}/{}k_O^+ kM, (k_O^+ S_\varepsilon - M)_+)_{\Omega_\varepsilon^{ER},t} - ({}^{1\!}/{}k_{CO}^+ kM, (k_{CO}^+ S_{C,\varepsilon} - M)_+)_{\Omega_\varepsilon^{ER},t} \\
& \qquad - \langle {}^{1\!}/{}k_O^- kM, (k_O^- S_{O,\varepsilon} - M)_+ \rangle_{\Gamma^1 \cap \partial \Omega_\varepsilon^{ER},t} \\
& \qquad\qquad - \langle {}^{1\!}/{}k_{CO}^- kM, (k_{CO}^- S_{CO,\varepsilon} - M)_+ \rangle_{\Gamma^1 \cap \partial \Omega_\varepsilon^{ER},t}.
\end{aligned}
$$

Now we distinguish two cases.

- There exists a non-nullset $V \subset \Omega_\varepsilon^{ER}$ such that $(k_O^+ S_\varepsilon - M)_+ > 0$ in $V$ or $(k_{CO}^+ S_{C,\varepsilon} - M)_+ > 0$ in $V$; or there exists a non-nullset $V \subset \Gamma^1 \cap \partial \Omega_\varepsilon^{ER}$ such that $(k_O^- S_{O,\varepsilon} - M)_+ > 0$ in $V$ or $(k_{CO}^- S_{CO,\varepsilon} - M)_+ > 0$ in $V$.
  Then there exists a $\delta > 0$ such that

$$
({}^{1\!}/{}k_O^+ M, (k_O^+ S_\varepsilon - M)_+)_{\Omega_\varepsilon^{ER},t} + ({}^{1\!}/{}k_{CO}^+ M, (k_{CO}^+ S_{C,\varepsilon} - M)_+)_{\Omega_\varepsilon^{ER},t} > \delta
$$

or

$$\langle 1/k_O^- M, (k_O^- S_{O,\varepsilon} - M)_+ \rangle_{\Gamma^1 \cap \partial\Omega_\varepsilon^{ER}, t}$$
$$+ \langle 1/k_{CO}^- M, (k_{CO}^- S_{CO,\varepsilon} - M)_+ \rangle_{\Gamma^1 \cap \partial\Omega_\varepsilon^{ER}, t} > \delta.$$

We choose $k$ to be $k\delta > c_1$, which is possible since $k$ and $\delta$ is growing with $k$, and we find

$$1/2k_O^+ \|(k_O^+ S_\varepsilon - M)_+\|_{\Omega_\varepsilon^{ER}}^2 + 1/2k_{CO}^+ \|(k_{CO}^+ S_{C,\varepsilon} - M)_+\|_{\Omega_\varepsilon^{ER}}^2$$
$$+ 1/2k_O^- \|(k_O^- S_{O,\varepsilon} - M)_+\|_{\Gamma^1 \cap \partial\Omega_\varepsilon^{ER}}^2$$
$$+ 1/2k_{CO}^- \|(k_{CO}^- S_{CO,\varepsilon} - M)_+\|_{\Gamma^1 \cap \partial\Omega_\varepsilon^{ER}}^2 \leq 0.$$

That contradicts the existence of such a subset $V$ and the proof is complete.

- Otherwise it holds that $(k_O^+ S_\varepsilon - M)_+ \leq 0$, $(k_{CO}^+ S_{C,\varepsilon} - M)_+ \leq 0$, $(k_O^- S_{O,\varepsilon} - M)_+ \leq 0$ and $(k_{CO}^- S_{CO,\varepsilon} - M)_+ \leq 0$ almost everywhere and we are finished.

$\square$

This lemma 4.17 yields that $I_{CRAC}$ is bounded on $\Gamma^1 \cap \partial\Omega_\varepsilon^{ER}$ using lemma 4.14. Now we show that $C_\varepsilon$ and $C_{e,\varepsilon}$ are $H^1$-functions to be able to use proposition 2.5 and lemma 3.4.

**Lemma 4.18 (Boundedness of $C_\varepsilon$ and $C_{e,\varepsilon}$ in $H^1$)**
*It holds that*

$$\|C_\varepsilon\|_{\Omega_\varepsilon^1}^2 + \|C_{e,\varepsilon}\|_{\Omega_\varepsilon^2}^2 + \|\nabla C_\varepsilon\|_{\Omega_\varepsilon^1, t}^2$$
$$+ \|\nabla C_{e,\varepsilon}\|_{\Omega_\varepsilon^2, t}^2 + \varepsilon\|C_\varepsilon - C_{e,\varepsilon}\|_{\Gamma_\varepsilon^{ER}, t}^2 \leq C$$

*for a constant $C > 0$, independent of $\varepsilon$.*

**Proof** We test the weak formulation for $C_\varepsilon$ with $C_\varepsilon$ and get

$$(\partial_t C_\varepsilon, C_\varepsilon)_{\Omega_\varepsilon^1} + D_C \|\nabla C_\varepsilon\|_{\Omega_\varepsilon^1}^2 + \varepsilon(L_0 + L_{IP3})\langle C_\varepsilon - C_{e,\varepsilon}, C_\varepsilon \rangle_{\Gamma_\varepsilon^{ER}}$$
$$+ \varepsilon\langle f_{SERCA}, C_\varepsilon \rangle_{\Gamma_\varepsilon^{ER}} + \langle \alpha I_{CRAC} + f_P + f_{NCX}, C_\varepsilon \rangle_{\Gamma^1 \cap \partial\Omega_\varepsilon^{ER}} = 0.$$

Also, we test the weak formulation for $C_{e,\varepsilon}$ with $C_{e,\varepsilon}$ and get

$$(\partial_t C_{e,\varepsilon}, C_{e,\varepsilon})_{\Omega_\varepsilon^2} + D_{\mathrm{ER}}\|\nabla C_{e,\varepsilon}\|_{\Omega_\varepsilon^2}^2$$
$$+ \varepsilon(L_0 + L_{\mathrm{IP3}})\langle C_{e,\varepsilon} - C_\varepsilon, C_{e,\varepsilon}\rangle_{\Gamma_\varepsilon^{\mathrm{ER}}} - \varepsilon\langle f_{\mathrm{SERCA}}, C_{e,\varepsilon}\rangle_{\Gamma_\varepsilon^{\mathrm{ER}}} = 0.$$

Adding the equations gives for any $\lambda > 0$

$$(\partial_t C_\varepsilon, C_\varepsilon)_{\Omega_\varepsilon^1} + D_C\|\nabla C_\varepsilon\|_{\Omega_\varepsilon^1}^2 + (\partial_t C_{e,\varepsilon}, C_{e,\varepsilon})_{\Omega_\varepsilon^2} + D_{\mathrm{ER}}\|\nabla C_{e,\varepsilon}\|_{\Omega_\varepsilon^2}^2$$
$$+ \varepsilon(L_0 + L_{\mathrm{IP3}})\|C_\varepsilon - C_{e,\varepsilon}\|_{\Gamma_\varepsilon^{\mathrm{ER}}}^2$$
$$= -\langle f_P + f_{\mathrm{NCX}} + \alpha I_{\mathrm{CRAC}}, C_\varepsilon\rangle_{\Gamma^1 \cap \partial\Omega_\varepsilon^1} + \varepsilon\langle f_{\mathrm{SERCA}}, C_{e,\varepsilon} - C_\varepsilon\rangle_{\Gamma_\varepsilon^{\mathrm{ER}}}$$
$$\leq \frac{\lambda}{2}\|f_P + f_{\mathrm{NCX}} + \alpha I_{\mathrm{CRAC}}\|_{\Gamma^1 \cap \partial\Omega_\varepsilon^1}^2 + \frac{1}{2\lambda}\|C_\varepsilon\|_{\Gamma^1 \cap \partial\Omega_\varepsilon^1}^2$$
$$+ \varepsilon\|f_{\mathrm{SERCA}}\|_{\Gamma_\varepsilon^{\mathrm{ER}}}^2 + \varepsilon\|C_\varepsilon\|_{\Gamma_\varepsilon^{\mathrm{ER}}}^2 + \varepsilon\|C_{e,\varepsilon}\|_{\Gamma_\varepsilon^{\mathrm{ER}}}^2$$
$$\leq \underbrace{\frac{\lambda}{2}\|f_P + f_{\mathrm{NCX}} + \alpha I_{\mathrm{CRAC}}\|_{\Gamma^1 \cap \partial\Omega_\varepsilon^1}^2 + \varepsilon\|f_{\mathrm{SERCA}}\|_{\Gamma_\varepsilon^2}^2}_{\leq c_1}$$
$$+ \frac{c_0}{2\lambda}\left(\|C_\varepsilon\|_{\Omega_\varepsilon^1}^2 + \|\nabla C_\varepsilon\|_{\Omega_\varepsilon^1}^2\right) + c_0\left(\|C_{e,\varepsilon}\|_{\Omega_\varepsilon^2}^2 + \varepsilon^2\|\nabla C_{e,\varepsilon}\|_{\Omega_\varepsilon^2}^2\right)$$
$$+ c_0\left(\|C_\varepsilon\|_{\Omega_\varepsilon^1}^2 + \varepsilon^2\|\nabla C_\varepsilon\|_{\Omega_\varepsilon^1}^2\right).$$

Note, that we used two different trace inequalities, see lemma 2.11 and lemma 4.13. We deduce

$$(\partial_t C_\varepsilon, C_\varepsilon)_{\Omega_\varepsilon^1} + \left(D_C - \frac{c_0}{2\lambda} - \varepsilon^2 c_0\right)\|\nabla C_\varepsilon\|_{\Omega_\varepsilon^1}^2 + (\partial_t C_{e,\varepsilon}, C_{e,\varepsilon})_{\Omega_\varepsilon^2}$$
$$+ \left(D_{\mathrm{ER}} - c_0\varepsilon^2\right)\|\nabla C_{e,\varepsilon}\|_{\Omega_\varepsilon^2}^2 + \varepsilon(L_0 + L_{\mathrm{IP3}})\|C_\varepsilon - C_{e,\varepsilon}\|_{\Gamma_\varepsilon^{\mathrm{ER}}}^2$$
$$\leq \lambda c_1 + c_2\|C_\varepsilon\|_{\Omega_\varepsilon^1}^2 + c_3\|C_{e,\varepsilon}\|_{\Omega_\varepsilon^2}^2$$

for some constants $c_1$, $c_2$ and $c_3$. By integration from 0 to $t$ we get

$$\frac{1}{2}\|C_\varepsilon\|_{\Omega_\varepsilon^1}^2 + \left(D_C - \frac{c_0}{2\lambda} - \varepsilon^2 c_0\right)\|\nabla C_\varepsilon\|_{\Omega_\varepsilon^1, t}^2 + \frac{1}{2}\|C_{e,\varepsilon}\|_{\Omega_\varepsilon^2}^2$$
$$+ \left(D_{\mathrm{ER}} - c_0\varepsilon^2\right)\|\nabla C_{e,\varepsilon}\|_{\Omega_\varepsilon^2, t}^2 + \varepsilon(L_0 + L_{\mathrm{IP3}})\|C_\varepsilon - C_{e,\varepsilon}\|_{\Gamma_\varepsilon^{\mathrm{ER}}, t}^2$$
$$\leq \lambda c_1 + c_2\|C_\varepsilon\|_{\Omega_\varepsilon^1, t}^2 + c_3\|C_{e,\varepsilon}\|_{\Omega_\varepsilon^2, t}^2 + \frac{1}{2}\|C_\varepsilon(0)\|_{\Omega_\varepsilon^1}^2 + \frac{1}{2}\|C_{e,\varepsilon}(0)\|_{\Omega_\varepsilon^2}^2.$$

For $\lambda$ big enough but finite, and small $\varepsilon$ we conclude with the lemma of Gronwall that

$$\|C_\varepsilon\|^2_{\Omega^1_\varepsilon} + \|\nabla C_\varepsilon\|^2_{\Omega^1_\varepsilon,t} + \|C_{e,\varepsilon}\|^2_{\Omega^2_\varepsilon}$$
$$+ \|\nabla C_{e,\varepsilon}\|^2_{\Omega^2_\varepsilon,t} + \varepsilon\|C_\varepsilon - C_{e,\varepsilon}\|^2_{\Gamma^{ER}_\varepsilon,t} \leq C$$

for a merged constant C. $\qquad\square$

Next, we show that $C_\varepsilon$ and $C_{e,\varepsilon}$ are also bounded in $L^\infty$. This assertion is needed to apply lemma 3.4 for strong convergence.

**Lemma 4.19 (Boundedness of $C_\varepsilon$ and $C_{e,\varepsilon}$ in $L^\infty$)**
*There exists a constant $C > 0$, independent of $\varepsilon$, such that*

$$\|C_\varepsilon\|_{L^\infty(\Omega^1_\varepsilon)} + \|C_{e,\varepsilon}\|_{L^\infty(\Omega^2_\varepsilon)} < C$$

*for almost every $t \in [0, T]$.*

**Proof** Let $k > 0$. We define the function $M(t) := \max\{\|C_\varepsilon(0)\|_{L^\infty}, \|C_{e,\varepsilon}(0)\|_{L^\infty}\}e^{kt}$ and test the equation for $C_\varepsilon$ with $(C_\varepsilon - M)_+$.

$$(\partial_t C_\varepsilon, (C_\varepsilon - M)_+)_{\Omega^1_\varepsilon} + D_C(\nabla C_\varepsilon, \nabla(C_\varepsilon - M)_+)_{\Omega^1_\varepsilon}$$
$$+ \varepsilon(L_0 + L_{IP3})\langle C_\varepsilon - C_{e,\varepsilon}, (C_\varepsilon - M)_+\rangle_{\Gamma^{ER}_\varepsilon}$$
$$= \underbrace{-\varepsilon\langle f_{SERCA}, (C_\varepsilon - M)_+\rangle_{\Gamma^{ER}_\varepsilon} - \langle f_P + f_{NCX}, (C_\varepsilon - M)_+\rangle_{\Gamma^1 \cap \partial\Omega^1_\varepsilon}}_{\leq 0}$$
$$- \alpha\langle I_{CRAC}, (C_\varepsilon - M)_+\rangle_{\Gamma^1 \cap \partial\Omega^1_\varepsilon}.$$

Considering the time and spatial derivatives of $(C_\varepsilon - M)_+$ gives

$$(\partial_t(C_\varepsilon - M)_+, (C_\varepsilon - M)_+)_{\Omega^1_\varepsilon} + D_C(\nabla(C_\varepsilon - M)_+, \nabla(C_\varepsilon - M)_+)_{\Omega^1_\varepsilon}$$
$$+ \varepsilon(L_0 + L_{IP3})\langle C_\varepsilon - C_{e,\varepsilon}, (C_\varepsilon - M)_+\rangle_{\Gamma^{ER}_\varepsilon}$$
$$\leq \alpha\langle |I_{CRAC}|, (C_\varepsilon - M)_+\rangle_{\Gamma^1 \cap \partial\Omega^1_\varepsilon} - (kM, (C_\varepsilon - M)_+)_{\Omega^1_\varepsilon}$$

Now we test the equation for $C_{e,\varepsilon}$ with $(C_{e,\varepsilon} - M)_+$ and get

$$(\partial_t(C_{e,\varepsilon} - M)_+, (C_{e,\varepsilon} - M)_+)_{\Omega^2_\varepsilon}$$
$$+ D_{ER}(\nabla(C_{e,\varepsilon} - M)_+, \nabla(C_{e,\varepsilon} - M)_+)_{\Omega^2_\varepsilon}$$
$$+ \varepsilon(L_0 + L_{IP3})\langle C_{e,\varepsilon} - C_\varepsilon, (C_{e,\varepsilon} - M)_+\rangle_{\Gamma^{ER}_\varepsilon} = -(kM, (C_{e,\varepsilon} - M)_+)_{\Omega^2_\varepsilon}.$$

We add the two results and integrate from 0 to $t$

$$\frac{1}{2}\|(C_\varepsilon - M)_+\|^2_{\Omega^1_\varepsilon} + D_C\|\nabla(C_\varepsilon - M)_+\|^2_{\Omega^1_\varepsilon,t}$$

$$+ \frac{1}{2}\|(C_{e,\varepsilon} - M)_+\|^2_{\Omega^2_\varepsilon} + D_{ER}\|\nabla(C_{e,\varepsilon} - M)_+\|^2_{\Omega^2_\varepsilon,t}$$

$$+ \varepsilon(L_0 + L_{IP3})\langle((C_\varepsilon - M) - (C_{e,\varepsilon} - M), (C_\varepsilon - M)_+ - (C_{e,\varepsilon} - M)_+\rangle_{\Gamma^{ER}_\varepsilon,t}$$

$$\leq \alpha\langle|I_{CRAC}|, (C_\varepsilon - M)_+\rangle_{\Gamma^1\cap\partial\Omega^1_\varepsilon,t} - (kM, (C_\varepsilon - M)_+)_{\Omega^1_\varepsilon,t}$$

$$- (kM, (C_{e,\varepsilon} - M)_+)_{\Omega^2_\varepsilon,t}.$$

Since

$$\langle(C_\varepsilon - M) - (C_{e,\varepsilon} - M), (C_\varepsilon - M)_+ - (C_{e,\varepsilon} - M)_+\rangle_{\Gamma^{ER}_\varepsilon,t}$$

$$= \|(C_\varepsilon - M)_+ - (C_{e,\varepsilon} - M)_+\|^2_{\Gamma^{ER}_\varepsilon,t} + \langle(C_\varepsilon - M)_-, (C_{e,\varepsilon} - M)_+\rangle_{\Gamma^{ER}_\varepsilon,t}$$

$$+ \langle(C_{e,\varepsilon} - M)_-, (C_\varepsilon - M)_+\rangle_{\Gamma^{ER}_\varepsilon,t}$$

$$\geq \|(C_\varepsilon - M)_+ - (C_{e,\varepsilon} - M)_+\|^2_{\Gamma^{ER}_\varepsilon,t}$$

we continue with

$$\frac{1}{2}\|(C_\varepsilon - M)_+\|^2_{\Omega^1_\varepsilon} + D_C\|\nabla(C_\varepsilon - M)_+\|^2_{\Omega^1_\varepsilon,t}$$

$$+ \frac{1}{2}\|(C_{e,\varepsilon} - M)_+\|^2_{\Omega^2_\varepsilon} + D_{ER}\|\nabla(C_{e,\varepsilon} - M)_+\|^2_{\Omega^2_\varepsilon,t}$$

$$+ \varepsilon(L_0 + L_{IP3})\|(C_\varepsilon - M)_+ - (C_{e,\varepsilon} - M)_+\|^2_{\Gamma^{ER}_\varepsilon,t}$$

$$\leq \alpha\|I_{CRAC}\|^2_{\Gamma^1\cap\partial\Omega^1_\varepsilon,t} + \underbrace{c_0\left(\|(C_\varepsilon - M)_+\|^2_{\Omega^1_\varepsilon,t} + \|\nabla(C_\varepsilon - M)_+\|^2_{\Omega^1_\varepsilon,t}\right)}_{\leq c_1}$$

$$- (kM, (C_\varepsilon - M)_+)_{\Omega^1_\varepsilon,t} - (kM, (C_{e,\varepsilon} - M)_+)_{\Omega^2_\varepsilon,t}$$

Now we distinguish two cases.

- There exists a non-nullset $V \subset \Omega^1_\varepsilon$ or $V \subset \Omega^2_\varepsilon$ with $(C_\varepsilon - M)_+ > 0$ or $(C_{e,\varepsilon} - M)_+ > 0$ in $V$, respectively.
  Then there exists a $\delta > 0$ such that $(M, (C_\varepsilon - M)_+)_{\Omega^1_\varepsilon,t} > \delta$ or $(M, (C_{e,\varepsilon} - M)_+)_{\Omega^2_\varepsilon,t} > \delta$, respectively, and we choose

$k\delta = c_1$. Then it follows that

$$
\frac{1}{2}\|(C_\varepsilon - M)_+\|^2_{\Omega^1_\varepsilon} + D_C\|\nabla(C_\varepsilon - M)_+\|^2_{\Omega^1_\varepsilon,t}
$$

$$
+ \frac{1}{2}\|(C_{e,\varepsilon} - M)_+\|^2_{\Omega^2_\varepsilon} + D_{ER}\|\nabla(C_{e,\varepsilon} - M)_+\|^2_{\Omega^2_\varepsilon,t}
$$

$$
+ \varepsilon(L_0 + L_{IP3})\|(C_\varepsilon - M)_+ - (C_{e,\varepsilon} - M)_+\|^2_{\Gamma^{ER}_\varepsilon,t}
$$

$$
\leq c_1 - (kM, (C_\varepsilon - M)_+)_{\Omega^1_\varepsilon,t} - (kM, (C_{e,\varepsilon} - M)_+)_{\Omega^2_\varepsilon,t} \leq 0.
$$

But this contradicts $(C_\varepsilon - M)_+ > 0$ or $(C_{e,\varepsilon} - M)_+ > 0$ in a non-nullset and we are finished.

- It holds that $(C_\varepsilon - M)_+ = 0$ and $(C_{e,\varepsilon} - M)_+ = 0$ almost everywhere.

From the above we conclude that $C_\varepsilon$ and $C_{e,\varepsilon}$ are bounded from above. Because $C_\varepsilon$ and $C_{e,\varepsilon}$ could be negative, we also show that they have a lower bound. Biologically it does not make sence for $C_\varepsilon$ or $C_{e,\varepsilon}$ to be negative, but the system is created such that mathematically we can not exclude it.

Therefore, we test the weak formulations with $(C_\varepsilon + M)_-$ and $(C_{e,\varepsilon} + M)_-$. We obtain

$$
(\partial_t(C_\varepsilon + M), (C_\varepsilon + M)_-)_{\Omega^1_\varepsilon} + D_C(\nabla(C_\varepsilon + M), \nabla(C_\varepsilon + M)_-)_{\Omega^1_\varepsilon}
$$

$$
+ \varepsilon(L_0 + L_{IP3})\langle C_\varepsilon - C_{e,\varepsilon}, (C_\varepsilon + M)_-\rangle_{\Gamma^{ER}_\varepsilon}
$$

$$
= -\langle f_P + f_{NCX} + \alpha I_{CRAC}, (C_\varepsilon + M)_-\rangle_{\Gamma^1 \cap \partial\Omega^1_\varepsilon}
$$

$$
- \varepsilon\langle f_{SERCA}, (C_\varepsilon + M)_-\rangle_{\Gamma^{ER}_\varepsilon} + (kM, (C_\varepsilon + M)_-)_{\Omega^1_\varepsilon}
$$

and

$$
(\partial_t(C_{e,\varepsilon} + M), (C_{e,\varepsilon} + M)_-)_{\Omega^2_\varepsilon} + D_{ER}(\nabla(C_{e,\varepsilon} + M), \nabla(C_{e,\varepsilon} + M)_-)_{\Omega^2_\varepsilon}
$$

$$
- \varepsilon(L_0 + L_{IP3})\langle C_\varepsilon - C_{e,\varepsilon}, (C_{e,\varepsilon} + M)_-\rangle_{\Gamma^{ER}_\varepsilon}
$$

$$
= \varepsilon\langle f_{SERCA}, (C_{e,\varepsilon} + M)_-\rangle_{\Gamma^{ER}_\varepsilon} + (kM, (C_{e,\varepsilon} + M)_-)_{\Omega^2_\varepsilon}.
$$

We add the equations and multiply by $-1$ on both sides

$$(\partial_t(C_\varepsilon + M)_-, (C_\varepsilon + M)_-)_{\Omega_\varepsilon^1} + D_C\|\nabla(C_\varepsilon + M)_-\|^2_{\Omega_\varepsilon^1}$$
$$+ (\partial_t(C_{e,\varepsilon} + M)_-, (C_{e,\varepsilon} + M)_-)_{\Omega_\varepsilon^2} + D_{ER}\|\nabla(C_{e,\varepsilon} + M)_-\|^2_{\Omega_\varepsilon^2}$$
$$+ \varepsilon(L_0 + L_{IP3})\|(C_\varepsilon + M)_- - (C_{e,\varepsilon} + M)_-\|^2_{\Gamma_\varepsilon^{ER}}$$
$$\leq \varepsilon\langle f_{SERCA}, (C_\varepsilon + M)_-\rangle_{\Gamma_\varepsilon^{ER}} - \varepsilon\langle f_{SERCA}, (C_{e,\varepsilon} + M)_-\rangle_{\Gamma_\varepsilon^{ER}}$$
$$+ \langle f_P + f_{NCX} + \alpha I_{CRAC}, (C_\varepsilon + M)_-\rangle_{\Gamma^1 \cap \partial\Omega_\varepsilon^1}$$
$$- (kM, (C_\varepsilon + M)_-)_{\Omega_\varepsilon^1} - (kM, (C_{e,\varepsilon} + M)_-)_{\Omega_\varepsilon^2}.$$

With standard estimations and integration from $0$ to $t$ we find

$$\frac{1}{2}\|(C_\varepsilon + M)_-\|^2_{\Omega_\varepsilon^1} + \frac{1}{2}\|(C_{e,\varepsilon} + M)_-\|^2_{\Omega_\varepsilon^2} + D_{ER}\|\nabla(C_{e,\varepsilon} + M)_-\|^2_{\Omega_\varepsilon^2,t}$$
$$+ \left(D_C - \varepsilon^2 c_0 - \frac{c_0}{\lambda}\right)\|\nabla(C_\varepsilon + M)_-\|^2_{\Omega_\varepsilon^1,t}$$
$$+ \varepsilon(L_0 + L_{IP3})\|C_\varepsilon + M)_- - (C_{e,\varepsilon} + M)_-\|^2_{\Gamma_\varepsilon^{ER},t}$$
$$\leq c_1 - (kM, (C_\varepsilon + M)_-)_{\Omega_\varepsilon^1,t} + c_2\|(C_\varepsilon + M)_-\|^2_{\Omega_\varepsilon^1,t}.$$

Note that we used two different trace inequalities, see lemma 2.11 and lemma 4.13. Then we do a similar separation of cases as in the first half of the proof and

- either find a non-nullset $V \subset \Omega_\varepsilon^1$ with $(C_\varepsilon + M)_- > 0$ on $V$. Then we can make $k$ big enough such that $c_1 - (kM, (C_\varepsilon + M)_-)_{\Omega_\varepsilon^1} < 0$. Then we find with Gronwall's lemma that $\|(C_\varepsilon + M)_-\|^2_{\Omega_\varepsilon^1} + \|(C_{e,\varepsilon} + M)_-\|^2_{\Omega_\varepsilon^2} < 0$ and we are finished.

- or we find that $C_\varepsilon + M > 0$ almost everywhere in $x \in \Omega_\varepsilon^1$. But then we have

$$(\partial_t(C_{e,\varepsilon} + M)_-, (C_{e,\varepsilon} + M)_-)_{\Omega_\varepsilon^2}$$
$$+ D_{ER}\|\nabla(C_{e,\varepsilon} + M)_-\|^2_{\Omega_\varepsilon^2,t} + \varepsilon(L_0 + L_{IP3})\|(C_{e,\varepsilon} + M)_-\|^2_{\Gamma_\varepsilon^{ER},t}$$
$$= -\varepsilon\langle f_{SERCA}, (C_{e,\varepsilon} + M)_-\rangle_{\Gamma_\varepsilon^{ER},t} - (kM, (C_{e,\varepsilon} + M)_-)_{\Omega_\varepsilon^2,t} \leq 0.$$

Integration from $0$ to $t$ leads to $\|(C_{e,\varepsilon} + M)_-\|^2_{\Omega_\varepsilon^2} < 0$ and we are also finished.

$\square$

Finally, we estimate the time derivatives. These are the last estimations we need to show strong convergences by applying lemma 3.4.

**Lemma 4.20 (Boundedness of $\partial_t C_\varepsilon$, $\partial_t C_{e,\varepsilon}$, $\partial_t S_\varepsilon$, $\partial_t S_{C,\varepsilon}$, $\partial_t S_{O,\varepsilon}$ and $\partial_t S_{CO,\varepsilon}$ in $H^{-1}$)**

1. *There exists a constant $C > 0$, independent of $\varepsilon$, such that*

$$\|\partial_t C_\varepsilon\|_{L^2([0,T],H^{-1}(\Omega_\varepsilon^1))} < C \quad \text{and} \quad \|\partial_t C_{e,\varepsilon}\|_{L^2([0,T],H^{-1}(\Omega_\varepsilon^2))} < C.$$

2. *There exists a constant $C > 0$, independent of $\varepsilon$, such that*

$$\|\partial_t S_\varepsilon\|_{L^2([0,T],H^{-1}(\Omega_\varepsilon^{ER}))} + \|\partial_t S_{C,\varepsilon}\|_{L^2([0,T],H^{-1}(\Omega_\varepsilon^{ER}))} < C.$$

3. *There exists a constant $C > 0$, independent of $\varepsilon$, such that*

$$\|\partial_t S_{O,\varepsilon}\|_{L^2([0,T],L^2(\Gamma^1 \cap \partial\Omega_\varepsilon^{ER})')} + \|\partial_t S_{CO,\varepsilon}\|_{L^2([0,T],L^2(\Gamma^1 \cap \partial\Omega_\varepsilon^{ER})')} < C.$$

**Proof**

1. We start with $\partial_t C_\varepsilon$ and the definition of the $H^{-1}$ norm. We drop the boundary terms, because test functions in $H_0^1$ are zero at the boundary.

$$
\begin{aligned}
\|\partial_t C_\varepsilon\|_{H^{-1}(\Omega_\varepsilon^1)} &= \sup_{\varphi \in H_0^1(\Omega_\varepsilon^1), \|\varphi\|=1} (\partial_t C_\varepsilon, \varphi)_{H_0^1(\Omega_\varepsilon^1)' \times H_0^1(\Omega_\varepsilon^1)} \\
&= \sup_{\varphi \in H_0^1(\Omega_\varepsilon^1), \|\varphi\|=1} (-D_C(\nabla C_\varepsilon, \nabla\varphi)_{H_0^1(\Omega_\varepsilon^1)' \times H_0^1(\Omega_\varepsilon^1)} \\
&\qquad \underbrace{- \varepsilon(L_0 + L_{IP3})\langle C_\varepsilon - C_{e,\varepsilon}, \varphi\rangle_{\Gamma_\varepsilon^{ER}}}_{=0} \\
&\qquad \underbrace{- \langle \alpha I_{CRAC} + f_P(C_\varepsilon) + f_{NCX}(C_\varepsilon), \varphi\rangle_{\Gamma^1 \cap \partial\Omega_\varepsilon^1}}_{=0} ) \\
&\leq \sup_{\varphi \in H_0^1(\Omega_\varepsilon^1), \|\varphi\|=1} (D_C\|\nabla C_\varepsilon\|_{L^2(\Omega_\varepsilon^1)}\|\nabla\varphi\|_{L^2(\Omega_\varepsilon^1)}) \\
&\leq c_1\|\nabla C_\varepsilon\|_{L^2(\Omega_\varepsilon^1)}.
\end{aligned}
$$

Integration from 0 to $T$ leads to $\|\partial_t C_\varepsilon\|_{L^2([0,T],H^{-1}(\Omega_\varepsilon^1))} \leq c_1 \|\nabla C_\varepsilon\|_{L^2([0,T]\times\Omega_\varepsilon^1)} \leq C$, see lemma 4.18. Analogously we estimate $\|\partial_t C_{e,\varepsilon}\|_{L^2([0,T],H^{-1}(\Omega_\varepsilon^2))}$.

2. We start with $\partial_t S_\varepsilon$ and the definition of the $H^{-1}$ norm.

$$\|\partial_t S_\varepsilon\|_{H^{-1}(\Omega_\varepsilon^{\mathrm{ER}})} = \sup_{\varphi \in H_0^1(\Omega_\varepsilon^{\mathrm{ER}}), \|\varphi\|=1} (\partial_t S_\varepsilon, \varphi)_{H_0^1(\Omega_\varepsilon^{\mathrm{ER}})' \times H_0^1(\Omega_\varepsilon^{\mathrm{ER}})}$$

$$= \sup_{\varphi \in H_0^1(\Omega_\varepsilon^{\mathrm{ER}}), \|\varphi\|=1} ((-D_S \nabla S_\varepsilon, \nabla \varphi)_{H_0^1(\Omega_\varepsilon^{\mathrm{ER}})' \times H_0^1(\Omega_\varepsilon^{\mathrm{ER}})}$$

$$- k_C^+ (f_e(C_{e,\varepsilon}) S_\varepsilon, \varphi)_{H_0^1(\Omega_\varepsilon^{\mathrm{ER}})' \times H_0^1(\Omega_\varepsilon^{\mathrm{ER}})}$$

$$+ k_C^- (S_{C,\varepsilon}, \varphi)_{H_0^1(\Omega_\varepsilon^{\mathrm{ER}})' \times H_0^1(\Omega_\varepsilon^{\mathrm{ER}})}$$

$$- \underbrace{\langle k_O^+ S_\varepsilon - k_O^- S_{O,\varepsilon}, \varphi \rangle_{\Gamma^1 \cap \Omega_\varepsilon^{\mathrm{ER}}}}_{=0})$$

$$\leq \sup_{\varphi \in H_0^1(\Omega_\varepsilon^{\mathrm{ER}}), \|\varphi\|=1} \left( D_S \|\nabla S_\varepsilon\|_{L^2(\Omega_\varepsilon^{\mathrm{ER}})} \|\nabla \varphi\|_{L^2(\Omega_\varepsilon^{\mathrm{ER}})} \right.$$

$$+ k_C^+ v_e \|S_\varepsilon\|_{L^2(\Omega_\varepsilon^{\mathrm{ER}})} \|\varphi\|_{L^2(\Omega_\varepsilon^{\mathrm{ER}})}$$

$$\left. + k_C^- \|S_{C,\varepsilon}\|_{L^2(\Omega_\varepsilon^{\mathrm{ER}})} \|\varphi\|_{L^2(\Omega_\varepsilon^{\mathrm{ER}})} \right)$$

$$\leq c_1 \left( \|\nabla S_\varepsilon\|_{L^2(\Omega_\varepsilon^{\mathrm{ER}})} + \|S_\varepsilon\|_{L^2(\Omega_\varepsilon^{\mathrm{ER}})} + \|S_{C,\varepsilon}\|_{L^2(\Omega_\varepsilon^{\mathrm{ER}})} \right).$$

Integration from 0 to $T$ yields

$$\|\partial_t S_\varepsilon\|_{L^2([0,T],H^{-1}(\Omega_\varepsilon^{\mathrm{ER}}))} \leq c_1 \left( \|\nabla S_\varepsilon\|_{L^2([0,T]\times\Omega_\varepsilon^{\mathrm{ER}})} \right.$$

$$\left. + \|S_\varepsilon\|_{L^2([0,T]\times\Omega_\varepsilon^{\mathrm{ER}})} + \|S_{C,\varepsilon}\|_{L^2([0,T]\times\Omega_\varepsilon^{\mathrm{ER}})} \right) \leq C,$$

see lemma 4.16. Analogously we estimate $\|\partial_t S_{C,\varepsilon}\|_{L^2([0,T],H^{-1}(\Omega_\varepsilon^{\mathrm{ER}}))}$.

3. Finally, we estimate $\|\partial_t S_{O,\varepsilon}\|_{L^2([0,T]\times\Gamma^1\cap\partial\Omega_\varepsilon^{\mathrm{ER}})}$.

$$\|\partial_t S_{O,\varepsilon}\|_{L^2([0,T]\times\Gamma^1\cap\partial\Omega_\varepsilon^{\mathrm{ER}})}$$

$$= \|k_O^+ S_\varepsilon - k_O^- S_{O,\varepsilon} - k_C^+ f_e(C_{e,\varepsilon}) S_{O,\varepsilon} + k_C^+ S_{CO,\varepsilon}\|_{L^2([0,T]\times\Gamma^1\cap\partial\Omega_\varepsilon^{\mathrm{ER}})},$$

which is bounded because of lemma 4.16. The estimation for $\|\partial_t S_{CO,\varepsilon}\|_{L^2([0,T] \times \Gamma^1 \cap \partial \Omega_\varepsilon^{ER})}$ is analogous.

$\square$

**Remark 4.21** With the estimations found in the lemmas of this section we know that $C_\varepsilon \in L^2([0,T], H^1(\Omega_\varepsilon^1)) \cap L^\infty([0,T] \times \Omega_\varepsilon^1) \cap H^1([0,T], H^{-1}(\Omega_\varepsilon^1))$. We apply theorem 3.1 and consider remarks 3.2 and 3.3 to deduce that there exists an extension $\tilde{C}_\varepsilon$ of function $C_\varepsilon$ such that $\tilde{C}_\varepsilon \in L^2([0,T], H^1(\Omega)) \cap L^\infty([0,T] \times \Omega) \cap H^1([0,T], H^{-1}(\Omega))$. Now we denote $\tilde{C}_\varepsilon$ again as $C_\varepsilon$ for convenience.
Analogously, we find that $C_{e,\varepsilon} \in L^2([0,T], H^1(\Omega)) \cap L^\infty([0,T] \times \Omega) \cap H^1([0,T], H^{-1}(\Omega))$, $S_\varepsilon \in L^2([0,T], H^1(\Omega)) \cap L^\infty([0,T] \times \Omega) \cap H^1([0,T], H^{-1}(\Omega))$ and $S_{C,\varepsilon} \in L^2([0,T], H^1(\Omega)) \cap L^\infty([0,T] \times \Omega) \cap H^1([0,T], H^{-1}(\Omega))$.
On the boundary $\Gamma^1$ we anticipate lemma 4.26 and deduce that $S_{O,\varepsilon} \in L^2([0,T], H^{1/2}(\Gamma^1 \cap \partial \Omega_\varepsilon^{ER}))$. Using also lemma 4.17 and lemma 4.20 we find with the extension theorem 3.1 that $S_{O,\varepsilon}$, $S_{CO,\varepsilon} \in L^2([0,T], H^{1/2}(\Gamma^1)) \cap L^\infty([0,T] \times \Gamma^1) \cap H^1([0,T], L^2(\Gamma^1))$.

**Remark 4.22** With lemma 3.4 and remark 4.21 we deduce strong convergence for the functions $C_\varepsilon$, $C_{e,\varepsilon}$, $S_\varepsilon$ and $S_{C,\varepsilon}$ in $L^2([0,T] \times \Omega)$ and strong convergence for the functions $S_{O,\varepsilon}$ and $S_{CO,\varepsilon}$ in $L^2([0,T] \times \Gamma^1)$.

**Remark 4.23** With remark 4.21 and theorem 4.9 we deduce also strong convergence on the outer boundary $\Gamma^1$ for the functions $C_\varepsilon$, $C_{e,\varepsilon}$, $S_\varepsilon$ and $S_{C,\varepsilon}$.

## 4.5 Existence of a solution

The purpose of this section is to show that there exists at least one solution of the weak formulation (4.14) for every $\varepsilon > 0$. For the main part we will use Schauder's fixed point theorem [58]. But we start with three lemmas to prepare the setting. The first lemma shows existence for the flux $I_{CRAC}$ living on the plasma membrane $\Gamma^1 \cap \partial \Omega_\varepsilon^1$ by use of Carathéodory's existence theorem, Theorem 1.1 of Chapter 2 in [14].

**Lemma 4.24 (Existence of $I_{CRAC}$)**

*Let $ext_\varepsilon S_{O,\varepsilon}$ and $C_\varepsilon$ be in $L^2([0,T] \times \Gamma^1 \cap \partial\Omega_\varepsilon^1)$. Then there exist solutions $a_\varepsilon \in C([0,T], L^2(\Gamma^1 \cap \partial\Omega_\varepsilon^1))$ and $V_\varepsilon \in C([0,T], L^2(\Gamma^1 \cap \partial\Omega_\varepsilon^1))$ of the differential equation (4.9). There also exists the function $I_{CRAC} \in L^1([0,T] \times \Gamma^1 \cap \partial\Omega_\varepsilon^1)$ as defined in (4.13).*

**Proof** We fix the equation for $V_\varepsilon$ and $a_\varepsilon$ in an arbitrary point $x$ almost everywhere in $\Gamma^1 \cap \partial\Omega_\varepsilon^1$ and show existence. We recall the ordinary differential equations for $a_\varepsilon$ and $V_\varepsilon$,

$$\frac{d}{dt}a_\varepsilon = \frac{C_\varepsilon^4}{\tau_{CAN}(C_\varepsilon^4 + K_{CAN}^4)} - \frac{a_\varepsilon}{\tau_{CAN}} =: j_a(a_\varepsilon, t),$$

$$\frac{d}{dt}V_\varepsilon = -\frac{g_K}{C_m}\frac{1}{1+e^{\frac{-(V_\varepsilon - V_n)}{K_n}}}(V_\varepsilon - V_K)$$

$$-\frac{g_{K(Ca)}}{C_m}\frac{C_\varepsilon^4}{C_\varepsilon^4 + K_{K(Ca)}^4}(V_\varepsilon - V_K) - \frac{g_{CRAC}}{C_m}(ext_\varepsilon S_{O,\varepsilon})^2(V_\varepsilon - V_{Ca})$$

$$-\frac{g_{CAN}}{C_m}a_\varepsilon(V_\varepsilon - V_{Na}) =: j_V(V_\varepsilon, t).$$

We need to check that

a) the right-hand sides $j_a$ and $j_V$ of the differential equations are defined on rectangles $[0,T] \times [\underline{a}, \overline{a}]$ and $[0,T] \times [\underline{V}, \overline{V}]$, respectively,

b) the right-hand sides of the differential equations are measurable in $t$ for all fixed $a_\varepsilon \in [\underline{a}, \overline{a}]$ and $V_\varepsilon \in [\underline{V}, \overline{V}]$, respectively,

c) the right-hand sides of the differential equations are continuous in $a_\varepsilon$ and $V_\varepsilon$ for all fixed $t \in [0,T]$, respectively,

d) there exists a Lebesgue-integrable functions $m_a(t)$ and $m_V(t)$, $t \in [0,T]$, such that $|j_a(a_\varepsilon, t)| \leq m_a(t)$ and $|j_V(V_\varepsilon, t)| \leq m_V(t)$ for all $(t, a_\varepsilon) \in [0,T] \times [\underline{a}, \overline{a}]$ and $(t, V_\varepsilon) \in [0,T] \times [\underline{V}, \overline{V}]$, respectively.

For a): See proof of lemma 4.14.

For b): The functions $j_a$ and $j_V$ are measurable because $S_{O,\varepsilon}$ and $C_\varepsilon$ are measurable in $t$ for fixed $a_\varepsilon \in [0, \overline{a}]$ and $V_\varepsilon \in [\underline{V}, \overline{V}]$, respectively.

For c): The functions $j_a$ and $j_V$ are continuous in $a_\varepsilon$ and $V_\varepsilon$, respectively, for all fixed $t \in [0, T]$.

For d): With $a_\varepsilon$ and $V_\varepsilon$ being bounded almost everywhere, also $j_a$ and $j_V$ are bounded almost everywhere. Using additionally b) and c), $j_a$ and $j_V$ are Lebesgue-integrable.

It follows that there exist solutions $V_\varepsilon(\cdot, x) \in C[0, T]$ and $a_\varepsilon(\cdot, x) \in C[0, T]$ for almost every $x \in \Gamma^1 \cap \partial\Omega_\varepsilon^1$ and the functions $V_\varepsilon$ and $a_\varepsilon$ are bounded.

The function

$$I_{\text{CRAC}}(S_{O,\varepsilon}, C_\varepsilon) = g_{\text{CRAC}}(\text{ext}_\varepsilon S_{O,\varepsilon})^2 (V_\varepsilon - V_{Ca})$$

also exists dependent on $C_\varepsilon$ and $S_{O,\varepsilon}$ and is an element of $L^1([0, T] \times \Gamma^1 \cap \partial\Omega_\varepsilon^1)$ since $\text{ext}_\varepsilon S_{O,\varepsilon} \in L^2([0, T] \times \Gamma^1 \cap \partial\Omega_\varepsilon^1)$. $\qquad \square$

The second lemma shows existence for the functions $S_{O,\varepsilon}$ and $S_{CO,\varepsilon}$ of the system of equations (4.14).

**Lemma 4.25 (Existence of $S_{O,\varepsilon}$ and $S_{CO,\varepsilon}$)**
*Let $C_{e,\varepsilon}$, $S_\varepsilon$ and $S_{C,\varepsilon}$ be in $L^2([0, T] \times \Gamma^1 \cap \partial\Omega_\varepsilon^{ER})$ and the initial values $S_{O,\varepsilon}(0), S_{CO,\varepsilon}(0)$ be in $L^2(\Gamma^1 \cap \partial\Omega_\varepsilon^{ER})$. Then there exists a solution $(S_{O,\varepsilon}, S_{CO,\varepsilon}) \in L^2([0, T] \times \Gamma^1 \cap \partial\Omega_\varepsilon^{ER})^2$ of the differential equation*

$$\frac{d}{dt}S_{O,\varepsilon} = k_O^+ S_\varepsilon - k_O^- S_{O,\varepsilon} - k_C^+ f_e(C_{e,\varepsilon})S_{O,\varepsilon} + k_C^- S_{CO,\varepsilon},$$

$$\frac{d}{dt}S_{CO,\varepsilon} = k_{CO}^+ S_{C,\varepsilon} - k_{CO}^- S_{CO,\varepsilon} + k_C^+ f_e(C_{e,\varepsilon})S_{O,\varepsilon} - k_C^- S_{CO,\varepsilon}$$

**Proof** We fix the equations in an arbitraty point $x$ almost everywhere in $\Gamma^1 \cap \partial\Omega_\varepsilon^{ER}$. Then, we claim that

$$\begin{pmatrix} S_{O,\varepsilon}(t) \\ S_{CO,\varepsilon}(t) \end{pmatrix} = e^{A(t)} \begin{pmatrix} S_{O,\varepsilon}(0) \\ S_{CO,\varepsilon}(0) \end{pmatrix} + e^{A(t)} \int_0^t e^{-A(\tau)} b(\tau) d\tau, \quad t \in [0, T]$$

solves the equation, where

$$A(t) := \begin{pmatrix} -tk_O^- - k_C^+ \int_0^t f_e(C_{e,\varepsilon}(\tau))d\tau & tk_C^- \\ k_C^+ \int_0^t f_e(C_{e,\varepsilon}(\tau))d\tau & -t(k_{CO}^- - k_C^-) \end{pmatrix}$$

and

$$b(t) := \begin{pmatrix} k_O^+ S_\varepsilon(t) \\ k_{CO}^+ S_{C,\varepsilon}(t) \end{pmatrix}.$$

Because every component of the matrix $A$ is the integral of bounded function over the bounded domain $[0, t]$, the components of $A$ are $L^\infty$ functions. Hence, $A \in L^\infty([0, T] \times \Gamma^1 \cap \partial\Omega_\varepsilon^{ER})^{2\times 2}$ and $b \in L^2([0, T] \times \Gamma^1 \cap \partial\Omega_\varepsilon^{ER})^2$. It holds that

$$\frac{\mathrm{d}}{\mathrm{d}t}\left( e^{A(t)} \begin{pmatrix} S_{O,\varepsilon}(0) \\ S_{CO,\varepsilon}(0) \end{pmatrix} + \int_0^t e^{A(t)-A(\tau)}b(\tau)\mathrm{d}\tau \right)$$

$$= \begin{pmatrix} -k_O^- - k_C^+ f_e(C_{e,\varepsilon}) & k_C^- \\ k_C^+ f_e(C_{e,\varepsilon}) & -k_{CO}^- - k_C^- \end{pmatrix}$$

$$\left( e^{A(t)} \begin{pmatrix} S_{O,\varepsilon}(0) \\ S_{CO,\varepsilon}(0) \end{pmatrix} + e^{A(t)} \int_0^t e^{-A(\tau)}b(\tau)\mathrm{d}\tau \right) + e^{A(t)}e^{-A(t)}b(t)$$

$$= \begin{pmatrix} -k_O^- - k_C^+ f_e(C_{e,\varepsilon}) & k_C^- \\ k_C^+ f_e(C_{e,\varepsilon}) & -k_{CO}^- - k_C^- \end{pmatrix} \begin{pmatrix} S_{O,\varepsilon}(t) \\ S_{CO,\varepsilon}(t) \end{pmatrix} + \begin{pmatrix} k_O^+ S_\varepsilon(t) \\ k_{CO}^+ S_{C,\varepsilon}(t) \end{pmatrix}.$$

Thus, we know

$$\begin{pmatrix} S_{O,\varepsilon}(t) \\ S_{CO,\varepsilon}(t) \end{pmatrix} = e^{A(t)} \begin{pmatrix} S_{O,\varepsilon}(0) \\ S_{CO,\varepsilon}(0) \end{pmatrix} + \int_0^t e^{A(t)-A(\tau)}b(\tau)\mathrm{d}\tau. \quad (4.15)$$

almost everywhere in $\Gamma^1 \cap \partial\Omega_\varepsilon^{ER}$. Because $e^{A(t)} \in L^\infty([0, T] \times \Gamma^1 \cap \partial\Omega_\varepsilon^{ER})$ and integration in the time variable is independent of the spatial variables, we deduce that the functions $S_{O,\varepsilon}$ and $S_{CO,\varepsilon}$ are elements of $L^2([0, T] \times \Gamma^1 \cap \partial\Omega_\varepsilon^{ER})$ and the claim holds true. □

**Lemma 4.26** *If* $\quad \|S_{O,\varepsilon}(0)\|^2_{H^{1/2-\delta}(\Gamma^1 \cap \partial\Omega_\varepsilon^{ER})} \quad < \quad r \quad$ *and* $\|S_{CO,\varepsilon}(0)\|^2_{H^{1/2-\delta}(\Gamma^1 \cap \partial\Omega_\varepsilon^{ER})} < r$ *for a* $r > 0$ *and* $S_\varepsilon, S_{C,\varepsilon}, C_{e,\varepsilon} \in L^2([0, \tau], H^{1-\delta}(\Omega_\varepsilon^{ER}))$ *for a* $\delta \in [0, 1/2)$ *and a* $\tau \in (0, T]$, *then it holds* *for $S_{O,\varepsilon}$, $S_{CO,\varepsilon}$ defined in (4.15) that*

$$\|S_{O,\varepsilon}\|^2_{L^2([0,\tau],H^{1/2-\delta}(\Gamma^1\cap\partial\Omega^{ER}_\varepsilon))} + \|S_{CO,\varepsilon}\|^2_{L^2([0,\tau],H^{1/2-\delta}(\Gamma^1\cap\partial\Omega^{ER}_\varepsilon))}$$
$$\leq c\left(r + \|S_\varepsilon\|^2_{L^2([0,\tau],H^{1-\delta}(\Omega^{ER}_\varepsilon))} + \|S_{C,\varepsilon}\|^2_{L^2([0,\tau],H^{1-\delta}(\Omega^{ER}_\varepsilon))}\right.$$
$$\left. + \|C_{e,\varepsilon}\|^2_{L^2([0,\tau],H^{1-\delta}(\Omega^{ER}_\varepsilon))}\right).$$

*for a constant $c > 0$.*

**Proof** *Step 1.* First we show that $S_{O,\varepsilon}$ and $S_{CO,\varepsilon}$ are bounded almost everywhere under the conditions of this lemma. Therefore, we test the equation of $S_{O,\varepsilon}$ of the system (4.14) with $(S_{O,\varepsilon} - M)_+$, where $M(t) = e^{kt}\|S_{O,\varepsilon}(0)\|_{L^\infty}$ for a $k > 0$.

$$\langle \partial_t S_{O,\varepsilon}, (S_{O,\varepsilon} - M)_+\rangle_{\Gamma^1\cap\partial\Omega^{ER}_\varepsilon}$$
$$= \langle k^+_O S_\varepsilon - k^-_O S_{O,\varepsilon} - k^+_C f_e(C_{e,\varepsilon})S_{O,\varepsilon} + k^+_C S_{CO,\varepsilon}, (S_{O,\varepsilon} - M)_+\rangle_{\Gamma^1\cap\partial\Omega^{ER}_\varepsilon}.$$

Considering the derivative with respect to time yields

$$\langle \partial_t(S_{O,\varepsilon} - M)_+, (S_{O,\varepsilon} - M)_+\rangle_{\Gamma^1\cap\partial\Omega^{ER}_\varepsilon}$$
$$= \langle k^+_O S_\varepsilon - k^-_O S_{O,\varepsilon} - k^+_C f_e(C_{e,\varepsilon})S_{O,\varepsilon} + k^+_C S_{CO,\varepsilon}, (S_{O,\varepsilon} - M)_+\rangle_{\Gamma^1\cap\partial\Omega^{ER}_\varepsilon}$$
$$- \langle kM, (S_{O,\varepsilon} - M)_+\rangle_{\Gamma^1\cap\partial\Omega^{ER}_\varepsilon}$$
$$\leq \|k^+_O S_\varepsilon - k^-_O S_{O,\varepsilon} - k^+_C f_e(C_{e,\varepsilon})S_{O,\varepsilon} + k^+_C S_{CO,\varepsilon}\|_{\Gamma^1\cap\partial\Omega^{ER}_\varepsilon}$$
$$\|(S_{O,\varepsilon} - M)_+\|_{\Gamma^1\cap\partial\Omega^{ER}_\varepsilon} - \langle kM, (S_{O,\varepsilon} - M)_+\rangle_{\Gamma^1\cap\partial\Omega^{ER}_\varepsilon}$$
$$\leq \frac{1}{2}\|k^+_O S_\varepsilon - k^-_O S_{O,\varepsilon} - k^+_C f_e(C_{e,\varepsilon})S_{O,\varepsilon} + k^+_C S_{CO,\varepsilon}\|^2_{\Gamma^1\cap\partial\Omega^{ER}_\varepsilon}$$
$$+ \frac{1}{2}\|S_{O,\varepsilon}\|^2_{\Gamma^1\cap\partial\Omega^{ER}_\varepsilon} - \langle kM, (S_{O,\varepsilon} - M)_+\rangle_{\Gamma^1\cap\partial\Omega^{ER}_\varepsilon}.$$

Integration with respect to time leads to

$$\frac{1}{2}\|(S_{O,\varepsilon} - M)_+\|^2_{\Gamma^1\cap\partial\Omega^{ER}_\varepsilon} = \overbrace{\frac{1}{2}\|(S_{O,\varepsilon}(0) - M)_+\|^2_{\Gamma^1\cap\partial\Omega^{ER}_\varepsilon,t}}^{=0}$$
$$+ c_1 - \langle kM, (S_{O,\varepsilon} - M)_+\rangle_{\Gamma^1\cap\partial\Omega^{ER}_\varepsilon,t}.$$

This is true, since $S_{O,\varepsilon}, S_\varepsilon, S_{CO,\varepsilon}$ are elements of $L^2(\Gamma^1 \cap \partial\Omega_\varepsilon^{ER})$ and we arrive at

$$\frac{1}{2}\|(S_{O,\varepsilon} - M)_+\|^2_{\Gamma^1\cap\partial\Omega_\varepsilon^{ER}} \leq c_1 - \langle kM, (S_{O,\varepsilon} - M)_+\rangle_{\Gamma^1\cap\partial\Omega_\varepsilon^{ER},t}$$

Now we consider two cases:

- Either there exists a non-nullset $V \subset \Gamma^1 \cap \partial\Omega_\varepsilon^{ER}$ where $(S_{O,\varepsilon} - M)_+ > 0$. Then, we define $\delta > 0$ as $\delta(k) = \langle M(k), (S_{O,\varepsilon} - M(k))_+\rangle_{\Gamma^1\cap\partial\Omega_\varepsilon^{ER},t}$ and choose $\delta(k)k \geq c_1$, where $k$ and $\delta(k)$ are growing with $k$. We conclude that

$$\frac{1}{2}\|(S_{O,\varepsilon} - M)_+\|^2_{\Gamma^1\cap\partial\Omega_\varepsilon^{ER}}$$
$$\leq c_1 - \langle kM, (S_{O,\varepsilon} - M)_+\rangle_{\Gamma^1\cap\partial\Omega_\varepsilon^{ER},t} = c_1 - k\delta < 0.$$

  This means $\|(S_{O,\varepsilon} - M)_+\|^2_{\Gamma^1\cap\partial\Omega_\varepsilon^{ER}} = 0$ and $(S_{O,\varepsilon} - M)_+ = 0$ almost everywhere in $\Gamma^1 \cap \partial\Omega_\varepsilon^{ER}$. But this contradicts to $(S_{O,\varepsilon} - M)_+ > 0$ on $V$.

- Or it holds that $(S_{O,\varepsilon} - M)_+ = 0$ almost everywhere and we deduce that $S_{O,\varepsilon} \leq M$ almost everywhere in $\Gamma^1 \cap \partial\Omega_\varepsilon^{ER}$.

To show that $S_{O,\varepsilon}$ is also bounded from below, we test the equation of $S_{O,\varepsilon}$ with $(S_{O,\varepsilon} + M)_-$ for another $k > 0$ and find with similar estimations that

$$\frac{1}{2}\|(S_{O,\varepsilon} + M)_-\|^2_{\Gamma^1\cap\partial\Omega_\varepsilon^{ER}} \leq c_1 - \langle kM, (S_{O,\varepsilon} + M)_-\rangle_{\Gamma^1\cap\partial\Omega_\varepsilon^{ER},t}$$

and with a similar case differentiation we find that $(S_{O,\varepsilon} + M)_- = 0$ almost everywhere in $\Gamma^1 \cap \partial\Omega_\varepsilon^{ER}$ and $S_{O,\varepsilon} \geq -M$ almost everywhere, which means that $S_{O,\varepsilon} \in L^\infty([0,\tau] \times \Gamma^1 \cap \partial\Omega_\varepsilon^{ER})$.

Analogously we find that $S_{CO,\varepsilon} \in L^\infty([0,\tau] \times \Gamma^1 \cap \partial\Omega_\varepsilon^{ER})$.

*Step 2.* In the second step we show that $f \cdot g \in L^\infty([0,\tau] \times \Gamma^1 \cap \partial\Omega_\varepsilon^{ER}) \cap L^2([0,\tau], H^{1/2-\delta}(\Gamma^1 \cap \partial\Omega_\varepsilon^{ER}))$ for two functions $f, g \in L^\infty([0,\tau] \times \Gamma^1 \cap \partial\Omega_\varepsilon^{ER}) \cap L^2([0,\tau], H^{1/2-\delta}(\Gamma^1 \cap \partial\Omega_\varepsilon^{ER}))$. It holds that $f \cdot g$ is bounded almost everywhere, if $f$ and $g$ are bounded almost everywhere.

Now we consider $\|f \cdot g\|^2_{H^{1/2-\delta}(\Gamma^1 \cap \partial\Omega^{ER}_\varepsilon)'}$

$$\|f \cdot g\|^2_{H^{1/2-\delta}(\Gamma^1 \cap \partial\Omega^{ER}_\varepsilon)}$$

$$= \int_{\Gamma^1 \cap \partial\Omega^{ER}_\varepsilon} \int_{\Gamma^1 \cap \partial\Omega^{ER}_\varepsilon} \frac{|f(x)g(x) - f(y)g(y)|^2}{|x-y|^{n-2\delta}} d\sigma_y d\sigma_x$$

$$\leq 2 \int_{\Gamma^1 \cap \partial\Omega^{ER}_\varepsilon} \int_{\Gamma^1 \cap \partial\Omega^{ER}_\varepsilon} f(x)^2 \frac{|g(x) - g(y)|^2}{|x-y|^{n-2\delta}} d\sigma_y d\sigma_x$$

$$+ 2 \int_{\Gamma^1 \cap \partial\Omega^{ER}_\varepsilon} \int_{\Gamma^1 \cap \partial\Omega^{ER}_\varepsilon} g(y)^2 \frac{|f(x) - f(y)|^2}{|x-y|^{n-2\delta}} d\sigma_y d\sigma_x$$

$$\leq 2\|f\|^2_{L^\infty} \int_{\Gamma^1 \cap \partial\Omega^{ER}_\varepsilon} \int_{\Gamma^1 \cap \partial\Omega^{ER}_\varepsilon} \frac{|g(x) - g(y)|^2}{|x-y|^{n-2\delta}} d\sigma_y d\sigma_x$$

$$+ 2\|g\|^2_{L^\infty} \int_{\Gamma^1 \cap \partial\Omega^{ER}_\varepsilon} \int_{\Gamma^1 \cap \partial\Omega^{ER}_\varepsilon} \frac{|f(x) - f(y)|^2}{|x-y|^{n-2\delta}} d\sigma_y d\sigma_x$$

$$= 2\|f\|^2_{L^\infty}\|g\|^2_{H^{1/2-\delta}(\Gamma^1 \cap \partial\Omega^{ER}_\varepsilon)} + 2\|g\|^2_{L^\infty}\|f\|^2_{H^{1/2-\delta}(\Gamma^1 \cap \partial\Omega^{ER}_\varepsilon)}$$

Integration from 0 to $\tau$ yields

$$\|f \cdot g\|^2_{L^2([0,\tau],H^{1/2-\delta}(\Gamma^1 \cap \partial\Omega^{ER}_\varepsilon))} \leq 2\|f\|^2_{\infty,\tau}\|g\|^2_{L^2([0,\tau],H^{1/2-\delta}(\Gamma^1 \cap \partial\Omega^{ER}_\varepsilon))}$$

$$+ 2\|g\|^2_{\infty,\tau}\|f\|^2_{L^2([0,\tau],H^{1/2-\delta}(\Gamma^1 \cap \partial\Omega^{ER}_\varepsilon))'}$$

which is bounded because of $g, f \in L^\infty([0,\tau] \times \Gamma^1 \cap \partial\Omega^{ER}_\varepsilon) \cap L^2([0,\tau], H^{1/2-\delta}(\Gamma^1 \cap \partial\Omega^{ER}_\varepsilon))$.

*Step 3.* With $f_e$ Lipschitz-continuous and $C_{e,\varepsilon} \in L^2([0,\tau], H^{1/2-\delta}(\Gamma^1 \cap \partial\Omega^{ER}_\varepsilon))$ we deduce that

$$\|f_e(C_{e,\varepsilon})\|^2_{H^{1/2-\delta}(\Gamma^1 \cap \partial\Omega^{ER}_\varepsilon)}$$

$$= \int_{\Gamma^1 \cap \partial\Omega^{ER}_\varepsilon} \int_{\Gamma^1 \cap \partial\Omega^{ER}_\varepsilon} \frac{|f_e(C_{e,\varepsilon}(x)) - f_e(C_{e,\varepsilon}(y))|^2}{|x-y|^{n-2\delta}} d\sigma_y d\sigma_x$$

$$= \int_{\Gamma^1 \cap \partial\Omega_\varepsilon^{ER}} \int_{\Gamma^1 \cap \partial\Omega_\varepsilon^{ER}} \frac{|f_e(C_{e,\varepsilon}(x)) - f_e(C_{e,\varepsilon}(y))|^2}{|C_{e,\varepsilon}(x) - C_{e,\varepsilon}(y)|^2}$$

$$\frac{|C_{e,\varepsilon}(x) - C_{e,\varepsilon}(y)|^2}{|x - y|^{n-2\delta}} d\sigma_y d\sigma_x$$

$$\leq c_1 \|f_e'(C_{e,\varepsilon})\|^2_{\Gamma^1 \cap \partial\Omega_\varepsilon^{ER}} \|C_{e,\varepsilon}\|^2_{H^{1/2-\delta}(\Gamma^1 \cap \partial\Omega_\varepsilon^{ER})}.$$

Integration with respect to time yields

$$\|f_e(C_{e,\varepsilon})\|^2_{L^2([0,\tau],H^{1/2-\delta}(\Gamma^1 \cap \partial\Omega_\varepsilon^{ER}))}$$

$$\leq c_1 \|f_e'(C_{e,\varepsilon})\|^2_{\Gamma^1 \cap \partial\Omega_\varepsilon^{ER},\tau} \|C_{e,\varepsilon}\|^2_{L^2([0,\tau],H^{1/2-\delta}(\Gamma^1 \cap \partial\Omega_\varepsilon^{ER}))},$$

which is bounded because $C_{e,\varepsilon} \in L^2([0,\tau], H^{1/2-\delta}(\Gamma^1 \cap \partial\Omega_\varepsilon^{ER}))$ and $f_e$ is Lipschitz-continuous.

*Step 4.* We consider the norm of the function $S_{O,\varepsilon}(t) = S_{O,\varepsilon}(0) + \int_0^t (k_O^+ S_\varepsilon + k_C^+ S_{CO,\varepsilon} - k_O^- S_{O,\varepsilon} - k_C^- f_e(C_{e,\varepsilon}) S_{O,\varepsilon}) dt$ for $0 \leq t \leq \tau$,

$$\|S_{O,\varepsilon}(t)\|^2_{H^{1/2-\delta}(\Gamma^1 \cap \partial\Omega_\varepsilon^{ER})} \leq \|S_{O,\varepsilon}(0)\|^2_{H^{1/2-\delta}(\Gamma^1 \cap \partial\Omega_\varepsilon^{ER})}$$

$$+ \left\| \int_0^t (k_O^+ S_\varepsilon + k_C^+ S_{CO,\varepsilon} \right.$$

$$\left. - k_O^- S_{O,\varepsilon} - k_C^- f_e(C_{e,\varepsilon}) S_{O,\varepsilon}) dt \right\|^2_{H^{1/2-\delta}(\Gamma^1 \cap \partial\Omega_\varepsilon^{ER})}$$

$$\leq \|S_{O,\varepsilon}(0)\|^2_{H^{1/2-\delta}(\Gamma^1 \cap \partial\Omega_\varepsilon^{ER})} + k_O^+ \left\| \int_0^t S_\varepsilon dt \right\|^2_{H^{1/2-\delta}(\Gamma^1 \cap \partial\Omega_\varepsilon^{ER})}$$

$$+ k_C^+ \left\| \int_0^t S_{CO,\varepsilon} dt \right\|^2_{H^{1/2-\delta}(\Gamma^1 \cap \partial\Omega_\varepsilon^{ER})}$$

$$+ k_O^- \left\| \int_0^t S_{O,\varepsilon} dt \right\|^2_{H^{1/2-\delta}(\Gamma^1 \cap \partial\Omega_\varepsilon^{ER})}$$

$$+ k_C^- \left\| \int_0^t f_e(C_{e,\varepsilon}) S_{O,\varepsilon} dt \right\|^2_{H^{1/2-\delta}(\Gamma^1 \cap \partial\Omega_\varepsilon^{ER})}$$

$$\leq \|S_{O,\varepsilon}(0)\|^2_{H^{1/2-\delta}(\Gamma^1 \cap \partial\Omega_\varepsilon^{ER})} + k_O^+ \|S_\varepsilon\|^2_{L^2([0,\tau],H^{1/2-\delta}(\Gamma^1 \cap \partial\Omega_\varepsilon^{ER}))}$$

$$+ k_C^+ \|S_{CO,\varepsilon}\|^2_{L^2([0,\tau],H^{1/2-\delta}(\Gamma^1 \cap \partial\Omega_\varepsilon^{ER}))}$$

$$+ k_O^- \|S_{O,\varepsilon}\|^2_{L^2([0,\tau],H^{1/2-\delta}(\Gamma^1 \cap \partial\Omega_\varepsilon^{ER}))}$$

$$+ k_C^- \|f_e(C_{e,\varepsilon}) S_{O,\varepsilon}\|^2_{L^2([0,\tau],H^{1/2-\delta}(\Gamma^1 \cap \partial\Omega_\varepsilon^{ER}))}.$$

Step 2, and Step 3 yield

$$
\begin{aligned}
\|S_{O,\varepsilon}(t)\|^2_{H^{1/2-\delta}(\Gamma^1\cap\partial\Omega_\varepsilon^{ER})} \leq{}& \|S_{O,\varepsilon}(0)\|^2_{H^{1/2-\delta}(\Gamma^1\cap\partial\Omega_\varepsilon^{ER})} \\
&+ k_O^+\|S_\varepsilon\|^2_{L^2([0,\tau],H^{1/2-\delta}(\Gamma^1\cap\partial\Omega_\varepsilon^{ER}))} \\
&+ k_C^+\|S_{CO,\varepsilon}\|^2_{L^2([0,\tau],H^{1/2-\delta}(\Gamma^1\cap\partial\Omega_\varepsilon^{ER}))} \\
&+ k_O^-\|S_{O,\varepsilon}\|^2_{L^2([0,\tau],H^{1/2-\delta}(\Gamma^1\cap\partial\Omega_\varepsilon^{ER}))} \\
&+ 2k_C^-\|f_e(C_{e,\varepsilon})\|^2_{\infty,\tau}\|S_{O,\varepsilon}\|^2_{L^2([0,\tau],H^{1/2-\delta}(\Gamma^1\cap\partial\Omega_\varepsilon^{ER}))} \\
&+ 2k_C^-\|S_{O,\varepsilon}\|^2_{\infty,\tau}\|f_e'(C_{e,\varepsilon})\|^2_{\Gamma^1\cap\partial\Omega_\varepsilon^{ER},\tau}\|C_{e,\varepsilon}\|^2_{L^2([0,\tau],H^{1/2-\delta}(\Gamma^1\cap\partial\Omega_\varepsilon^{ER}))}.
\end{aligned}
$$

Using the upper bound for some terms we find

$$
\begin{aligned}
\|S_{O,\varepsilon}(t)\|^2_{H^{1/2-\delta}(\Gamma^1\cap\partial\Omega_\varepsilon^{ER})} \leq{}& \|S_{O,\varepsilon}(0)\|^2_{H^{1/2-\delta}(\Gamma^1\cap\partial\Omega_\varepsilon^{ER})} \\
&+ c_1\|S_\varepsilon\|^2_{L^2([0,\tau],H^{1/2-\delta}(\Gamma^1\cap\partial\Omega_\varepsilon^{ER}))} \\
&+ c_2\|S_{CO,\varepsilon}\|^2_{L^2([0,\tau],H^{1/2-\delta}(\Gamma^1\cap\partial\Omega_\varepsilon^{ER}))} \\
&+ c_3\|S_{O,\varepsilon}\|^2_{L^2([0,\tau],H^{1/2-\delta}(\Gamma^1\cap\partial\Omega_\varepsilon^{ER}))} \\
&+ c_4\|C_{e,\varepsilon}\|^2_{L^2([0,\tau],H^{1/2-\delta}(\Gamma^1\cap\partial\Omega_\varepsilon^{ER}))}.
\end{aligned}
$$

Analogously we find for $S_{CO,\varepsilon}(t)$ that

$$
\begin{aligned}
\|S_{CO,\varepsilon}(t)\|^2_{H^{1/2-\delta}(\Gamma^1\cap\partial\Omega_\varepsilon^{ER})} \leq{}& \|S_{CO,\varepsilon}(0)\|^2_{H^{1/2-\delta}(\Gamma^1\cap\partial\Omega_\varepsilon^{ER})} \\
&+ c_1\|S_{C,\varepsilon}\|^2_{L^2([0,\tau],H^{1/2-\delta}(\Gamma^1\cap\partial\Omega_\varepsilon^{ER}))} \\
&+ c_2\|S_{CO,\varepsilon}\|^2_{L^2([0,\tau],H^{1/2-\delta}(\Gamma^1\cap\partial\Omega_\varepsilon^{ER}))} \\
&+ c_3\|S_{O,\varepsilon}\|^2_{L^2([0,\tau],H^{1/2-\delta}(\Gamma^1\cap\partial\Omega_\varepsilon^{ER}))} \\
&+ c_4\|C_{e,\varepsilon}\|^2_{L^2([0,\tau],H^{1/2-\delta}(\Gamma^1\cap\partial\Omega_\varepsilon^{ER}))}.
\end{aligned}
$$

We add the results and with Gronwall's lemma we find

$$\|S_{O,\varepsilon}(t)\|^2_{H^{1/2-\delta}(\Gamma^1 \cap \partial\Omega^{ER}_\varepsilon)} + \|S_{CO,\varepsilon}(t)\|^2_{H^{1/2-\delta}(\Gamma^1 \cap \partial\Omega^{ER}_\varepsilon)}$$

$$\leq c_1 \left( \|S_{O,\varepsilon}(0)\|^2_{H^{1/2-\delta}(\Gamma^1 \cap \partial\Omega^{ER}_\varepsilon)} + \|S_{CO,\varepsilon}(0)\|^2_{H^{1/2-\delta}(\Gamma^1 \cap \partial\Omega^{ER}_\varepsilon)} \right.$$

$$+ \|S_\varepsilon\|^2_{L^2([0,\tau],H^{1/2-\delta}(\Gamma^1 \cap \partial\Omega^{ER}_\varepsilon))} + \|S_{C,\varepsilon}\|^2_{L^2([0,\tau],H^{1/2-\delta}(\Gamma^1 \cap \partial\Omega^{ER}_\varepsilon))}$$

$$\left. + \|C_{e,\varepsilon}\|^2_{L^2([0,\tau],H^{1/2-\delta}(\Gamma^1 \cap \partial\Omega^{ER}_\varepsilon))} \right).$$

We use that
$\|S_{O,\varepsilon}(0)\|^2_{H^{1/2-\delta}(\Gamma^1 \cap \partial\Omega^{ER}_\varepsilon)} < r$ and $\|S_{CO,\varepsilon}(0)\|^2_{H^{1/2-\delta}(\Gamma^1 \cap \partial\Omega^{ER}_\varepsilon)} < r$
and the trace inequality $\|\cdot\|_{H^{1/2-\delta}(\Gamma^1 \cap \partial\Omega^{ER}_\varepsilon)} \leq c_0 \|\cdot\|_{H^{1-\delta}(\Omega^{ER}_\varepsilon)}$ and
conclude with

$$\|S_{O,\varepsilon}(t)\|^2_{H^{1/2-\delta}(\Gamma^1 \cap \partial\Omega^{ER}_\varepsilon)} + \|S_{CO,\varepsilon}(t)\|^2_{H^{1/2-\delta}(\Gamma^1 \cap \partial\Omega^{ER}_\varepsilon)}$$

$$\leq c_1 r + c_1 c_0 \left( \|S_\varepsilon\|^2_{L^2([0,\tau],H^{1-\delta}(\Omega^{ER}_\varepsilon))} \right.$$

$$\left. + \|S_{C,\varepsilon}\|^2_{L^2([0,\tau],H^{1-\delta}(\Omega^{ER}_\varepsilon))} + \|C_{e,\varepsilon}\|^2_{L^2([0,\tau],H^{1-\delta}(\Omega^{ER}_\varepsilon))} \right).$$

Finally, integration from 0 to $\tau$ and $\tau \leq T$ yields

$$\|S_{O,\varepsilon}\|^2_{L^2([0,\tau],H^{1/2-\delta}(\Gamma^1 \cap \partial\Omega^{ER}_\varepsilon))} + \|S_{CO,\varepsilon}\|^2_{L^2([0,\tau],H^{1/2-\delta}(\Gamma^1 \cap \partial\Omega^{ER}_\varepsilon))}$$

$$\leq T c_1 r + T c_1 c_0 \left( \|S_\varepsilon\|^2_{L^2([0,\tau],H^{1-\delta}(\Omega^{ER}_\varepsilon))} \right.$$

$$\left. + \|S_{C,\varepsilon}\|^2_{L^2([0,\tau],H^{1-\delta}(\Omega^{ER}_\varepsilon))} + \|C_{e,\varepsilon}\|^2_{L^2([0,\tau],H^{1-\delta}(\Omega^{ER}_\varepsilon))} \right).$$

$\square$

We recall the definitions of the functions $f_{\text{SERCA}}$, $f_{\text{P}}$, $f_{\text{NCX}}$, $f_e$ :
$\mathbb{R} \to \mathbb{R}$,

$$f_{\text{SERCA}}(u) = v_{\text{SERCA}} \frac{u^2}{K^2_{\text{SERCA}} + u^2} \qquad f_{\text{P}}(u) = v_{\text{P}} \frac{u^2}{K^2_{\text{P}} + u^2}$$

$$f_{\text{NCX}}(u) = v_{\text{NCX}} \frac{u^4}{K^4_{\text{NCX}} + u^4} \qquad f_e(u) = v_e \frac{u^4}{K^4_e + u^4}.$$

The following lemma 4.27 shows that the functions

$$F_{\text{SERCA}} : L^2([0,T] \times \Gamma_\varepsilon^{\text{ER}}) \to L^2([0,T] \times \Gamma_\varepsilon^{\text{ER}}),$$
$$F_{\text{P}} : L^2([0,T] \times \Gamma^1 \cap \partial\Omega_\varepsilon^1) \to L^2([0,T] \times \Gamma^1 \cap \partial\Omega_\varepsilon^1),$$
$$F_{\text{NCX}} : L^2([0,T] \times \Gamma^1 \cap \partial\Omega_\varepsilon^1) \to L^2([0,T] \times \Gamma^1 \cap \partial\Omega_\varepsilon^1),$$
$$F_e : L^2([0,T] \times \Gamma^1 \cap \partial\Omega_\varepsilon^{\text{ER}}) \to L^2([0,T] \times \Gamma^1 \cap \partial\Omega_\varepsilon^{\text{ER}})$$

or

$$F_e : L^2([0,T] \times \Omega_\varepsilon^{\text{ER}}) \to L^2([0,T] \times \Omega_\varepsilon^{\text{ER}})$$

defined by

$$F_X(u)(x) = f_X(u(x))$$

with $u \in L^2$ and $f_X$ for $X \in \{\text{SERCA}, \text{P}, \text{NCX}, e\}$ are continuous and bounded.

**Lemma 4.27** *For $X \in \{\text{SERCA, P, NCX, } e\}$ the functions $F_X$ are continuous and bounded.*

**Proof** For $X \in \{\text{SERCA}, \text{P}, \text{NCX}, e\}$ it holds that $f_X$ is continuous and bounded. Further,

$$|f_X(x)| = \left| x \frac{x v_X}{x^2 + K_X^2} \right| \le |x|^{\frac{2}{2}} \frac{v_X}{2K_X}$$

for $X \in \{\text{SERCA}, \text{P}\}$ and $\frac{v_X}{2K_X} > 0$. Analogously we find

$$|f_X(x)| = \left| x \frac{x^3 v_X}{x^4 + K_X^4} \right| \le |x|^{\frac{2}{2}} \frac{3v_X}{K_X^3}$$

for $X \in \{\text{NCX}, e\}$ and $\frac{3v_X}{K_X^3} > 0$. Then, by theorem 3.6 we deduce that the functions $F_X : L^2 \to L^2$ are continuous and bounded for $X \in \{\text{SERCA, P, NCX, } e\}$. $\square$

We also deduce from the proof of lemma 4.27 that there are constants $L_{\text{SERCA}}$, $L_P$, $L_{\text{NCX}}$ and $L_e$ such that $(F_{\text{SERCA}}(C_\varepsilon))(t) \le L_{\text{SERCA}} C_\varepsilon(t)$, $(F_P(C_\varepsilon))(t) \le L_P C_\varepsilon(t)$ and $(F_{\text{NCX}}(C_\varepsilon))(t) \le L_{\text{NCX}} C_\varepsilon(t)$ and $(F_e(C_{e,\varepsilon}))(t) \le L_e C_{e,\varepsilon}(t)$.

Now we apply Schauder's fixed point theorem 3.40 to ensure a solution of the complete system of differential equations (4.14).

For $I_{CRAC}$, depending on $C_\varepsilon$ and $S_{O,\varepsilon}$, we use $I_{CRAC}(C_\varepsilon, S_{O,\varepsilon}) = g_{CRAC}(\text{ext}_\varepsilon S_{O,\varepsilon}) b(\text{ext}_\varepsilon S_{O,\varepsilon})(V_\varepsilon - V_{Ca})$ with

$$b(z) := \begin{cases} z, & \text{for } z \leq b_{max}, \\ b_{max}, & \text{for } z > b_{max} \end{cases}$$

for a constant $b_{max} > 0$. We deduce that $I_{CRAC} \leq L_{CRAC}(\text{ext}_\varepsilon S_{O,\varepsilon})$ for a constant $L_{CRAC} > 0$.
In lemma 4.17 we saw that $S_{O,\varepsilon}$ is a bounded function, i.e $S_{O,\varepsilon} < \tilde{c}$ almost everywhere, for an existing solution and so, the auxiliary function $b$ is just used for the proof of existence. We choose $b_{max} = \tilde{c}$, and the function $b$ never reaches the bound.

**Theorem 4.28 (Existence)**
*The system of differential equations (4.14) has at least one solution $(C_\varepsilon, C_{e,\varepsilon}, S_\varepsilon, S_{C,\varepsilon})$ in $\mathcal{V}(\Omega_\varepsilon^1) \times \mathcal{V}(\Omega_\varepsilon^2) \times \mathcal{V}(\Omega_\varepsilon^{ER}) \times \mathcal{V}(\Omega_\varepsilon^{ER})$.*

**Proof** We show existence on a small time interval $[0, \tau]$. The existing solutions must be patched together bit by bit.
For a $\delta \in [0, 1/2)$ we define the spaces $V_1 := L^2([0, \tau], H^{1-\delta}(\Omega_\varepsilon^1))$, $V_2 := L^2([0, \tau], H^{1-\delta}(\Omega_\varepsilon^2))$ and $V^{ER} := L^2([0, \tau], H^{1-\delta}(\Omega_\varepsilon^{ER}))$.

Further, we define the function

$$T : V^1 \times V^2 \times \left(V^{ER}\right)^2$$
$$\rightarrow \{u \in L^2([0, \tau], H^1(\Omega_\varepsilon^1)) | \, \partial_t u \in L^2([0, \tau], H^1(\Omega_\varepsilon^1)')\}$$
$$\times \{u \in L^2([0, \tau], H^1(\Omega_\varepsilon^2)) | \, \partial_t u \in L^2([0, \tau], H^1(\Omega_\varepsilon^2)')\}$$
$$\times \{u \in L^2([0, \tau], H^1(\Omega_\varepsilon^{ER})) | \, \partial_t u \in L^2([0, \tau], H^1(\Omega_\varepsilon^{ER})')\}^2$$

with

$$T(\tilde{C}_\varepsilon, \tilde{C}_{e,\varepsilon}, \tilde{S}_\varepsilon, \tilde{S}_{C,\varepsilon}) := (C_\varepsilon, C_{e,\varepsilon}, S_\varepsilon, S_{C,\varepsilon}),$$

given by

$$
\begin{aligned}
\partial_t C_\varepsilon - D_C \Delta C_\varepsilon &= 0 \\
-D_C \nabla C_\varepsilon \cdot n &= \varepsilon (L_0 + L_{IP3})(\tilde{C}_\varepsilon - \tilde{C}_{e,\varepsilon}) + \varepsilon F_{SERCA}(\tilde{C}_\varepsilon) \\
-D_C \nabla C_\varepsilon \cdot n &= \alpha I_{CRAC}(\tilde{C}_\varepsilon, S_{O,\varepsilon}) + F_P(\tilde{C}_\varepsilon) + F_{NCX}(\tilde{C}_\varepsilon) \\
-D_C \nabla C_\varepsilon \cdot n &= 0
\end{aligned}
$$

$$
\begin{aligned}
\partial_t C_{e,\varepsilon} - D_{ER} \Delta C_{e,\varepsilon} &= 0 \\
-D_{ER} \nabla C_{e,\varepsilon} \cdot n &= \varepsilon (L_0 + L_{IP3})(\tilde{C}_{e,\varepsilon} - \tilde{C}_\varepsilon) - \varepsilon F_{SERCA}(\tilde{C}_\varepsilon) \\
-D_{ER} \nabla C_{e,\varepsilon} \cdot n &= 0
\end{aligned}
$$

$$
\begin{aligned}
\partial_t S_\varepsilon - D_S \Delta S_\varepsilon &= -k_C^+ f_e(\tilde{C}_{e,\varepsilon}) \tilde{S}_\varepsilon + k_C^- \tilde{S}_{C,\varepsilon} \\
-D_S \nabla S_\varepsilon \cdot n &= k_O^+ \tilde{S}_\varepsilon - k_O^- S_{O,\varepsilon}(\tilde{C}_{e,\varepsilon}, \tilde{S}_\varepsilon, \tilde{S}_{C,\varepsilon}) \\
-D_S \nabla S_\varepsilon \cdot n &= 0
\end{aligned}
$$

$$
\begin{aligned}
\partial_t S_{C,\varepsilon} - D_{SC} \Delta S_{C,\varepsilon} &= k_C^+ f_e(\tilde{C}_{e,\varepsilon}) \tilde{S}_\varepsilon - k_C^- \tilde{S}_{C,\varepsilon} \\
-D_{SC} \nabla S_{C,\varepsilon} \cdot n &= k_{CO}^+ \tilde{S}_{C,\varepsilon} - k_{CO}^- S_{CO,\varepsilon}(\tilde{C}_{e,\varepsilon}, \tilde{S}_\varepsilon, \tilde{S}_{C,\varepsilon}) \\
-D_{SC} \nabla S_{C,\varepsilon} \cdot n &= 0
\end{aligned}
$$

The solution of this system is unique and the operator $T$ is continuous, where we find $I_{CRAC}(\tilde{C}_\varepsilon, \tilde{S}_{O,\varepsilon})$ in lemma 4.24 and $S_{O,\varepsilon}(\tilde{C}_{e,\varepsilon}, \tilde{S}_\varepsilon, \tilde{S}_{C,\varepsilon})$, $S_{CO,\varepsilon}(\tilde{C}_{e,\varepsilon}, \tilde{S}_\varepsilon, \tilde{S}_{C,\varepsilon})$ in lemma 4.25.

The space $\{u \in L^2([0,\tau], H^1(\Omega_\varepsilon^1)) \mid \partial_t u \in L^2([0,\tau], H^1(\Omega_\varepsilon^1)')\}$ is compactly embedded in $V_1$, $\{u \in L^2([0,\tau], H^1(\Omega_\varepsilon^2)) \mid \partial_t u \in L^2([0,\tau], H^1(\Omega_\varepsilon^2)')\}$ is compactly embedded in $V_2$ and $\{u \in L^2([0,\tau], H^1(\Omega_\varepsilon^{ER})) \mid \partial_t u \in L^2([0,\tau], H^1(\Omega_\varepsilon^{ER})')\}$ is compactly embedded in $V^{ER}$ (lemma of Lion-Aubin 3.5 and Rellich-Kondrachov theorem [21]), and we denote the embedding with $I$. We deduce that the fixed-point operator that maps $(\tilde{C}_\varepsilon, \tilde{C}_{e,\varepsilon}, \tilde{S}_\varepsilon, \tilde{S}_{C,\varepsilon}) \in V^1 \times V^2 \times (V^{ER})^2$ to $(C_\varepsilon, C_{e,\varepsilon}, S_\varepsilon, S_{C,\varepsilon}) \in V^1 \times V^2 \times (V^{ER})^2$ is continuous and compact.

It is left to show that for the initial value $y_0 = (C_\varepsilon(0), C_{e,\varepsilon}(0), S_\varepsilon(0), S_{C,\varepsilon}(0), S_{O,\varepsilon}(0), S_{CO,\varepsilon}(0))$ it holds that

$$
(\tilde{C}_\varepsilon, \tilde{C}_{e,\varepsilon}, \tilde{S}_\varepsilon, \tilde{S}_{C,\varepsilon}) \in B_{y_0}(r)
$$

implies

$$(I \circ T)\left(\tilde{C}_\varepsilon, \tilde{C}_{e,\varepsilon}, \tilde{S}_\varepsilon, \tilde{S}_{C,\varepsilon}\right) \in B_{y_0}(r).$$

This means that $\|\tilde{C}\|_{V^1}^2 + \|\tilde{C}_{e,\varepsilon}\|_{V^2}^2 + \|\tilde{S}_\varepsilon\|_{VER}^2 + \|\tilde{S}_{C,\varepsilon}\|_{VER}^2 \leq r$ should imply $\|C_\varepsilon\|_{V^1}^2 + \|C_{e,\varepsilon}\|_{V^2}^2 + \|S_\varepsilon\|_{VER}^2 + \|S_{C,\varepsilon}\|_{VER}^2 \leq r$ for some $r > 0$, where we may assume that the initial conditions are smaller than $r$.

We test the weak formulation of the equation for $C_\varepsilon$ with $C_\varepsilon$ and integrate from 0 to $t \leq \tau$.

$$
\begin{aligned}
\frac{1}{2}\|C_\varepsilon\|_{\Omega_\varepsilon^1}^2 &+ D_C\|\nabla C_\varepsilon\|_{\Omega_\varepsilon,t}^2 \\
&\leq c_1\varepsilon\|\tilde{C}_\varepsilon\|_{\Gamma_\varepsilon^{\mathrm{ER}},t}^2 + c_2\varepsilon\|\tilde{C}_{e,\varepsilon}\|_{\Gamma_\varepsilon^{\mathrm{ER}},t}^2 + c_3\varepsilon\|C_\varepsilon\|_{\Gamma_\varepsilon^{\mathrm{ER}},t}^2 \\
&\quad + c_4\frac{1}{\lambda}\|C_\varepsilon\|_{\Gamma^1 \cap \partial\Omega_\varepsilon^1,t}^2 + c_4\lambda\|\mathrm{ext}_\varepsilon\, S_{O,\varepsilon}\|_{\Gamma^1 \cap \partial\Omega_\varepsilon^1,t}^2 \\
&\quad + c_5\frac{1}{\lambda}\|C_\varepsilon\|_{\Gamma^1 \cap \partial\Omega_\varepsilon^1,t}^2 + c_5\lambda\|\tilde{C}_\varepsilon\|_{\Gamma^1 \cap \partial\Omega_\varepsilon^1,t}^2 + r \\
&\leq c_1\varepsilon\|\tilde{C}_\varepsilon\|_{\Gamma_\varepsilon^{\mathrm{ER}},t}^2 + \varepsilon^{n-2}c_1|\tilde{C}_\varepsilon|_{L^2([0,t],H^{1/2-\delta}(\Gamma_\varepsilon^{\mathrm{ER}}))}^2 + c_2\varepsilon\|\tilde{C}_{e,\varepsilon}\|_{\Gamma_\varepsilon^{\mathrm{ER}},t}^2 \\
&\quad + \varepsilon^{n-2}c_2|\tilde{C}_{e,\varepsilon}|_{L^2([0,t],H^{1/2-\delta}(\Gamma_\varepsilon^{\mathrm{ER}}))}^2 + c_3\varepsilon\|C_\varepsilon\|_{\Gamma_\varepsilon^{\mathrm{ER}},t}^2 \\
&\quad + c_4\frac{1}{\lambda}\|C_\varepsilon\|_{L^2([0,t],H^{1/2}(\Gamma^1 \cap \partial\Omega_\varepsilon^1))}^2 + c_5\lambda\|\tilde{S}_\varepsilon\|_{L^2([0,t],H^{1-\delta}(\Omega_\varepsilon^{\mathrm{ER}}))}^2 \\
&\quad + c_5\lambda\|\tilde{S}_{C,\varepsilon}\|_{L^2([0,t],H^{1-\delta}(\Omega_\varepsilon^{\mathrm{ER}}))}^2 c_6\frac{1}{\lambda}\|C_\varepsilon\|_{\Gamma^1 \cap \partial\Omega_\varepsilon^1,t}^2 \\
&\quad + c_7\lambda\|\tilde{C}_\varepsilon\|_{L^2([0,t],H^{1/2-\delta}(\Gamma^1 \cap \partial\Omega_\varepsilon^1))}^2 + c_8 r.
\end{aligned}
$$

We deduce this estimate from lemma 4.26. With the trace inequality and lemma 3.24 we continue with

$$
\frac{1}{2}\|C_\varepsilon\|^2_{\Omega^1_\varepsilon} + D_C\|\nabla C_\varepsilon\|^2_{\Omega_{\varepsilon,t}}
$$

$$
\leq c_1\|\tilde{C}_\varepsilon\|^2_{\Omega^1_{\varepsilon,t}} + \varepsilon^{-n}c_1|\tilde{C}_\varepsilon|^2_{L^2([0,t],H^{1-\delta}(\Omega^1_\varepsilon))} + c_2\|\tilde{C}_{e,\varepsilon}\|^2_{\Omega^2_{\varepsilon,t}}
$$

$$
+ \varepsilon^{-n}c_2|\tilde{C}_{e,\varepsilon}|^2_{L^2([0,t],H^{1-\delta}(\Omega^2_\varepsilon))} + c_3\|C_\varepsilon\|^2_{\Omega^1_{\varepsilon,t}} + c_3\varepsilon^2\|\nabla C_\varepsilon\|^2_{\Omega^1_{\varepsilon,t}}
$$

$$
+ c_4\frac{1}{\lambda}\|C_\varepsilon\|^2_{L^2([0,t],H^1(\Omega^1_\varepsilon))} + c_5\lambda\|\tilde{S}_\varepsilon\|^2_{L^2([0,t],H^{1-\delta}(\Omega^{\mathrm{ER}}_\varepsilon))}
$$

$$
+ c_5\lambda\|\tilde{S}_{C,\varepsilon}\|^2_{L^2([0,t],H^{1-\delta}(\Omega^{\mathrm{ER}}_\varepsilon))} + c_6\frac{1}{\lambda}\|C_\varepsilon\|^2_{L^2([0,t],H^1(\Omega^1_\varepsilon))}
$$

$$
+ c_7\lambda\|\tilde{C}_\varepsilon\|^2_{L^2([0,t],H^{1-\delta}(\Omega^1_\varepsilon))} + c_8 r.
$$

This leads to

$$
\|C_\varepsilon\|^2_{\Omega^1_\varepsilon} + \left(D_C - c_3\varepsilon^2 - \frac{1}{\lambda}c_4 - \frac{1}{\lambda}c_5\right)\|\nabla C_\varepsilon\|^2_{\Omega^1_{\varepsilon,t}}
$$

$$
\leq c_1\frac{1}{\lambda}\|C_\varepsilon\|^2_{\Omega^1_{\varepsilon,t}} + c_2\lambda\varepsilon^{-n}\underbrace{\|\tilde{C}_\varepsilon\|^2_{L^2([0,t],H^{1-\delta}(\Omega^1_\varepsilon))}}_{\leq r}
$$

$$
+ c_3\varepsilon^{-n}\underbrace{\|\tilde{C}_{e\varepsilon}\|^2_{L^2([0,t],H^{1-\delta}(\Omega^2_\varepsilon))}}_{\leq r} + c_4\lambda\underbrace{\|\tilde{S}_\varepsilon\|^2_{L^2([0,t],H^{1-\delta}(\Omega^{\mathrm{ER}}_\varepsilon))}}_{\leq r}
$$

$$
+ c_4\lambda\underbrace{\|\tilde{S}_{C,\varepsilon}\|^2_{L^2([0,t],H^{1-\delta}(\Omega^{\mathrm{ER}}_\varepsilon))}}_{\leq r} + c_5 r
$$

with $c_2$, $c_3$ large but finite for $\varepsilon > 0$. For $\varepsilon$ small enough and $\lambda$ big enough we find with Gronwall's lemma that

$$
\|C_\varepsilon\|^2_{\Omega^1_\varepsilon} + \|\nabla C_\varepsilon\|^2_{\Omega^1_{\varepsilon,t}} \leq c_1 r.
$$

This inequality yields $\|C_\varepsilon\|^2_{L^2([0,\tau],H^1(\Omega^1_\varepsilon))} \leq c_1 r$ and $\|C_\varepsilon\|^2_{L^2([0,\tau]\times\Omega^1_\varepsilon)} \leq \tau c_2 r$. With the interpolation inequality in [1] we find

$$
\|C_\varepsilon\|^2_{V^1} \leq c_3\|C_\varepsilon\|^{2-2\delta}_{L^2([0,\tau],H^1(\Omega^1_\varepsilon))}\|C_\varepsilon\|^{2\delta}_{L^2([0,\tau]\times\Omega^1_\varepsilon)}
$$

$$
\leq c_3(c_1 r)^{1-\delta}(\tau c_2 r)^\delta = cr\tau^\delta.
$$

With similar transformations we also get the corresponding inequality for the equations for $C_{e,\varepsilon}$. Next we estimate $S_\varepsilon$ by testing the equation with $S_\varepsilon$.

$$\int_{\Omega_\varepsilon^{\mathrm{ER}}} \partial_t S_\varepsilon S_\varepsilon \mathrm{d}x + D_S \int_{\Omega_\varepsilon^{\mathrm{ER}}} (\nabla S_\varepsilon)^2 \mathrm{d}x$$

$$= -k_C^+ \int_{\Omega_\varepsilon^{\mathrm{ER}}} f_e(\tilde{C}_{e,\varepsilon}) \tilde{S}_\varepsilon S_\varepsilon \mathrm{d}x + k_C^- \int_{\Omega_\varepsilon^{\mathrm{ER}}} \tilde{S}_{C,\varepsilon} S_\varepsilon \mathrm{d}x$$

$$- \int_{\Gamma^1 \cap \partial\Omega_\varepsilon^{\mathrm{ER}}} (k_O^+ \tilde{S}_\varepsilon S_\varepsilon - k_O^+ S_{O,\varepsilon}(\tilde{C}_{e,\varepsilon}, \tilde{S}_\varepsilon, \tilde{S}_{C,\varepsilon}) S_\varepsilon) \mathrm{d}\sigma_x.$$

Integration from 0 to $t \le \tau$ leads to

$$\frac{1}{2}\|S_\varepsilon\|^2_{\Omega_\varepsilon^{\mathrm{ER}}} + D_S\|\nabla S_\varepsilon\|^2_{\Omega_\varepsilon^{\mathrm{ER}}}$$

$$\le k_C^+ v_e \frac{1}{2}\left(\|\tilde{S}_\varepsilon\|^2_{\Omega_\varepsilon^{\mathrm{ER}},t} + \|S_\varepsilon\|^2_{\Omega_\varepsilon^{\mathrm{ER}},t}\right)$$

$$+ k_C^- \frac{1}{2}\left(\|\tilde{S}_{C,\varepsilon}\|^2_{\Omega_\varepsilon^{\mathrm{ER}},t} + \|S_\varepsilon\|^2_{\Omega_\varepsilon^{\mathrm{ER}},t}\right)$$

$$+ k_O^+ \frac{c_0 \lambda}{2}\|\tilde{S}_\varepsilon\|^2_{L^2([0,t],H^{1-\delta}(\Omega_\varepsilon^{\mathrm{ER}}))}$$

$$+ k_O^+ \frac{c_0}{2\lambda}\left(\|S_\varepsilon\|^2_{\Omega_\varepsilon^{\mathrm{ER}},t} + \|\nabla S_\varepsilon\|^2_{\Omega_\varepsilon^{\mathrm{ER}},t}\right)$$

$$+ k_O^- \frac{\lambda c_0}{2}\left(\|\tilde{S}_\varepsilon\|^2_{L^2([0,t],H^{1-\delta}(\Omega_\varepsilon^{\mathrm{ER}}))} + \|\tilde{S}_{C,\varepsilon}\|^2_{L^2([0,t],H^{1-\delta}(\Omega_\varepsilon^{\mathrm{ER}}))}\right)$$

$$+ k_O^- \frac{c_0}{2\lambda}\left(\|S_\varepsilon\|^2_{\Omega_\varepsilon^{\mathrm{ER}},t} + \|\nabla S_\varepsilon\|^2_{\Omega_\varepsilon^{\mathrm{ER}},t}\right) + c_1 r.$$

This leads to

$$\|S_\varepsilon\|^2_{\Omega_\varepsilon^{\mathrm{ER}}} + \left(D_S - k_O^+ \frac{c_0}{2\lambda} - k_O^- \frac{c_0}{2\lambda}\right)\|\nabla S_\varepsilon\|^2_{\Omega_\varepsilon^{\mathrm{ER}},t}$$

$$\le c_1\|S_\varepsilon\|^2_{\Omega_\varepsilon^{\mathrm{ER}},t} + c_2\lambda\|\tilde{S}_\varepsilon\|^2_{L^2([0,t],H^{1-\delta}(\Omega_\varepsilon^{\mathrm{ER}}))}$$

$$+ c_3\lambda\|\tilde{S}_{C,\varepsilon}\|^2_{L^2([0,t],H^{1-\delta}(\Omega_\varepsilon^{\mathrm{ER}}))} + c_4 r$$

$$\le c_1\|S_\varepsilon\|^2_{\Omega_\varepsilon^{\mathrm{ER}},t} + \lambda c_2 r.$$

With Gronwall's lemma and large, but finite $\lambda$ we find

$$\|S_\varepsilon\|^2_{\Omega_\varepsilon^{\mathrm{ER}}} + \|\nabla S_\varepsilon\|^2_{\Omega_\varepsilon^{\mathrm{ER}},t} \le c_1 r.$$

Again, the interpolation inequality and integration from 0 to $\tau$ yields

$$\|S_\varepsilon\|_{VER}^2 \leq c_3 \|S_\varepsilon\|_{L^2([0,\tau],H^1(\Omega_\varepsilon^{ER}))}^{2-2\delta} \|S_\varepsilon\|_{L^2([0,\tau]\times\Omega_\varepsilon^{ER})}^{2\delta}$$

$$\leq c_3(c_1 r)^{1-\delta}(\tau c_2 r)^\delta = cr\tau^\delta.$$

With similar estimations we find that $\|S_{C,\varepsilon}\|_{VER}^2 \leq cr\tau^\delta$.
In the end, we choose $\tau$ such that

$$\tau < \frac{1}{\sqrt[\delta]{4c}}$$

and get

$$\|C_\varepsilon\|_{V^1}^2 + \|C_{e,\varepsilon}\|_{V^2}^2 + \|S_\varepsilon\|_{VER}^2 + \|S_{C,\varepsilon}\|_{VER}^2 \leq r,$$

and the proof is complete. $\qquad\qquad\qquad\qquad\qquad\qquad\square$

# 4.6 Identification of the calcium-stim1 limit model

In this section we determine the limit equation of the system (4.14) for $\varepsilon$ tending to zero. For homogenization of the spatial derivatives we use proposition 2.5.
We define $\chi^1(y)$, $\chi^2(y)$, $\chi^{CRAC}(y)$ and $\chi^{ER}(y)$ with $y = x/\varepsilon$, which is 1 in $\Omega_\varepsilon^1$, $\Omega_\varepsilon^2$, $\Omega_\varepsilon^{CRAC}$ and $\Omega_\varepsilon^{ER}$, respectively, and 0 otherwise.

**Nonlinear terms and terms on the Robin boundary**
To handle the nonlinear terms we apply lemma 3.4 and lemmas 4.18, 4.19, 4.16, 4.17, 4.20 (or remark 4.22) and see that the functions $C_\varepsilon$, $C_{e,\varepsilon}$, $S_\varepsilon$ and $S_{C,\varepsilon}$ each have a strongly converging subsequence in $L^2([0,T], L^2(\Omega))$.
With remark 4.22 we find that $S_{O,\varepsilon}$ and $S_{CO,\varepsilon}$ converge strongly on $L^2([0,T], L^2(\Gamma^1))$, up to a subsequence.

- For the function $f_{SERCA}$ on $\Gamma_\varepsilon^{ER}$ we use theorem 3.7 to easily get that

$$\lim_{\varepsilon \to 0} \varepsilon \langle f_{SERCA}(C_\varepsilon), \varphi_\varepsilon \rangle_{L^2(\Gamma_\varepsilon^{ER}),t} = (|\Gamma^{ER}| f_{SERCA}(C_0), \varphi_0)_{L^2(\Omega),t}$$

with $C_\varepsilon$ strongly converging to $C_0$.

- In the domain $\Omega_\varepsilon^{ER}$ we find the nonlinear function $f_e(C_{e,\varepsilon})$ in the equations for $S_\varepsilon$ and $S_{C,\varepsilon}$. Since $C_{e,\varepsilon}$ and $S_\varepsilon$ converge strongly in $\Omega_\varepsilon^{ER}$, we derive that

$$\lim_{\varepsilon \to 0} \int_{\Omega \times (0,t)} \chi^{ER}\left(\frac{x}{\varepsilon}\right) k_C^+ f_e(C_{e,\varepsilon}) S_\varepsilon \varphi_\varepsilon \, dx \, dt$$
$$= \int_{\Omega \times (0,t)} \int_{Y^{ER}} k_C^+ f_e(C_{e,0}) S_0 \varphi_0 \, dy \, dx \, dt$$

where $\chi^{ER}$ is equal to 1 for $x/\varepsilon$ in $\Omega_\varepsilon^{ER}$ and 0 otherwise and $C_{e,\varepsilon}$ converges strongly to $C_{e,0}$.

- To find the limit of the term containing the $I_{CRAC}$ we use the periodic unfolding operator,

$$\int_{\Gamma^1} \chi^{CRAC}\left(\frac{x}{\varepsilon}\right)(\text{ext}_\varepsilon S_{O,\varepsilon}(x))^2(V_\varepsilon(x) - V_{Ca})\varphi_\varepsilon(x) \, d\sigma_x$$
$$= \frac{1}{|\partial Y|} \int_{\Gamma^1} \int_{\partial Y^{CRAC}} \mathcal{T}_\varepsilon(\text{ext}_\varepsilon S_{O,\varepsilon})^2(x,y)$$
$$\mathcal{T}_\varepsilon(V_\varepsilon - V_{Ca})(x,y)\varphi(x,y) \, d\sigma_y \, d\sigma_x.$$

We denote the extension operating on the unit cell $Y = [0,1]^n$ by $\text{ext}_1$, which is independent of $\varepsilon$, and find that

$$\frac{1}{|\partial Y|} \int_{\Gamma^1} \int_{\partial Y^{CRAC}}$$
$$\mathcal{T}_\varepsilon(\text{ext}_\varepsilon S_{O,\varepsilon})^2(x,y)\mathcal{T}_\varepsilon(V_\varepsilon - V_{Ca})(x,y)\varphi(x,y) \, d\sigma_y \, d\sigma_x$$
$$= \frac{1}{|\partial Y|} \int_{\Gamma^1} \int_{\partial Y^{CRAC}} (\text{ext}_1(\mathcal{T}_\varepsilon(S_{O,\varepsilon})))^2(x,y)$$
$$\mathcal{T}_\varepsilon(V_\varepsilon - V_{Ca})(x,y)\varphi(x,y) \, d\sigma_y \, d\sigma_x$$

Since the extension $\text{ext}_1$ is continuous and bounded and $S_{O,\varepsilon}$ converges strongly (see remark 4.22), we deduce

$$\lim_{\varepsilon \to 0} \frac{1}{|\partial Y|} \int_{\Gamma^1} \int_{\partial Y^{CRAC}} (\text{ext}_1(\mathcal{T}_\varepsilon(S_{O,\varepsilon})))^2(x,y)$$
$$\mathcal{T}_\varepsilon(V_\varepsilon - V_{Ca})(x,y)\varphi(x,y) \, d\sigma_y \, d\sigma_x$$
$$= \int_{\Gamma^1} \int_{\partial Y^{CRAC}} (\text{ext}_1 S_{O,0}(x,y))^2(V_0(x,y) - V_{Ca})\varphi(x,y) \, d\sigma_y \, d\sigma_x,$$

for all $\varphi \in C^\infty(\Omega, C_\#^\infty(Y))$, because in our case $|\partial Y| = 1$.

- For the boundary term in the equation for $C_\varepsilon$ we use theorem 4.9 b) to deduce that $C_\varepsilon$ converges strongly in $L^2([0,T] \times \Gamma^1)$ and that

$$\lim_{\varepsilon \to 0} \int_{\Gamma^1 \times (0,t)} \chi^1\left(\frac{x}{\varepsilon}\right) (f_P(C_\varepsilon) + f_{NCX}(C_\varepsilon))\varphi_\varepsilon d\sigma_x dt$$

$$= |\partial Y^1| \int_{\Gamma^1 \times (0,t)} (f_P(C_0) + f_{NCX}(C_0))\varphi_0 d\sigma_x dt$$

for every $\varphi_0 \in C^\infty(\Omega)$.

- We use again theorem 4.9 for the Robin boundary term in the equations for $S_\varepsilon$,

$$\lim_{\varepsilon \to 0} \int_{\Gamma^1 \times (0,t)} \chi^{ER}\left(\frac{x}{\varepsilon}\right) k_O^+ S_\varepsilon \varphi_\varepsilon d\sigma_x dt$$

$$= |\partial Y^{ER}| \int_{\Gamma^1 \times (0,t)} k_O^+ S_0 \varphi_0 d\sigma_x dt$$

for all $\varphi_0 \in C^\infty(\Omega)$.

- Analogously we find

$$\lim_{\varepsilon \to 0} \int_{\Gamma^1 \times (0,t)} \chi^{ER}\left(\frac{x}{\varepsilon}\right) k_{CO}^+ S_{C,\varepsilon} \varphi_\varepsilon d\sigma_x dt$$

$$= |\partial Y^{ER}| \int_{\Gamma^1 \times (0,t)} k_{CO}^+ S_{C,0} \varphi_0 d\sigma_x dt$$

for all $\varphi_0 \in C^\infty(\Omega)$.

- With theorem 4.9 we deduce that $C_{e,\varepsilon}$ converges strongly in $L^2([0,T] \times \Gamma^1)$. Further, with theorem 2.4 we find that

$$\lim_{\varepsilon \to 0} \int_{\Gamma^1 \times (0,t)} \chi^{ER}\left(\frac{x}{\varepsilon}\right) k_C^+ f_e(C_{e,\varepsilon}) S_{O,\varepsilon} \varphi_\varepsilon dx dt$$

$$= \int_{\Gamma^1 \times (0,t)} \int_{\partial Y^{ER}} k_C^+ f_e(C_{e,0}) S_{O,0} \varphi_0 d\sigma_y d\sigma_x dt$$

for all $\varphi_0 \in C^\infty(\Omega)$.

For the following homogenization process we use the just derived limits of the nonlinear terms and on the boundaries. Furthermore, we apply proposition 2.5, lemma 2.3, lemma 2.9, remark 3.30 and the estimates found in lemma 4.16 and 4.18. As test functions $\varphi_\varepsilon \in C^\infty(\Omega, C_\#^\infty(Y))$ we choose functions of the form

$$\varphi_\varepsilon\left(x, \frac{x}{\varepsilon}\right) = \varphi_0(x) + \varepsilon\varphi_1\left(x, \frac{x}{\varepsilon}\right)$$

with $(\varphi_0, \varphi_1) \in C^\infty(\Omega) \times C^\infty(\Omega, C_\#^\infty(Y))$.

**Limit equation for $C_\varepsilon$**
We have the equation

$$(\chi^1 \partial_t C_\varepsilon, \varphi_\varepsilon)_{\Omega_\varepsilon^1, t} + D_C(\chi^1 \nabla C_\varepsilon, \nabla \varphi_\varepsilon)_{\Omega_\varepsilon^1, t}$$
$$+ \varepsilon\langle (L_0 + L_{IP3})(C_\varepsilon - C_{e,\varepsilon}) + f_{SERCA}(C_\varepsilon), \varphi_\varepsilon\rangle_{\Gamma_\varepsilon^{ER}, t}$$
$$+ \langle \alpha I_{CRAC} + f_P(C_\varepsilon) + f_{NCX}(C_\varepsilon), \varphi_\varepsilon\rangle_{\Gamma^1 \cap \partial\Omega_\varepsilon^1, t} = 0$$

for all admissible test functions $\varphi_\varepsilon \in C^\infty(\Omega, C_\#^\infty(Y))$. This is equivalent to

$$\int_{\Omega\times(0,t)} \chi^1\left(\frac{x}{\varepsilon}\right) \partial_t C_\varepsilon \varphi_\varepsilon dxdt + D_C \int_{\Omega\times(0,t)} \chi^1\left(\frac{x}{\varepsilon}\right) \nabla C_\varepsilon \nabla\varphi_\varepsilon dxdt$$
$$+ \varepsilon \int_{\Gamma_\varepsilon^{ER}\times(0,t)} ((L_0 + L_{IP3})(C_\varepsilon - C_{e,\varepsilon}) + f_{SERCA}(C_\varepsilon))\varphi_\varepsilon d\sigma_x dt$$
$$+ \int_{\Gamma^1\times(0,t)} \chi^{CRAC}\left(\frac{x}{\varepsilon}\right) \alpha g_{CRAC}(ext_\varepsilon S_{O,\varepsilon})^2 (V_\varepsilon - V_{Ca})\varphi_\varepsilon d\sigma_x dt$$
$$+ \int_{\Gamma^1\times(0,t)} \chi^1\left(\frac{x}{\varepsilon}\right) (f_P(C_\varepsilon) + f_{NCX}(C_\varepsilon))\varphi_\varepsilon d\sigma_x dt = 0$$

for all $\varphi_\varepsilon \in C^\infty(\Omega, C^\infty_\#(Y))$. For $\varepsilon \to 0$ we get

$$\int_{\Omega \times (0,t)} \int_{Y^1} \partial_t C_0 \varphi_0 dy dx dt$$

$$+ D_C \int_{\Omega \times (0,t)} \int_{Y^1} [\nabla_x C_0 + \nabla_y C_1][\nabla_x \varphi_0 + \nabla_y \varphi_1] dy dx dt$$

$$+ \int_{\Omega \times (0,t)} \int_{\Gamma_{ER}} ((L_0 + L_{IP3})(C_0 - C_{e,0}) + f_{SERCA}(C_0)) \varphi_0 dy dx dt$$

$$+ \int_{\Gamma^1 \times (0,t)} \int_{\partial Y_{CRAC}} \alpha g_{CRAC} (\text{ext}_1(S_{O,0}))^2 (V_0 - V_{Ca}) \varphi_0 d\sigma_y d\sigma_x dt$$

$$+ |\partial Y^1| \int_{\Gamma^1 \times (0,t)} (f_P(C_0) + f_{NCX}(C_0)) \varphi_0 d\sigma_x dt = 0$$

for all $(\varphi_0, \varphi_1) \in C^\infty(\Omega) \times C^\infty(\Omega, C^\infty_\#(Y))$, where $C_0 \in H^1(\Omega)$ and $C_1 \in L^2(\Omega, H^1_\#(Y))$.

**Limit equation for $C_{e,\varepsilon}$**
For the equation

$$\int_{\Omega \times (0,t)} \chi^2 \left(\frac{x}{\varepsilon}\right) \partial_t C_{e,\varepsilon} \varphi_\varepsilon dx dt + D_{ER} \int_{\Omega \times (0,t)} \chi^2 \left(\frac{x}{\varepsilon}\right) \nabla C_{e,\varepsilon} \nabla \varphi_\varepsilon dx dt$$

$$+ \varepsilon \int_{\Gamma^{ER}_\varepsilon \times (0,t)} ((L_0 + L_{IP3})(C_{e,\varepsilon} - C_\varepsilon) - f_{SERCA}(C_\varepsilon)) \varphi_\varepsilon d\sigma_x dt = 0$$

we find for $\varepsilon \to 0$

$$\int_{\Omega \times (0,t)} \int_{Y^2} \partial_t C_{e,0} \varphi_0 dy dx dt$$

$$+ D_{ER} \int_{\Omega \times (0,t)} \int_{Y^2} [\nabla_x C_{e,0} + \nabla_y C_{e,1}][\nabla_x \varphi_0 + \nabla_y \varphi_1] dy dx dt$$

$$+ \int_{\Omega \times (0,t)} \int_{\Gamma_{ER}} ((L_0 + L_{IP3})(C_{e,0} - C_0) - f_{SERCA}(C_0)) \varphi_0 dy dx dt = 0$$

for all $(\varphi_0, \varphi_1) \in C^\infty(\Omega) \times C^\infty(\Omega, C^\infty_\#(Y))$, where $C_{e,0} \in H^1(\Omega)$ and $C_{e,1} \in L^2(\Omega, H^1_\#(Y))$.

**Limit equation for $S_\varepsilon$**
We have

$$\int_{\Omega\times(0,t)} \chi^{ER}\left(\frac{x}{\varepsilon}\right) \partial_t S_\varepsilon \varphi_\varepsilon dxdt + D_S \int_{\Omega\times(0,t)} \chi^{ER}\left(\frac{x}{\varepsilon}\right) \nabla S_\varepsilon \nabla \varphi_\varepsilon dxdt$$

$$+ \int_{\Omega\times(0,t)} \chi^{ER}\left(\frac{x}{\varepsilon}\right) (k_C^+ f_e(C_{e,\varepsilon}) S_\varepsilon - k_C^- S_{C,\varepsilon}) \varphi_\varepsilon dxdt$$

$$+ \int_{\Gamma^1\times(0,t)} \chi^{ER}\left(\frac{x}{\varepsilon}\right) (k_O^+ S_\varepsilon - k_O^- S_{O,\varepsilon}) \varphi_\varepsilon d\sigma_x dt = 0$$

for all $\varphi_\varepsilon \in C^\infty(\Omega, C_\#^\infty(Y))$. For $\varepsilon \to 0$ we get

$$\int_{\Omega\times(0,t)} \int_{Y^{ER}} \partial_t S_0 \varphi_0 dydxdt$$

$$+ D_S \int_{\Omega\times(0,t)} \int_{Y^{ER}} [\nabla_x S_0 + \nabla_y S_1][\nabla_x \varphi_0 + \nabla_y \varphi_1] dydxdt$$

$$\int_{\Omega\times(0,t)} \int_{Y^{ER}} (k_C^+ f_e(C_{e,0}) S_0 - k_C^- S_{C,0}) \varphi_0 dydxdt$$

$$+ \int_{\Gamma^1\times(0,t)} \int_{\partial Y^{ER}} k_O^+ S_0 \varphi_0 d\sigma_y d\sigma_x dt$$

$$- \int_{\Gamma^1\times(0,t)} \int_{\partial Y^{ER}} k_O^- S_{O,0} \varphi_0 d\sigma_y d\sigma_x dt = 0$$

for all $(\varphi_0, \varphi_1) \in C^\infty(\Omega) \times C^\infty(\Omega, C_\#^\infty(Y))$, where $S_0 \in H^1(\Omega)$ and $S_1 \in L^2(\Omega, H_\#^1(Y))$.

**Limit equation for $S_{C,\varepsilon}$**
The equation for $S_{C,\varepsilon}$ is given by

$$\int_{\Omega\times(0,t)} \chi^{ER}\left(\frac{x}{\varepsilon}\right) \partial_t S_{C,\varepsilon} \varphi_\varepsilon dxdt$$

$$+ D_{SC} \int_{\Omega\times(0,t)} \chi^{ER}\left(\frac{x}{\varepsilon}\right) \nabla S_{C,\varepsilon} \nabla \varphi_\varepsilon dxdt$$

$$+ \int_{\Omega\times(0,t)} \chi^{ER}\left(\frac{x}{\varepsilon}\right) (k_C^- S_{C,\varepsilon} - k_C^+ f_e(C_{e,\varepsilon}) S_\varepsilon) \varphi_\varepsilon dxdt$$

$$+ \int_{\Gamma^1\times(0,t)} \chi^{ER}\left(\frac{x}{\varepsilon}\right) (k_{CO}^+ S_{C,\varepsilon} - k_{CO}^- S_{CO,\varepsilon}) \varphi_\varepsilon d\sigma_x dt = 0.$$

For $\varepsilon \to 0$ we arrive at

$$\int_{\Omega \times (0,t)} \int_{Y^{\mathrm{ER}}} \partial_t S_{C,0} \varphi_0 dy dx dt$$

$$+ D_{SC} \int_{\Omega \times (0,t)} \int_{Y^{\mathrm{ER}}} [\nabla_x S_{C,0} + \nabla_y S_{C,1}][\nabla_x \varphi_0 + \nabla_y \varphi_1] dy dx dt$$

$$\int_{\Omega \times (0,t)} \int_{Y^{\mathrm{ER}}} (k_C^- S_{C,0} - k_C^+ f_e(C_{e,0}) S_0) \varphi_0 dy dx dt$$

$$+ \int_{\Gamma^1 \times (0,t)} \int_{\partial Y^{\mathrm{ER}}} k_{CO}^+ S_{C,0} \varphi_0 d\sigma_y d\sigma_x dt$$

$$- \int_{\Gamma^1 \times (0,t)} \int_{\partial Y^{\mathrm{ER}}} k_{CO}^- S_{CO,0} \varphi_0 d\sigma_y d\sigma_x dt = 0$$

for all $(\varphi_0, \varphi_1) \in C^\infty(\Omega) \times C^\infty(\Omega, C^\infty_\#(Y))$, where $S_{C,0} \in H^1(\Omega)$ and $S_{C,1} \in L^2(\Omega, H^1_\#(Y))$.

Now we consider the functions $S_{O,\varepsilon}$ and $S_{CO,\varepsilon}$ that only live on the boundary $\Gamma^1 \cap \partial \Omega_\varepsilon^{\mathrm{ER}}$ and homogenization takes place in one dimension less.

**Limit equation for $S_{O,\varepsilon}$**
The equation for $S_{O,\varepsilon}$ is given by

$$\int_{\Gamma^1 \times (0,t)} \chi^{\mathrm{ER}} \left( \frac{x}{\varepsilon} \right) \partial_t S_{O,\varepsilon} \varphi_\varepsilon d\sigma_x dt + \int_{\Gamma^1 \times (0,t)} \chi^{\mathrm{ER}} \left( \frac{x}{\varepsilon} \right) (k_O^- S_{O,\varepsilon}$$

$$- k_O^+ S_\varepsilon - k_C^- S_{CO,\varepsilon} + k_C^+ f_e(C_{e,\varepsilon}) S_{O,\varepsilon}) \varphi_\varepsilon d\sigma_x dt = 0$$

for all $\varphi_\varepsilon \in C^\infty(\Gamma^1, C^\infty_\#(\partial Y))$. For $\varepsilon \to 0$ we get

$$\int_{\Gamma^1 \times (0,t)} \int_{\partial Y^{\mathrm{ER}}} \partial_t S_{O,0} \varphi_0 d\sigma_y d\sigma_x dt + \int_{\Gamma^1 \times (0,t)} \int_{\partial Y^{\mathrm{ER}}} (k_O^- S_{O,0}$$

$$- k_O^+ S_0 - k_C^- S_{CO,0} + k_C^+ f_e(C_{e,0}) S_{O,0}) \varphi_0 d\sigma_y d\sigma_x dt = 0$$

for all $\varphi_0 \in C^\infty(\Gamma^1, C^\infty_\#(\partial Y))$, where $S_{O,0} \in L^2(\Gamma^1 \times \partial Y^{\mathrm{ER}})$.

**Limit equation for $S_{CO,\varepsilon}$**
The equation for $S_{CO,\varepsilon}$ is given by

$$\int_{\Gamma^1 \times (0,t)} \chi^{\mathrm{ER}} \left( \frac{x}{\varepsilon} \right) \partial_t S_{CO,\varepsilon} \varphi_\varepsilon d\sigma_x dt + \int_{\Gamma^1 \times (0,t)} \chi^{\mathrm{ER}} \left( \frac{x}{\varepsilon} \right) (k_{CO}^- S_{CO,\varepsilon}$$

$$- k_{CO}^+ S_{C,\varepsilon} + k_C^- S_{CO,\varepsilon} - k_C^+ f_e(C_{e,\varepsilon}) S_{O,\varepsilon}) \varphi_\varepsilon d\sigma_x dt = 0.$$

For $\varepsilon \to 0$ we find

$$\int_{\Gamma^1 \times (0,t)} \int_{\partial Y^{ER}} \partial_t S_{CO,0} \varphi_0 d\sigma_y d\sigma_x dt + \int_{\Gamma^1 \times (0,t)} \int_{\partial Y^{ER}} (k_{CO}^- S_{CO,0}$$
$$- k_{CO}^+ S_{C,0} + k_C^- S_{CO,0} - k_C^+ f_e(C_{e,0}) S_{O,0}) \varphi_0 d\sigma_y d\sigma_x dt = 0$$

for $\varepsilon \to 0$ for all $\varphi_0 \in C^\infty(\Gamma^1, C_\#^\infty(\partial Y))$, where $S_{CO,0} \in L^2(\Gamma^1 \times \partial Y^{ER})$.

We do not need to consider the limit formation of the equations for $V_\varepsilon$ and $a_\varepsilon$ given in (4.14), since these functions are ordinary differential equations and the homogenization process is analogous to $S_{O,\varepsilon}$ or $S_{CO,\varepsilon}$, because $C_\varepsilon$ and $S_{O,\varepsilon}$ converge strongly on $\Gamma^1 \cap \partial \Omega_\varepsilon^1$. The limit functions $V_0$ and $a_0$ are given by

$$\partial_t V_0 = -\frac{I_K + I_{K(Ca)} + I_{CRAC} + I_{CAN}}{C_m},$$
$$I_K = g_K \frac{1}{1 + e^{\frac{-(V_0 - V_n)}{K_n}}} (V_0 - V_K),$$
$$I_{K(Ca)} = g_{K(Ca)} \frac{C_0^4}{C_0^4 + K_{K(Ca)}^4} (V_0 - V_K),$$
$$I_{CRAC} = g_{CRAC} (\text{ext}_1 S_{O,0})^2 (V_0 - V_{Ca}),$$
$$I_{CAN} = g_{CAN} a_0 (V_0 - V_{Na}),$$
$$\partial_t a_0 = \frac{C_0^4}{\tau_{CAN}(C_0^4 + K_{CAN}^4)} - \frac{a_0}{\tau_{CAN}}.$$

## Weak formulation of the homogeneous model

Now we regard the $y$-dependence of the functions and shorten some terms. Because the functions $C_{e,0}$, $S_0$ and $S_{C,0}$ and the initial conditions $S_{O,0}(0)$ and $S_{CO,0}(0)$ are $y$-independent and $S_{O,0}, S_{CO,0}$ are given by ordinary differential equations, also $S_{O,0}$ and $S_{CO,0}$ are $y$-independent and we simplify the just found equations to the following weak system of equations.

Let $(C_0, C_{e,0}, S_0, S_{C,0}, S_{O,0}, S_{CO,0}) \in \mathcal{V}(\Omega)^4 \times \mathcal{V}(\Gamma^1)^2$ and $(C_1, C_{e,1}, S_1, S_{C,1}) \in \mathcal{V}(\Omega, Y)$ such that

$$|Y^1|(\partial_t C_0, \varphi_0)_\Omega + D_C(\nabla_x C_0 + \nabla_y C_1, \nabla_x \varphi_0 + \nabla_y \varphi_1)_{\Omega \times Y^1}$$
$$+ |\Gamma^{ER}|((L_0 + L_{IP3})(C_0 - C_{e,0}) + f_{SERCA}(C_0), \varphi_0)_\Omega$$
$$+ (\alpha I_{CRAC}, \varphi_0)_{\Gamma^1 \times \partial Y^1} + |\partial Y^1|(f_P(C_0) + f_{NCX}(C_0), \varphi_0)_{\Gamma^1} = 0,$$

$$|Y^2|(\partial_t C_{e,0}, \varphi_0)_\Omega + D_{ER}(\nabla_x C_{e,0} + \nabla_y C_{e,1}, \nabla_x \varphi_0 + \nabla_y \varphi_1)_{\Omega \times Y^2}$$
$$+ |\Gamma^{ER}|((L_0 + L_{IP3})(C_{e,0} - C_0) - f_{SERCA}(C_0), \varphi_0)_\Omega = 0,$$

$$|Y^{ER}|(\partial_t S_0, \varphi_0)_\Omega + D_S(\nabla_x S_0 + \nabla_y S_1, \nabla_x \varphi_0 + \nabla_y \varphi_1)_{\Omega \times Y^{ER}}$$
$$+ |Y^{ER}|(k_C^+ f_e(C_{e,0}) S_0 - k_C^- S_{C,0}, \varphi_0)_\Omega$$
$$+ |\partial Y^{ER}|(k_O^+ S_0 - k_O^- S_{O,0}, \varphi_0)_{\Gamma^1} = 0,$$

$$|Y^{ER}|(\partial_t S_{C,0}, \varphi_0)_\Omega + D_S(\nabla_x S_{C,0} + \nabla_y S_{C,1}, \nabla_x \varphi_0 + \nabla_y \varphi_1)_{\Omega \times Y^{ER}}$$
$$+ |Y^{ER}|(k_C^- S_{C,0} - k_C^+ f_e(C_{e,0}) S_0, \varphi_0)_\Omega$$
$$+ |\partial Y^{ER}|(k_{CO}^+ S_{C,0} - k_{CO}^- S_{CO,0}, \varphi_0)_{\Gamma^1} = 0,$$

$$(\partial_t S_{O,0}, \varphi_0)_{\Gamma^1}$$
$$+ (k_O^- S_{O,0} - k_O^+ S_0 - k_C^- S_{CO,0} + k_C^+ f_e(C_{e,0}) S_{O,0}, \varphi_0)_{\Gamma^1} = 0,$$

$$(\partial_t S_{CO,0}, \varphi_0)_{\Gamma^1}$$
$$+ (k_{CO}^- S_{CO,0} - k_{CO}^+ S_{C,0} + k_C^- S_{CO,0} - k_C^+ f_e(C_{e,0}) S_{O,0}, \varphi_0)_{\Gamma^1} = 0,$$

for all $\varphi_0 \in C^\infty(\Omega)$ and $\varphi_1 \in C^\infty(\Omega, C_\#^\infty(Y))$.

The next step is to shrink the blown up membrane $Y^{ER}$ back to $\Gamma^{ER}$.

From now on, we rename the functions $(C_0, C_{e,0}, S_0, S_{C,0}, S_{O,0}, S_{CO,0}, V_0, a_0)$ by $(C, C_e, S, S_C, S_O, S_{CO}, V, a)$ to avoid confusion.

# 4.7 Delta-Limit of the homogeneous model

It is our aim to let $Y^{ER}$ tend to $\Gamma^{ER}$ as described in section 4.1. We use the two-step convergence and theorem 4.1 for the functions $S$ and $S_C$.

The condition that $\Gamma^{ER}$ is a smooth manifold and that

245

$Y^{\text{ER}} = \{p + d \cdot n_p | \ p \in \Gamma^{\text{ER}}, d \in (-\delta, \delta)\}$ needs to be satisfied, where $n_p$ is the outer normal in $p \in \Gamma^{\text{ER}}$. This also implies $\partial Y^{\text{ER}} = \{p + d \cdot n_p | \ p \in \partial\Gamma^{\text{ER}}, d \in (-\delta, \delta)\}$.

First we consider the behavior of the functions $S$ and $S_C$. The functions $S_O$ and $S_{CO}$ are hardly influenced by the $\delta$-limit formation since we divided the corresponding equations by $|\partial Y^{\text{ER}}|$. Nevertheless, we need to check if the $\delta$-limit formation works for the extension of the function $S_O$.

Then, we consider the impact of the limit formation for $\delta$ tending to zero for the functions $C$ and $C_e$.

### $\delta$-limit for the equations for $S$ and $S_C$

We are able to easily use the two-step convergence and the theorem 4.1, because the equations for $S$ and $S_C$ have the same form as used in theorem 4.1. For equations of this form, boundedness independently of $\delta$ and limit progression are proven and done in the proof of theorem 4.1 and we deduce that

$$
|\Gamma^{\text{ER}}| \int_\Omega \partial_t S \varphi_0 dx
$$
$$
+ \int_\Omega \sum_{i,j} \partial_{x_j} S \, D_S \underbrace{\int_{\Gamma^{\text{ER}}} (P_\Gamma(e_j + \nabla_\Gamma \mu_j^S))_i d\sigma_y}_{=P_{ij}^S} \partial_{x_i} \varphi_0 dx
$$
$$
+ |\Gamma^{\text{ER}}| \int_\Omega (k_C^+ f_e(C_e)S - k_C^- S_C)\varphi_0 dx
$$
$$
+ |\partial\Gamma^{\text{ER}}| \int_{\Gamma^1} (k_O^+ S - k_O^- S_O)\varphi_0 d\sigma_x = 0
$$

with

$$
\nabla_\Gamma \cdot \left(P_\Gamma(e_j + \nabla_\Gamma \mu_j^S)\right) = 0 \qquad \text{in } \Gamma^{\text{ER}}
$$
$$
P_\Gamma(e_j + \nabla_\Gamma \mu_j^S) \cdot n = 0 \qquad \text{on } \partial\Gamma^{\text{ER}},
$$

and $\mu_j^S$ being $Y$-periodic. Then we have

$$
S_1 = \sum_{i=1}^n \nabla_{x_i} S \mu_i^S.
$$

Hence, the equation for $S$ is

$$|\Gamma^{ER}|(\partial_t S, \varphi_0)_\Omega + (P^S \nabla S, \nabla \varphi_0)_\Omega$$
$$+ |\Gamma^{ER}|(k_C^+ f_e(C_e) S - k_C^- S_C, \varphi_0)_\Omega + |\partial \Gamma^{ER}|(k_O^+ S - k_O^- S_O, \varphi_0)_{\Gamma^1} = 0.$$

We have exactly the same case in the $S_C$ equation and find

$$|\Gamma^{ER}|(\partial_t S_C, \varphi_0)_\Omega + (P^{SC} \nabla S_C, \nabla \varphi_0)_\Omega$$
$$|\Gamma^{ER}|(k_C^- S_C - k_C^+ f_e(C_e) S, \varphi_0)_\Omega + |\partial \Gamma^{ER}|(k_{CO}^+ S_C - k_{CO}^- S_{CO}, \varphi_0)_{\Gamma^1} = 0$$

with the tensor $P^{SC} = (P_{ij}^{SC})_{i,j=1,\ldots,n}$,

$$P_{ij}^{SC} = D_{SC} \int_{\Gamma^{ER}} \left( P_\Gamma(e_j + \nabla_\Gamma \mu_j^S) \right)_i d\sigma_y.$$

**$\delta$-limit for $I_{CRAC}$ and $C$**
We have a special case for the $I_{CRAC}$, because the support of the extension $\text{ext}_1 S_O$ is $\partial Y^{CRAC}$ instead of $\partial Y^1$. We have

$$(\alpha I_{CRAC}, \varphi_0)_{\Gamma^1 \times \partial Y^1} = (\alpha g_{CRAC}(\text{ext}_1 S_O)^2(V - V_{Ca}), \varphi_0)_{\Gamma^1 \times \partial Y^{CRAC}}.$$

We need to check what happens to the extension $\text{ext}_1$, when $\delta$ tends to zero. Therefore, we take a look at the heart of the extension 4.10 and see that for $\delta \to 0$ the extension $\text{ext}_1$ tends to

$$\text{ext}_1 : L^2(\partial \Gamma^{ER}) \hookrightarrow L^2(\partial \Gamma^{ER} \cup \partial Y^{CRAC}),$$

$$\text{ext}_1 \, g_j(y) := \left\{ \begin{array}{ll} g_j(y), & y_2 = 0 \\ g_j(y_1, 0), & y_2 < 0 \end{array} \right\} = g_j(y_1, 0) \quad \forall y_2.$$

This means that functions defined on $L^2(\partial \Gamma^{ER})$ are constantly extended to $L^2(\partial Y^{CRAC})$ and the extension is well-defined for $\delta \to 0$. Because $S_O$ and $V$ is independent of $y$ we deduce that

$$(\alpha I_{CRAC}, \varphi_0)_{\Gamma^1 \times \partial Y^1} = |\partial Y^{CRAC}|(\alpha g_{CRAC} S_O^2(V - V_{Ca}), \varphi_0)_{\Gamma^1}.$$

To identify the cell problem of the equation for $C$ we set $\varphi_0 = 0$ and find the cell problem as we did in equation (2.22) of section

2.6. We find that $C_1 = \sum_{j=1}^{n} \partial_{x_j} C(x,t) \mu_j^C(y)$ with $\mu_j^C$ satisfying the cell problem

$$
\begin{aligned}
\nabla_y \cdot D_C(e_j + \nabla_y \mu_j^C) &= 0 && \text{in } Y^1, \\
D_{ER}(e_j + \nabla_y \mu_j^C) \cdot n &= 0 && \text{on } \Gamma^{ER},
\end{aligned}
$$

and $\mu_j^C$ must be $Y$-periodic for $j = 1, \ldots, n$. Further, we define the diffusion tensor

$$
P_{ij}^C := \int_{Y^1} D_C(\delta_{ij} + \partial_{y_i} \mu_j^C) dy.
$$

Finally we find the weak formulation of the function $C$ as

$$
\begin{aligned}
|Y^1|(\partial_t C, \varphi_0)_\Omega &+ (P^C \nabla C, \nabla \varphi_0)_\Omega \\
&+ |\Gamma^{ER}|((L_0 + L_{IP3})(C - C_e) + f_{SERCA}(C), \varphi_0)_\Omega \\
+ |\partial Y^{CRAC}|(\alpha I_{CRAC}, \varphi_0)_{\Gamma^1} &+ |\partial Y^1|(f_P(C) + f_{NCX}(C), \varphi_0)_{\Gamma^1} = 0.
\end{aligned}
$$

### $\delta$-limit for $C_e$, $S_O$ and $S_{CO}$

The equations for $C_e$, $S_O$ and $S_{CO}$ are independent of $\delta$ and hence stay untouched for $\delta$ tending to zero. Hence, it remains to calculate the diffusion tensor $P^e$ by the cell problem as we did in equation (2.22). We find $C_{e,1} = \sum_{j=1}^{n} \partial_{x_j} C_e(x,t) \mu_j^e(y)$ with $\mu_j^e$ satisfying the cell problem

$$
\begin{aligned}
\nabla_y \cdot D_{ER}(e_j + \nabla_y \mu_j^e) &= 0 && \text{in } Y^2, \\
D_{ER}(e_j + \nabla_y \mu_j^e) \cdot n &= 0 && \text{on } \Gamma^{ER},
\end{aligned}
$$

and $\mu_j^e$ must be $Y$-periodic for $j = 1, \ldots, n$. Further, we define the diffusion tensor

$$
P_{ij}^{ER} := \int_{Y^2} D_{ER}(\delta_{ij} + \partial_{y_i} \mu_j^e) dy.
$$

Finally we find the equations

$$
\begin{aligned}
|Y^2|(\partial_t C_e, \varphi)_\Omega &+ (P^e \nabla C_e, \nabla \varphi)_\Omega \\
&+ |\Gamma^{ER}|((L_0 + L_{IP3})(C_e - C) - f_{SERCA}(C), \varphi)_\Omega = 0
\end{aligned}
$$

for all $\varphi \in C^\infty(\Omega)$,

$$(\partial_t S_O, \varphi)_{\Gamma^1} + (k_O^- S_O - k_O^+ S - k_C^- S_{CO} + k_C^+ f_e(C_e) S_O, \varphi)_{\Gamma^1} = 0$$

and

$$(\partial_t S_{CO}, \varphi)_{\Gamma^1} + (k_{CO}^- S_{CO} - k_{CO}^+ S_C + k_C^- S_{CO} - k_C^+ f_e(C_e) S_O, \varphi)_{\Gamma^1} = 0$$

for all $\varphi \in C^\infty(\Gamma^1)$.

## Weak formulation of equation (4.14) after homogenization and $\delta$-limit formation

Let $(C, C_e, S, S_C, S_O, S_{CO}) \in \mathcal{V}(\Omega)^4 \times \mathcal{V}(\Gamma^1)^2$ be such that

$$|Y^1|(\partial_t C, \varphi)_\Omega + (P^C \nabla C, \nabla \varphi)_\Omega$$
$$+ |\Gamma^{ER}|((L_0 + L_{IP3})(C - C_e) + f_{SERCA}(C), \varphi)_\Omega$$
$$+ |\partial Y^{CRAC}|(\alpha I_{CRAC}, \varphi)_{\Gamma^1} + |\partial Y^1|(f_P(C) + f_{NCX}(C), \varphi)_{\Gamma^1} = 0,$$

$$|Y^2|(\partial_t C_e, \varphi)_\Omega + (P^e \nabla C_e, \nabla \varphi)_\Omega$$
$$+ |\Gamma^{ER}|((L_0 + L_{IP3})(C_e - C) - f_{SERCA}(C), \varphi)_\Omega = 0,$$

$$|\Gamma^{ER}|(\partial_t S, \varphi)_\Omega + (P^S \nabla S, \nabla \varphi)_\Omega$$
$$+ |\Gamma^{ER}|(k_C^+ f_e(C_e) S - k_C^- S_C, \varphi)_\Omega + |\partial \Gamma^{ER}|(k_O^+ S - k_O^- S_O, \varphi)_{\Gamma^1} = 0,$$

$$|\Gamma^{ER}|(\partial_t S_C, \varphi)_\Omega + (P^{SC} \nabla S_C, \nabla \varphi)_\Omega$$
$$+ |\Gamma^{ER}|(k_C^- S_C - k_C^+ f_e(C_e) S, \varphi)_\Omega + |\partial \Gamma^{ER}|(k_{CO}^+ S_C - k_{CO}^- S_{CO}, \varphi)_{\Gamma^1} = 0,$$

$$(\partial_t S_O, \varphi)_{\Gamma^1} + (k_O^- S_O - k_O^+ S - k_C^- S_{CO} + k_C^+ f_e(C_e) S_O, \varphi)_{\Gamma^1} = 0,$$

$$(\partial_t S_{CO}, \varphi)_{\Gamma^1} + (k_{CO}^- S_{CO} - k_{CO}^+ S_C + k_C^- S_{CO} - k_C^+ f_e(C_e) S_O, \varphi)_{\Gamma^1} = 0,$$
$$(4.16)$$

for all $\varphi \in C^\infty(\Omega)$.

## Continuity of $C_e$

Additionally we are able to show that the function $C_e$ is continuous in the spatial variable, if $\Omega \subset \mathbb{R}^3$. We use the Sobolev embedding theorem, see [1], and find that a function $u$ is continuous if $u \in W^{m,p}(\Omega)$ with $\Omega \in \mathbb{R}^n$ and

$$m - \frac{n}{p} > 0.$$

For us this means that $C_e$ should be in $L^\infty([0,T], H^2(\Omega))$, since $[0,T] \subset \mathbb{R}$ and $\Omega \subset \mathbb{R}^3$. Then we have continuity in space and boundedness almost everywhere in time.

**Theorem 4.29 (Boundedness of $C_e$ in $L^\infty([0,T], H^2(\Omega))$)**
Let $\Omega \subset \mathbb{R}^3$. The function $C_e$ in system (4.16) is an element of $L^\infty([0,T], H^2(\Omega))$.

**Proof** It suffices to show that

$$\|C_e\|^2_{L^2(\Omega)} + \|\nabla C_e\|^2_{L^2(\Omega)} + \|\Delta C_e\|^2_{L^2(\Omega)} \le c$$

for a constant $c > 0$ and for almost every $t \in [0,T]$.
We test the equation for $C_e$ with $\varphi = \partial_t \Delta C_e$, which can be made rigorous as seen in lemma 7.23 in [25]

$$|Y^2|(\partial_t C_e, \partial_t \Delta C_e)_\Omega - (P^e \Delta C_e, \partial_t \Delta C_e)_\Omega$$
$$= |\Gamma^{ER}|(L_0 + L_{IP3})(C - C_e, \partial_t \Delta C_e)_\Omega + |\Gamma^{ER}|(f_{SERCA}(C), \partial_t \Delta C_e)_\Omega.$$

We integrate by parts and multiply the result by $-1$ to get

$$|Y^2| \|\partial_t \nabla C_e\|^2_\Omega + P^e \frac{1}{2} \partial_t \|\Delta C_e\|^2_\Omega$$
$$= |\Gamma^{ER}|(L_0 + L_{IP3})((\nabla C, \partial_t \nabla C_e)_\Omega - (\nabla C_e, \partial_t \nabla C_e)_\Omega$$
$$- (C - C_e, \underbrace{\partial_t \nabla C_e \cdot n}_{=0})_{\Gamma^1}) + |\Gamma^{ER}|((f'_{SERCA}(C) \nabla C, \partial_t \nabla C_e)_\Omega$$
$$- (f_{SERCA}(C), \underbrace{\partial_t \nabla C_e \cdot n}_{=0})_{\Gamma^1}).$$

The binomial theorem leads to

$$|Y^2| \|\partial_t \nabla C_e\|_\Omega^2 + P^e \frac{1}{2} \partial_t \|\Delta C_e\|_\Omega^2 + |\Gamma^{ER}|(L_0 + L_{IP3}) \frac{1}{2} \partial_t \|\nabla C_e\|_\Omega^2$$

$$\leq |\Gamma^{ER}|(L_0 + L_{IP3}) \left( \frac{\lambda}{2} \|\nabla C\|_\Omega^2 + \frac{1}{2\lambda} \|\partial_t \nabla C_e\|_\Omega^2 \right)$$

$$+ |\Gamma^{ER}| \frac{\lambda}{2} \|\underbrace{f'_{SERCA}}_{\text{bounded}} \nabla C\|_\Omega^2 + |\Gamma^{ER}| \frac{1}{2\lambda} \|\partial_t \nabla C_e\|_\Omega^2.$$

By integration from 0 to $t$ we find

$$\left( |Y^2| - |\Gamma^{ER}|(L_0 + L_{IP3}) \frac{1}{2\lambda} - |\Gamma^{ER}| \frac{1}{2\lambda} \right) \|\partial_t \nabla C_e\|_{\Omega,t}^2$$

$$+ P^e \frac{1}{2} \|\Delta C_e\|_\Omega^2 + |\Gamma^{ER}|(L_0 + L_{IP3}) \frac{1}{2} \|\nabla C_e\|_\Omega^2$$

$$\leq \lambda c_1 \|\nabla C\|_{\Omega,t}^2 + c_2 \|\Delta C_e(0)\|_\Omega^2 + c_3 \|\nabla C_e(0)\|_\Omega^2.$$

With lemma 4.18 we see that $\|\nabla C\|_{\Omega,t}^2$ is bounded. Sufficiently smooth initial conditions yield boundedness of the last two terms. We choose $\lambda$ big enough, but finite, and find

$$\|\partial_t \nabla C_e\|_{\Omega,t}^2 + \|\Delta C_e\|_\Omega^2 + \|\nabla C_e\|_\Omega^2 \leq c$$

for almost every $t \in [0, T]$. Again, we find in lemma 4.18 that $\|C_e\|_\Omega^2 \leq c$ for a constant $c \in \mathbb{R}$ and we conclude that there exists a $c > 0$ such that

$$\|C_e\|_\Omega^2 + \|\nabla C_e\|_\Omega^2 + \|\Delta C_e\|_\Omega^2 \leq c$$

for almost every $t \in [0, T]$, which completes the proof. $\square$

# 4.8 Uniqueness of the Limit Model

We show that there exists just one solution of the limit model (4.16).

**Theorem 4.30 (Uniqueness)**
*There exists at most one solution for the limit model (4.16).*

**Proof** First we show that the tensors $P^C$, $P^e$ and $P^S$ are unique. We start with the tensors $P^C$ and $P^e$. The functions $\mu^C$ and $\mu^e$ live in different domains but satisfy the same differential equation, hence it is sufficient to show that there is at most one solution for

$$\nabla \cdot ((e_j + \nabla \mu_j)) = 0 \qquad \text{in } Y,$$
$$(e_j + \nabla \mu_j) \cdot n = 0 \qquad \text{on } \partial Y,$$

for $j = 1, \ldots, n$ and $Y$ a Lipschitz domain. Therefore, we assume that $\mu^1$ and $\mu^2$ are two solutions and subtract the weak formulations, which leads to

$$\int_Y (\nabla \mu_j^1 - \nabla \mu_j^2) \nabla \varphi \mathrm{d}x = 0.$$

We test with the function $\varphi = \mu_j^1 - \mu_j^2$ and obtain

$$\|\nabla (\mu_j^1 - \mu_j^2)\|_Y^2 = 0.$$

We conclude

$$\nabla \mu_j^1 = \nabla \mu_j^2 \qquad \text{which yields} \qquad \partial_{y_i} \mu_j^1(y, t) = \partial_{y_i} \mu_j^1(y, t)$$

for $i, j = 1, \ldots, n$ and almost every $y \in Y, t \in [0, T]$. This yields

$$P_{ij} = \int_Y (\delta_{ij} + \partial_{y_i} \mu_j^1(y, t)) \mathrm{d}y = \int_Y (\delta_{ij} + \partial_{y_i} \mu_j^2(y, t)) \mathrm{d}y$$

for $i, j = 1, \ldots, n$. We deduce that the solutions $\mu^1$ and $\mu^2$ are equal and $P^C$ and $P^e$ are unique.

Next we also show uniqueness for the tensor $P^S$. We assume that $\mu^1$ and $\mu^2$ are two solutions of the pde

$$\nabla_\Gamma \cdot (P_\Gamma(e_j + \nabla_\Gamma \mu_j)) = 0 \qquad \text{in } \Gamma^{\text{ER}},$$
$$P_\Gamma(e_j + \nabla_\Gamma \mu_j) \cdot n = 0 \qquad \text{on } \partial \Gamma^{\text{ER}},$$

for $j = 1, \ldots, n$. We subtract the weak formulations of the two solutions and test with the function $\varphi = \mu_j^1 - \mu_j^2$,

$$\int_{\Gamma^{\text{ER}}} \left( P_\Gamma(e_j + \nabla_\Gamma \mu_j^1) - P_\Gamma(e_j + \nabla_\Gamma \mu_j^2) \right) \nabla_\Gamma (\mu_j^1 - \mu_j^2) \mathrm{d}x = 0.$$

We substract $P_\Gamma e_j$, and obtain because of $P_\Gamma \nabla_\Gamma \mu_j = \nabla_\Gamma \mu_j$ that

$$\|\nabla_\Gamma (\mu_j^1 - \mu_j^2)\|_{\Gamma_{ER}}^2 = 0.$$

We deduce that

$$P_{ij}^S = \int_{\Gamma_{ER}} (P_\Gamma (e_j + \nabla_\Gamma \mu_j^1)))_i \mathrm{d}\sigma_y = \int_{\Gamma_{ER}} (P_\Gamma (e_j + \nabla_\Gamma \mu_j^2)))_i \mathrm{d}\sigma_y$$

for $i, j = 1, \ldots, n$ and there is at most one $P^S$.

Now we come to uniqueness for the solutions themself. We assume there are two solutions
$(C_1, C_{e,1}, S_1, S_{C,1}, S_{O,1}, S_{CO,1}, V_1, a_1)$ and
$(C_2, C_{e,2}, S_2, S_{C,2}, S_{O,2}, S_{CO,2}, V_2, a_2)$ of the system of equations
(4.16) with the same initial values. Starting with the equation for
$C_1$ and $C_2$, we test the weak formulations with $\varphi = C_1 - C_2$ and
subtract the two results.

$$|Y^1|(\partial_t (C_1 - C_2), (C_1 - C_2))_\Omega + P^C \|\nabla C_1 - \nabla C_2\|_\Omega^2$$
$$+ |\Gamma^{ER}|(L_0 + L_{IP3}) \underbrace{\|C_1 - C_2\|_\Omega^2}_{\geq 0}$$
$$- |\Gamma^{ER}|(L_0 + L_{IP3})(C_{e,1} - C_{e,2}, C_1 - C_2)_\Omega$$
$$+ \underbrace{|\Gamma^{ER}|(f_{SERCA}(C_1) - f_{SERCA}(C_2), C_1 - C_2)_\Omega}_{\geq 0, \text{ since } f_{SERCA} \text{ monotone, increasing}}$$
$$+ \alpha g_{CRAC}|\partial Y^{CRAC}|(S_{O,1}^2(V_1 - V_{Ca}) - S_{O,2}^2(V_2 - V_{Ca}), C_1 - C_2)_{\Gamma^1}$$
$$+ \underbrace{|\partial Y^1|(f_P(C_1) + f_{NCX}(C_1) - f_P(C_2) - f_{NCX}(C_2), C_1 - C_2)_{\Gamma^1}}_{\geq 0, \text{ since } f_p, f_{NCX} \text{ monotone, increasing}} = 0.$$

We use that $x_1, x_2, y_1, y_2$ satisfy

$$x_1^2 y_1 - x_2^2 y_2 = y_1(x_1 + x_2)(x_1 - x_2) + x_2^2(y_1 - y_2)$$

for $x_1 = S_{O,1}$, $x_2 = S_{O,2}$, $y_1 = V_1 - V_{Ca}$, $y_2 = V_2 - V_{Ca}$. Further we know that $S_{O,1}, S_{O,2}$ and $V_1, V_2$ are bounded almost everywhere by $\|S_O\|_{L^\infty}$ and $\|V\|_{L^\infty}$, because $S_{O,\varepsilon}$ and $V_\varepsilon$ were bounded almost everywhere independent of $\varepsilon$. Integrating from 0 to $t$ gives

$$\frac{1}{2}|Y^1| \|C_1 - C_2\|_\Omega^2 + P^C \|\nabla(C_1 - C_2)\|_{\Omega,t}^2$$
$$\leq |\Gamma^{ER}|(L_0 + L_{IP3}) |(C_{e,1} - C_{e,2}, C_1 - C_2)_{\Omega,t}|$$
$$+ |\alpha| g_{CRAC} |\partial Y^{CRAC}| \|V - V_{Ca}\|_{L^\infty} 2\|S_0\|_{L^\infty} |(S_{O,1} - S_{O,2}, C_1 - C_2)_{\Gamma^1,t}|$$
$$+ |\alpha| g_{CRAC} |\partial Y^{CRAC}| \|S_0\|_{L^\infty}^2 |(V_1 - V_2, C_1 - C_2)_{\Gamma^1,t}|.$$

The initial conditions for $C_1$ and $C_2$ cancel each other. Next, we use the binomial theorem with a factor $\lambda$ and the trace inequality. We merge the constants.

$$\frac{1}{2}|Y^1| \|C_1 - C_2\|_\Omega^2 + (P^C - c_5\lambda)\|\nabla(C_1 - C_2)\|_{\Omega,t}^2$$
$$\leq c_1 \|C_{e,1} - C_{e,2}\|_{\Omega,t}^2 + c_2\|C_1 - C_2\|_{\Omega,t}^2$$
$$+ c_3\|S_{O,1} - S_{O,2}\|_{\Gamma^1,t}^2 + c_4\|V_1 - V_2\|_{\Gamma^1,t}^2.$$

We perform a similar estimation for $C_{e,1}$ and $C_{e,2}$. Because $f_{SERCA}$ is Lipschitz-continuous, we easily find

$$\frac{1}{2}|Y^2| \|C_{e,1} - C_{e,2}\|_\Omega^2 + P^e \|\nabla(C_{e,1} - C_{e,2})\|_{\Omega,t}^2$$
$$\leq c_1 \|C_{e,1} - C_{e,2}\|_{\Omega,t}^2 + c_2\|C_1 - C_2\|_{\Omega,t}^2.$$

Further, we know that $f_e(C_{e,1}), f_e(C_{e,2}) \leq v_e$ and $S_1, S_2 \leq \|S\|_{L^\infty}$ almost everywhere and $f_e(C_{e,1}) - f_e(C_{e,2}) \leq L_e(C_{e,1} - C_{e,2})$. Hence, with similar transformations we find for the equation for $S_1$ and $S_2$ that

$$\frac{1}{2}|\Gamma^{ER}| \|S_1 - S_2\|_\Omega^2 + (P^S - c_5\lambda)\|\nabla(S_1 - S_2)\|_{\Omega,t}^2$$
$$\leq c_1 \|C_{e,1} - C_{e,2}\|_{\Omega,t}^2 + c_2\|S_1 - S_2\|_{\Omega,t}^2$$
$$+ c_3\|S_{C,1} - S_{C,2}\|_{\Omega,t}^2 + c_4\|S_{O,1} - S_{O,2}\|_{\Gamma^1,t}^2.$$

Analogously, we get for $S_C$, $S_O$ and $S_{CO}$

$$\frac{1}{2}|\Gamma^{ER}|\|S_{C,1} - S_{C,2}\|_{\Omega}^2 + (P^{SC} - c_5\lambda)\|\nabla(S_{C,1} - S_{C,2})\|_{\Omega,t}^2$$
$$\leq c_1\|C_{e,1} - C_{e,2}\|_{\Omega,t}^2 + c_2\|S_1 - S_2\|_{\Omega,t}^2$$
$$+ c_3\|S_{C,1} - S_{C,2}\|_{\Omega,t}^2 + c_4\|S_{CO,1} - S_{CO,2}\|_{\Gamma^1,t}^2$$

and

$$\frac{1}{2}\|S_{O,1} - S_{O,2}\|_{\Gamma^1}^2$$
$$\leq c_1\|S_{O,1} - S_{O,2}\|_{\Gamma^1,t}^2 + c_2\|S_{CO,1} - S_{CO,2}\|_{\Gamma^1,t}^2 + c_3\|S_1 - S_2\|_{\Omega,t}^2$$
$$+ c_4\|\nabla(S_1 - S_2)\|_{\Omega,t}^2 + c_5\|C_{e,1} - C_{e,2}\|_{\Omega,t}^2 + c_6\|\nabla(C_{e,1} - C_{e,2})\|_{\Omega,t}^2$$

and

$$\frac{1}{2}\|S_{CO,1} - S_{CO,2}\|_{\Gamma^1}^2$$
$$\leq c_1\|S_{O,1} - S_{O,2}\|_{\Gamma^1,t}^2 + c_2\|S_{CO,1} - S_{CO,2}\|_{\Gamma^1,t}^2$$
$$+ c_3\|S_{C,1} - S_{C,2}\|_{\Omega,t}^2 + c_4\lambda\|\nabla(S_{C,1} - S_{C,2})\|_{\Omega,t}^2$$
$$+ c_5\|C_{e,1} - C_{e,2}\|_{\Omega,t}^2 + c_6\lambda\|\nabla(C_{e,1} - C_{e,2})\|_{\Omega,t}^2.$$

It needs to be proved that the currents $I_K$, $I_{K(Ca)}$, $I_{CRAC}$ and $I_{CAN}$ are Lipschitz-continuous in order to find a similar estiamte for the potential $V$.

We immediately see that $I_K$ is continuous, differentiable and that its derivative is

$$I_K'(V) = \underbrace{\frac{g_K}{1 + e^{-\left(\frac{V-V_N}{K_N}\right)}}}_{>0, < g_K} + \underbrace{\frac{g_K(V - V_K)e^{-\left(\frac{V-V_N}{K_N}\right)}}{K_N\left(1 + e^{-\left(\frac{V-V_N}{K_N}\right)}\right)^2}}_{\to 0 \text{ for } V \to \pm\infty}.$$

Hence, it holds that

$$|I_K(V_1) - I_K(V_2)| \leq L_K|V_1 - V_2|$$

255

for a constant $L_K > 0$. We deduce for $I_{K(Ca)}$ that

$$I_{K(Ca)}(V_1, C_1) - I_{K(Ca)}(V_2, C_2)$$

$$= g_{K(Ca)} \frac{C_1^4}{C_1^4 + K_{K(Ca)}^4}(V_1 - V_2)$$

$$+ g_{K(Ca)}(V_2 - V_K)\left(\frac{C_1^4}{C_1^4 + K_{K(Ca)}^4} - \frac{C_2^4}{C_2^4 + K_{K(Ca)}^4}\right)$$

$$= g_{K(Ca)} \underbrace{\frac{C_1^4}{C_1^4 + K_{K(Ca)}^4}}_{\geq 0, \text{ bounded}}(V_1 - V_2)$$

$$+ g_{K(Ca)}(V_2 - V_K)K_{K(Ca)}^4(C_1 - C_2)\underbrace{\frac{(C_1^2 + C_2^2)(C_1 + C_2)}{(C_1^4 + K_{K(Ca)}^4)(C_2^4 + K_{K(Ca)}^4)}}_{\geq 0, \text{ bounded}}.$$

Hence,

$$|I_{K(Ca)}(V_1, C_1) - I_{K(Ca)}(V_2, C_2)| \leq L_{K(Ca)}\left(|V_1 - V_2| + |C_1 - C_2|\right)$$

for a constant $L_{K(Ca)} > 0$. Further, we estimate for the current $I_{CRAC}$

$$I_{CRAC}(V_1, S_{O,1}) - I_{CRAC}(V_2, S_{O,2})$$

$$= g_{CRAC}S_{O,1}^2(V_1 - V_2) + g_{CRAC}(V_2 - V_{Ca})(S_{O,1}^2 - S_{O,2}^2)$$

$$= \underbrace{g_{CRAC}S_{O,1}^2}_{\geq 0, \text{ bounded}}(V_1 - V_2)$$

$$+ \underbrace{g_{CRAC}(V_2 - V_{Ca})(S_{O,1} + S_{O,2})}_{\text{bounded}}(S_{O,1} - S_{O,2}).$$

It follows that

$$|I_{CRAC}(V_1, S_{O,1}) - I_{CRAC}(V_2, S_{O,2})|$$

$$\leq L_{CRAC}\left(|V_1 - V_2| + |S_{O,1} - S_{O,2}|\right)$$

for $L_{CRAC} > 0$. Finally, we consider $I_{CAN}$ with

$$
\begin{aligned}
I_{CAN}(V_1, a_1) &- I_{CAN}(V_2, a_2) \\
&= \underbrace{g_{CAN} a_1}_{\text{bounded}} (V_1 - V_2) + \underbrace{g_{CAN}(V_2 - V_{NA})}_{\text{bounded}} (a_1 - a_2).
\end{aligned}
$$

Hence,

$$
|I_{CAN}(V_1, a_1) - I_{CAN}(V_2, a_2)| \leq L_{CAN} \left( |V_1 - V_2| + |a_1 - a_2| \right)
$$

for $L_{CAN} > 0$. Now, it is left to show a similar estimate for $a$. We easily get

$$
|\partial_t a_1 - \partial_t a_2| = L_a \left( |C_1 - C_2| + |a_1 - a_2| \right)
$$

We deduce that

$$
\begin{aligned}
\|V_1 - V_2\|^2_{\Gamma^1} \leq{}& c_1 \|V_1 - V_2\|^2_{\Gamma^1, t} + c_2 \|C_1 - C_2\|^2_{\Omega, t} + c_3 \lambda \|\nabla(C_1 - C_2)\|^2_{\Omega, t} \\
&+ c_4 \|S_{O,1} - S_{O,2}\|^2_{\Gamma^1, t} + c_5 \|a_1 - a_2\|^2_{\Gamma^1, t}
\end{aligned}
$$

and

$$
\begin{aligned}
\|a_1 - a_2\|^2_{\Gamma^1} \leq{}& c_1 \|a_1 - a_2\|^2_{\Gamma^1, t} + c_2 \|C_1 - C_2\|^2_{\Omega, t} + \lambda c_3 \|\nabla(C_1 - C_2)\|^2_{\Omega, t}.
\end{aligned}
$$

Now we add all eight results, where the norms of the gradients can be omitted for small $\lambda$. Then we find that

$$
\begin{aligned}
\|C_1 &- C_2\|^2_{\Omega} + \|C_{e,1} - C_{e,2}\|^2_{\Omega} + \|S_1 - S_2\|^2_{\Omega} + \|S_{C,1} - S_{C,2}\|^2_{\Omega} \\
&+ \|S_{O,1} - S_{O,2}\|^2_{\Gamma^1} + \|S_{CO,1} - S_{CO,2}\|^2_{\Gamma^1} + \|V_1 - V_2\|^2_{\Gamma^1} + \|a_1 - a_2\|^2_{\Gamma^1} \\
\leq{}& c \left( \|C_1 - C_2\|^2_{\Omega, t} + \|C_{e,1} - C_{e,2}\|^2_{\Omega, t} + \|S_1 - S_2\|^2_{\Omega, t} \right. \\
&\quad + \|S_{C,1} - S_{C,2}\|^2_{\Omega, t} + \|S_{O,1} - S_{O,2}\|^2_{\Gamma^1, t} \\
&\quad \left. + \|S_{CO,1} - S_{CO,2}\|^2_{\Gamma^1, t} + \|V_1 - V_2\|^2_{\Gamma^1, t} + \|a_1 - a_2\|^2_{\Gamma^1, t} \right).
\end{aligned}
$$

With Gronwall's lemma we deduce that

$$\|C_1 - C_2\|_\Omega^2 + \|C_{e,1} - C_{e,2}\|_\Omega^2 + \|S_1 - S_2\|_\Omega^2$$
$$+ \|S_{C,1} - S_{C,2}\|_\Omega^2 + \|S_{O,1} - S_{O,2}\|_{\Gamma^1}^2 + \|S_{CO,1} - S_{CO,2}\|_{\Gamma^1}^2$$
$$+ \|V_1 - V_2\|_{\Gamma^1}^2 + \|a_1 - a_2\|_{\Gamma^1}^2 \leq 0$$

and uniqueness of the solution of systme (4.16) holds. ☐

## 4.9 Implementation and discussion

Patrick Fletcher implemented in [23] a numerical solver for the partial differential equation (4.16). He also used the following parameters for implementation. For numerical simulations we use his program and parameters. Patrick Fletcher assumed a radially symmetric cell. This allowes him to use polar coordinates and to exploit the independence of the angle to project the 3-dimensional problem to a 1-dimensional problem, where the distance $r$ from the cell's center is the only remaining variable. New to Fletcher's implementation are the volume factors obtained from homogenization.

We consider a circular cell (radius $5\mu m$) with big nucleus (radius $3\mu m$) that is located in the middle of the cell. One time unit corresponds to one second.

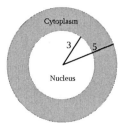

**Remark 4.31** To perform homogenization in terms of the outer boundary with Robin conditions, the domain $\Omega$ must be a union of squares, see section 4.2. Hence, the domain $\Omega$, illustarting the human cell and used for simulations of the calcium-stim1 project,

258

also ought to be composed of squares.
Nevertheless, we use the circle defined above due to the following reasons.
A circle is the favored shape of a human cell and presents the real live most likely. Futhermore, we perform homogenization in spirit for a rectangular shape and obtain the flux conditions over the outer boundary.
Since the flux condition is just a relative measure of the mass-compensation outside and inside the domain, we may modify the outer boundary to a circular shape and use the same flux condition as for the rectangular domain.

**Parameter values**
The values of the parameters are

| $k_O^+ = 0.005\frac{1}{s}$ | $k_{CO}^+ = 0$ | $k_c^+ = 6.25 \cdot 10^{-9}\frac{1}{s}$ |
|---|---|---|
| $k_O^- = 0$ | $k_{CO}^- = 0.07\frac{1}{s}$ | $k_c^- = 10\frac{1}{s}$ |
| $D_c = 15\frac{\mu m}{s}$ | $D_S = 0.05\frac{\mu m}{s}$ | |
| $D_{ER} = 5\frac{\mu m}{s}$ | $D_{SC} = 0.1\frac{\mu m}{s}$ | |

For the functions $f_{SERCA}$, $f_P$, $f_{NCX}$, $f_e$ and for $L_0$ and $L_{IP3}$ we use

| $v_{SERCA} = 5000$ | $v_P = 100$ | $v_{NCX} = 200$ |
|---|---|---|
| $K_{SERCA} = 0.2$ | $K_P = 0.1$ | $K_{NCX} = 3$ |
| $v_e = 10^{10}$ | $L_0 = 3$ | |
| $K_e = 10^{2.5}$ | $L_{IP3} \approx 0.25$ | |

The parameters for the current $V$ used for the $I_{CRAC}$ are

| $C_m = 0.0031$ | $g_K = 5$ | $g_{K(Ca)} = 5$ | $g_{CAN} = 0.5$ |
|---|---|---|---|
| $g_{CRAC} = 0.1$ | $V_N = -45$ | $V_K = -80$ | $V_{NA} = 60$ |
| $V_{Ca} = 100$ | $K_N = 15$ | $K_{K(Ca)} = 0.75$ | $K_{CAN} = 0.75$ |
| | $V_K = -80$ | | $\tau_{CAN} = 60$ |

The initial conditions are constant functions given by $C = 0.05\mu M$, $C_e = 150\mu M$, $S = 0\mu M$, $S_C = 0\mu M$, $S_O = 0.0637\mu M$, $S_{CO} = 0\mu M$, $V = -80mV$ and $a = 0$. We let the system run for 500 time units until an equilibrium is reached. Then the real experiment starts.

**Specific geometry**
We build a specific geometry of the unit cell $Y$ to calculate the required volume factors found by homogenization. The left figure shows the domain of $Y^1$ in the unit cell $Y$, the figure in the middle shows the domain $Y^2$ and the right one shows the domain $\Gamma^{ER}$.

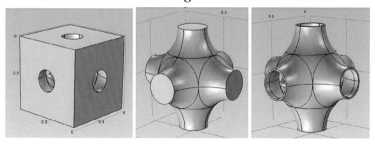

It is important for the domains $\Omega_\varepsilon^1$, $\Omega_\varepsilon^2$ and $\Gamma_\varepsilon^{ER}$ to be connected to receive global diffusion. The three profiles of the unit cell $[0,1]^2 \times \{\frac{1}{2}\} \subset \mathbb{R}^3$, $[0,1] \times \{\frac{1}{2}\} \times [0,1] \subset \mathbb{R}^3$ and $\{\frac{1}{2}\} \times [0,1]^2 \subset \mathbb{R}^3$ are equal and illustrated in the following figure on the left-hand side. The unit cell $[0,1]^2 \subset \mathbb{R}^2$ of the boundary $\Gamma^1$ is illustrated on the figure on the right-hand side.

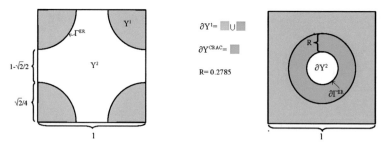

For this choice of geometry, the numbers needed in the equations are

$$|Y^1| = 0.7248253161 \qquad |\partial Y^1| = 0.9326 \qquad P^C = 0.5929$$
$$|Y^2| = 0.2751525797 \qquad |\partial Y^{CRAC}| = 0.5 \qquad P^e = 0.13609410$$
$$|\Gamma^{ER}| = 2.0925616 \qquad |\partial \Gamma^{ER}| = 0.92015 \qquad P^S = 1.39504.$$

$$(4.17)$$

## Numerical results

### 3-dimensional plots of $C$, $C_e$, $S$ and $S_C$

Using these parameter values and volume factors, we get the numerical calculated solutions $C$, $C_e$, $S$ and $S_C$ in the cell. Here, the light color means a high concentration and a dark color means a low concentration.

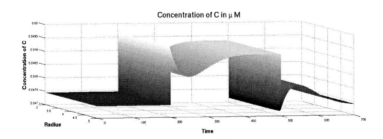

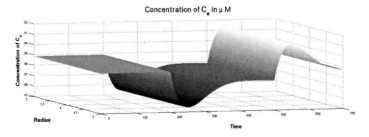

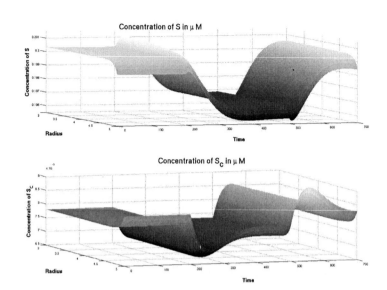

One can see that the concentration of the molecules is almost uniform throughout the cell for a fixed time. The concentration of $S$ molecules at the plasma membrane is smaller - the concentration of $S_C$ molecules at the plasma membrane is greater than in the interior of the cell. But the shapes of the concentrations are very similar.

Now knowing that the radius does not play an important role, we just go into details concerning the time during the following explanations. First we look at the calcium dynamics, illustrated on the first two figures.

Until $t = 200$ there is an equilibrium with $C \approx 0.0475\mu M$ in cytosol and $C_e \approx 89\mu M$ in the lumen of the ER. At $t = 200$ IP3 molecules appear and the osmosis coefficient through the surface of the ER is set manually higher (from $L = 3$ to $L \approx 3.25$). As we see, the calcium concentration in cytosol immediately increases to about $C \approx 0.495\mu M$ and the calcium concentration in the ER decreases. After the peak of calcium in cytosol it again decreases a bit. Since the calcium in ER is also still decreasing, we lose the calcium through the plasma membrane with the pumps PMCA and NCX.

At time $t \approx 250$ a local minimum is reached in $C$ (at $C \approx 0.0485\mu M$) and $C_e$ (at $C_e \approx 85.5\mu M$). Then the calcium concentration in cytosol and ER slowly starts increasing. We get more calcium in total in the cell. This means the CRAC channels have opened to let calcium into the cytosol that balances out with the calcium in ER.

At time $t \approx 400$ an equilibrium is reached at $C \approx 0.4925\mu M$ and $C_e \approx 87.5\mu M$.

Then at $t = 500$ the IP3 molecules leave the cell and the osmosis coefficient between cytosol and lumen of the ER is manually set back to $L = 3$. Immediately the calcium concentration in cytosol drops down to $C \approx 0.0475\mu M$, because calcium is sucked back into the ER. Then it increases again a little bit (to $C \approx 0.048\mu M$), since the CRAC channels still are open. Then until $t \approx 650$ the CRAC channels close and at $t \approx 700$ the equilibrium value $C \approx 0.0475\mu M$ is reached.

Calcium in ER is increasing much after $t = 500$ to a global maximum at $C_e \approx 91\mu M$, because calcium moves back into the ER. Then it decreases until it reaches its equilibrium value $C_e \approx 89\mu M$ at time $t \approx 700$.

**Stim1 dynamics**

Now we consider the dynamics of the Stim1 molecules. They form a complex interaction and we need to look more closely at what exactly is happening. The behavior in equilibrium is displayed in the following figure. On the arrows we find the components of the derivatives. For example the light-blue arrow from $S_C$ to $S$ means that in the equilibrium $-k_C^+ f_e(C_e)S + k_C^- S$ is positive and more molecules change from $S_C$ to $S$ than from $S$ to $S_C$.

The sum of the concentrations for $S$, $S_C$, $S_O$ and $S_{CO}$ stays constant. Because we consider the behavior of $S$ and $S_C$ on the plasma membrane, we can conclude that in equilibrium $S$ molecules steadily diffuse from the interior of the cell to the plasma membrane and $S_C$ molecules steadily diffuse from the plasma membrane in the interior of the cell.

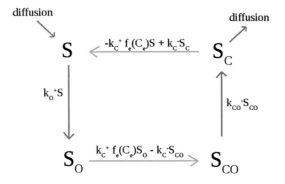

Note that the upper blue arrow points in the left direction and the lower blue arrow in the right direction, though the factors $k_C^+ f_e(C_e)$ and $k_C^-$ are the same. This means that the ratio $s/s_C$ must be smaller than $s_O/s_{CO}$.

We regard the four functions $S$, $S_C$, $S_O$ and $S_{CO}$ as mostly decoupled from the calcium dynamics. We regard the function $f_e(C_e)$ as a catalyst function for changing $S$ in $S_C$ and $S_O$ in $S_{CO}$.

**Plots and explanations to Stim1 $S_O$ bound with Orai1**

We start with the explanation for $S_O$, the Stim1 molecules bound with Orai1 at the plasma membrane. In the following figure we see in the top plot the function $S_O$ over time. In the middle plot there are two summands of the derivative of $S_O$. In the equilibrium the blue and the pink functions cancel each other. The light-blue function represents the interactions between $S_O$ and $S_{CO}$, namely $-k_C^+ f_e(C_e) S_O + k_C^- S_{CO}$. The pink function is the term $k_O^+ S$. We see that the pink function almost does not change over time. The dynamics in $S$ only have very little impact on $S_O$. In the third plot we see the derivative of $S_O$, the sum of the blue and the pink functions. It has almost the same form as the light-blue function above. This means, the behavior of $S_O$ is mostly determined by the interaction between $S_O$ and $S_{CO}$. The increase in $S_O$ for $t > 200$ originates in the lower catalyst function $f_e(C_e)$, because then fewer $S_O$ molecules change into $S_{CO}$ molecules. The decrease at time $t = 500$ also comes from the greater catalyst function $f_e(C_e)$ because more $S_O$ molecules change into $S_{CO}$ molecules.

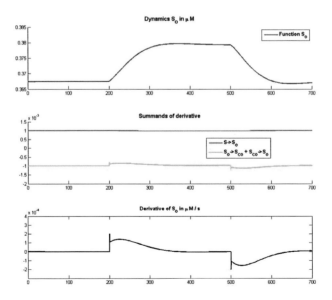

**Plots and explanations to Stim1 $S_{CO}$ bound with Orai1 and calcium**

We want to understand the dynamics of the function $S_{CO}$, the Stim1 molecules bound to Orai1 at the plasma membrane and to calcium. Again, we see in the following, top plot the function $S_{CO}$. In the middle one, we see both summands of the derivative of $S_{CO}$. The third plot contains the derivative of $S_{CO}$.

In the equilibrium, $t < 200$, the blue and the pink functions cancel each other. The light-blue line illustrates the function $k_C^+ f_e(C_e) S_O - k_C^- S_{CO}$. The reactions in the derivative of $S_{CO}$ for $t > 200$ is mainly determined by the light-blue function. We lose molecules to $S_O$.

But here also the pink function plays a role, that represents the term $-k_{CO}^- S_{CO}$ which means losing molecules to $S_C$. Clearly we see the increase and decrease in the pink graph. When the dynamics start at $t = 200$, $S_O$ decays and we lose fewer molecules to the function $S_C$. It counteracts to the blue function and induces the derivative to increase again for $t > 250$. Then also the catalyst function $C_e$ increases again and the effects cease. For $t > 500$ we see a converse behavior.

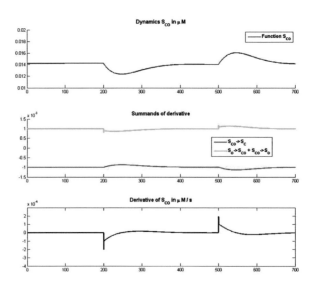

**Plots and explanations to Stim1 $S_C$ bound with calcium**

Next we consider the behavior of the function $S_C$, the Stim1 molecules in the cytosol bound to calcium. We see in the following picture in the top plot the function $S_C$. In the middle plot we show the summands $k_C^+ f_e(C_e) - k_C^- S_C$ in blue and the diffusion term added to $k_{CO}^- S_{CO}$ in pink. The third plot is the derivative of $S_C$.

In the equilibrium for $t < 200$ the blue and the pink functions cancel each other. Here we see that the light-blue graph coming from the term $k_C^+ f_e(C_e)S - k_C^- S_C$ is a negative function. For $t \geq 200$ the pink function decreases, because we get fewer molecules from $S_{CO}$.

The derivative of $S_C$ itself is a complex interaction between the light-blue and the pink function. Sometimes they counteract and sometimes they strengthen each other. Anyway, the sum of the light-blue and the pink function as shown in the third window is negative after $t = 200$. At $t \approx 250$ it becomes positive and then decays back to zero. For $t > 500$, the converse behavior occurs.

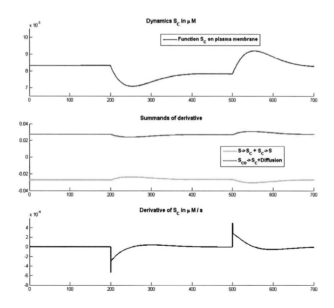

**Plots and explanations to Stim1 $S$ unbound**

Finally we look at the function $S$, the unbound Stim1 molecules in the cytoplasm. Again, in the first plot we see the function $S$. In the middle one we have the term $-k_C^+ f_e(C_e)S + k_C^- S_C$ in light-blue and the diffusion term minus the term $-k_O^+ S$ in pink. The third plot shows the sum of the light-blue and the pink function, the derivative of $S$.

In the equilibrium these two functions cancel each other. Again, the blue function plays the main role. The shape of the derivative of $S$ is qualitatively similar to the light-blue one. We have at first a strong peak at $t = 200$. It originates directly from the decrease of the catalyst function $f_e(C_e)$ and fewer $S$ molecules change into $S_C$ molecules.

The following strong decrease arises also from the light-blue function. Because after $t \approx 250$ the catalyst function $f_e(C_e)$ is increasing again, we see the derivative of $S$ decays back to zero.

For $t \geq 500$ we have the converse behavior.

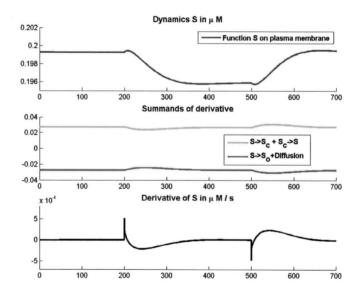

**Summary and CRAC channels**

By looking at the functions $S$, $S_C$, $S_O$ and $S_{CO}$, we could think that just unbound $S$ molecules from the interior of the cell bind to the Orai1 molecules at the plasma membrane, which would be a possible explanaition of the decrease in $S$ and the increase in $S_O$. But, as we understood now by considering the derivatives, the $S$ molecules go the indirect route by changing to $S_C$ first, then to $S_{CO}$ and only then to $S_O$. Hence, the process of creating the current $I_{CRAC} = g_{CRAC} S_O^2 (V - V_{Ca})$ is dexterous and elaborate. We see the current $I_{CRAC}$ given in $mV$ in the following figure.

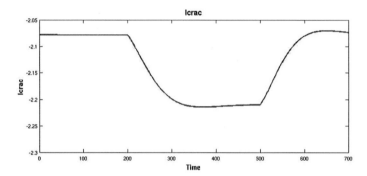

Since the current is negative, it becomes more negative as $S_O$ increases. As we saw in the figure for the concentration of the calcium in cytosol $C$, we realize the effect of the $I_{CRAC}$ at time $t \approx 250$ when the calcium is increasing again after the local minimum.

The channels start to open immediately after depletion of the ER and reach a saturation at time $\approx 350s$, this is $150s$ after depletion of the ER.

# Conclusion

Now we come back to the question 4.10 we asked in the introduction to the signaling in Lymphocytes. Do the Stim1 molecules move to the plasma membrane just by diffusion or do they need an additional driving force?
To answer this question we refer to the following figure taken from article [39], figure 2B.

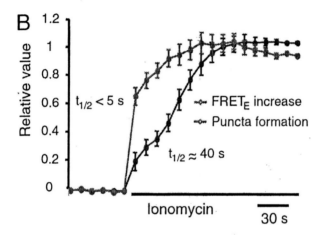

Figure 4.2: Puncta formation over time

In this figure one sees the Stim1 molecules bound to the plasma membrane, called "Puncta formation", over time. We also see a so-called "$FRET_E$ increase", which is of no relevance in this context. The figure is made by biological experimets; to know more about the setup of the experiments, see article [39]. The calcium depletion of the lumen of the ER happens at time 0.
Translated to our setting, the Puncta formation stands for the concentration of $S_O$, which are the Stim1 molecules bound to the plasma membrane. Hence, to compare the occurences with our results we consider the concentration of $S_O$ over time, illustrated in the following figure.

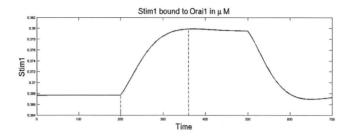

Figure 4.3: $S_O$ over time

Here, the depletion of the lumen of the ER happens at time $t = 200s$.

In figure 4.2, one measures an increase in Stim1 molecules at the plasma membrane only after about 60 seconds after depletion of the ER. Then, a strong increase occures, which saturates at about $135s$ after depletion of the ER. After this maximum, the number of Stim1 molecules is slightly decreasing. About $240s$ after depletion the experiment stops.

In the results of our mathematical examinations, in figure 4.3, the increase of $S_O$ starts immediately after depletion of the ER at time $t = 200s$. About $160s$ after depletion, the concentration of Stim1 molecules bound to the plasma membrane reaches a maximal value, and after that it decreases again a bit.

By comparing these results, we notice that the shape of the two functions is very similar. The saturation values are reached after about the same time, $135s - 160s$ after depletion; the time difference is only about 25 seconds.

Maybe an increase in Stim1 molecules is hard to measure in the experiment for small values and therefore, one measures the growth only after some time passed; this would explain the sudden strong increase in figure 4.2.

As far as we can measure up the situation, diffusion and no other force is sufficient for Stim1 molecules to reach the plasma membrane in the given time frame.

271

# BIBLIOGRAPHY

[1] R. A. Adams and J. J. F. Fournier. *Sobolev Spaces*. Academic Press, 2 edition, 2003.

[2] G. Allaire. Homogenization and two-scale convergence. *SIAM J. Math. Anal.*, 23(6):1482–1518, 1992.

[3] G. Allaire, A. Damlamian, and U. Hornung. Two-scale convergence on periodic surfaces and applications. In A. P. Bourgeat, C. Carasso, S. Luckhaus, and A. Mikelić, editors, *Proceedings of the international conference on mathematical modelling of flow through porous media*, pages 15–25. World Scientific, 1995.

[4] H. W. Alt. *Lineare Funktionalanalysis*. Springer DE, 2006.

[5] T. Arbogast, J. Douglas, and U. Hornung. Derivation of the double porosity model of single phase flow via homogenization theory. *SIAM J. Math. Anal.*, 21:823–836, 1990.

[6] A. Bensoussan, J.-L. Lions, and G. Papanicolaou. *Asymptotic analysis for periodic structures*. North-Holland, 1978.

[7] D. O. Besong. Mathematical modelling and numerical solution of chemical reactions and diffusion of carcinogenic compounds in cells. 2004. TRITA NA E04152.

[8] A. Bourgeat, G. Chechkin, and A. Piatnitski. Singular double porosity model. *Applicable Analysis*, 82, 2003.

[9] A. Bourgeat, L. Pankratov, and M. Panfilov. Study of the double porosity model versus the fissures thickness. *Asymptotic Analysis*, 38, 2004.

[10] D. Cioranescu, A. Damlamian, and G. Griso. Periodic unfolding and homogenization. *Comptes Rendus Mathematique*, 335(1):99–104, 2002.

[11] D. Cioranescu, A. Damlamian, and G. Griso. The periodic unfolding method in homogenization. *SIAM J. Math. Anal.*, 40(4):1585–1620, 2008.

[12] D. Cioranescu and P. Donato. Homogénéisation du problème de neumann non homogène dans des ouverts perforés. *Asymptotic Anal.*, 1:115–138, 1988.

[13] D. Cioranescu, P. Donato, and R. Zaki. The periodic unfolding method in perforated domains. *Portugaliae Mathematica*, 63(4):467–496, 2006.

[14] E. A. Coddington and N. Levinson. *Theory of ordinary differential equations*. McGraw-Hill, 1955.

[15] C. Conca. On the aplication of the homogenization theory to a class of problems arising in fluid mechanics. *J. Math. Pures et Appl.*, 64:31–75, 1985.

[16] C. Conca, J. I. Diaz, and C. Timofte. Effective chemical processes in porous media. *Math. Mod. Meth. Appl. Sci.*, 13(10):1437–1462, 2003.

[17] A. M. Czochra and M. Ptashnyk. Derivation of a macroscopic receptor-based model using homogenization techniques. *SIAM J. Math. Anal.*, 40(1):215–237, 2008.

[18] E. B. Davies. *Heat Kernels and Spectral Theory*. Cambridge University Press, 1989.

[19] M. P. do Carmo. *Riemannian Geometry*. Springer, 1992.

[20] M. Domijan, R. Murray, and J. Sneyd. Dynamical probing of the mechanisms underlying calcium oscillations. *Journal of Nonlinear Science*, 16(5):483–506, 2006.

[21] L. C. Evans. *Partial Differential Equations*. American Mathematical Society, 2010.

[22] J. Feldman and G. Uhlmann. *Inverse Problems*. www.math.ubc.ca/~feldman/ibook/, UBC, Version of Dec. 4, 2004.

[23] P. Fletcher and Y.-X. Li. A model of Ca2+ oscillations in T-lymphocytes due to Ca2+ store-operated redistribution of STIM1. 2011. (in preparation).

[24] H. V. Gelboin. Benzo[a]parene metabolism, activation, and carcinogenesis: Role and regulation of mixed-function oxidases and related enzymes. *Physiological Reviews*, 60(4):1107–1155, 1980.

[25] D. Gilbarg and N.S. Trudinger. *Elliptic Partial Differential Equations of Second Order*. Springer, 2000.

[26] P. Goel, J. Sneyd, and A. Friedman. Homogenization of the cell cytoplasm: The calcium bidomain equations. *Multiscale Model. Simul.*, 5(4):1045–1062, 2006.

[27] A. Gossauer. *Struktur und Reaktivität der Biomoleküle*. John Wiley & Sons, 2003.

[28] W. Hackbusch. *Elliptic Differential Equations*. Springer, 2003.

[29] M. Hanke and M. Cabauatan-Villaneuva. Analytical and numerical approximation of effective diffusivities in the cytoplasm of biological cells. 2007. TRITA CSC NA 2007:06.

[30] D. Haroske and H. Triebel. *Distributions, Sobolev Spaces, Elliptic Equations*. European Mathematical Society, 2008.

[31] A. M. Hofer, C. Fasolato, and T. Pozzan. Capacitative Ca2+ entry is closely linked to the filling state of internal Ca2+ stores: A study using simultaneous measurements of icrac ans intraluminal [Ca2+]. *Journal of Cell Biology*, 140(2):325–334, 1998.

[32] P. G. Hogan, R. S. Lewis, and A. Rao. Molecular basis of calcium signaling in lymphocytes: STIM and ORAI. *Annual Review of Immunology*, 28:491–533, 2010.

[33] P. J. Hoover and R. S. Lewis. Stoichiometric requirements for trapping and gating of Ca2+ release-activated Ca2+ (CRAC) channels by stromal interaction molecule 1 (STIM1). *Proceedings of the National Academy of Sciance*, 108(32):13299–13304, 2011.

[34] U. Hornung and W. Jäger. Diffusion, convection, adsorption, and reaction of chemicals in porous media. *J. Diff. Eq.*, 92:199–225, 1991.

[35] H. Jiang, S. L. Gelhaus, D. Mangal, R. G. Harvey, I. A. Blair, and T. M. Penning. Metabolism of benzo[a]pyrene in human bronchoalveolar h358 cells using liquid chromatography-mass spectrometry. *Chem. Res. Toxicol.*, 20(9):1331–1341, 2007.

[36] H. Kleinig and U. Maier. *Zellbiologie*. Gustav Fischer Verlag, 1999.

[37] C. Kreisbeck. Another approach to the thin-film Gamma-limit of the micromagnetic free energy in the regime of small samples. *Quarterly of Applied Mathematics*, 2012. (in press).

[38] P. Li. *Seminar on Differential Geometry*. Princeton University Press, 1982.

[39] J. Liou, M. Fivaz, T. Inoue, and T Meyer. Live-cell imaging reveals sequential oligomerization and local plasma membrane targeting of stromal interaction molecule 1 after Ca2+ store depletion. *Proceedings of the National Academy of Sciance*, 104(22):9301–9306, 2007.

[40] D. Lukkassen, G. Nguetseng, and P. Wall. Two-scale convergence. *International Journal of Pure and Applied Mathematics*, 2(1):35–86, 2002.

[41] M. Mabrouk and S. Hassan. Homogenization of a composite medium with a thermal barrier. *Mathematical Methods in the applied Sciences*, 27:405 – 425, 2004.

[42] S. Monsurrò. Homogenization of a two-component composite with interfacial thermal barrier. *Adv. Math. Sci. Appl.*, 13(1):44–63, 2003.

[43] M. Neuss-Radu. Some extensions of two-scale convergence. *C. R. Acad. Sci. Paris, Ser. I*, 322:899–904, 1996.

[44] G. Nguetseng. A general convergence result for a functional related to the theory of homogenization. *SIAM J. Math. Anal.*, 20(3):608–629, 1989.

[45] O. Pelkonen and D. W. Nebert. Metabolism of polycyclic aromatic hydrocarbons: Etiologic role in carcinogenesis. *Pharmacological Reviews*, 34(2), 1982.

[46] M. A. Peter. Coupled reaction-diffusion processes inducing an evolution of the microstructure: Analysis and homogenization. *Nonlinear Analysis*, 70:806–821, 2009.

[47] M. A. Peter and M. Böhm. Different choises of scaling in homogenization of diffusion and interfacial exchange in a porous medium. *Mathematical Methods in the Applieds Sciences*, 31:1257–1282, 2008.

[48] M. A. Peter and M. Böhm. Multiscale modelling of chemical degradation mechanisms in porous media with evolving microstructure. *Multiscale Model. Simul.*, 7(4):1643–1668, 2009.

[49] M. A. Peter, J. Sneyd, and I. Graf. Homogenization of the two-compartment model of calcium dynamics in biological cells. 2012. (in preparation).

[50] D. H. Phillips. Fifty years of benzo[a]pyrene. *Nature*, 303(9):468–472, 1983.

[51] E. Sanchez-Palencia. *Non-homogeneous media and vibration theory.* Springer, 1980.

[52] R. E. Showalter. Microstructure models of porous media. In U. Hornung, editor, *Homogenization and Porous Media*, pages 183–202. Springer, 1997.

[53] R. E. Showalter. *Monotone operators in Banach space and nonlinear partial differential equations.* American Mathematical Society, 1997.

[54] P. Sims, P. L. Grover, A. Swaisland, K. Pal, and A. Hewer. Metabolic activation of benzo(a)pyrene proceeds by a diol-epoxide. *Nature*, 252:326–328, 1974.

[55] P. B. Stathopulos, G.-Y. Li, M. J. Plevin, J. B. Ames, and M. Ikura. Stored Ca2+ depletion-induced oligomerization of stromal interaction molecule 1 (STIM1) via the EF-SAM region. *Journal of Biological Chemistry*, 281(47):35855–35862, 2006.

[56] M. Valadier. Admissible functions in two-scale convergence. *Portugaliae Mathematica*, 54(2):147–164, 1997.

[57] J. Wloka. *Partial differential equations*. Cambridge University Press, 1987.

[58] E. Zeidler. *Nonlinear functional analysis and its applications I - fixed-point theorems*. Springer, 1986.

# Curriculum Vitae

**Personal data**

| | |
|---|---|
| first name: | Isabella Maria, |
| surname: | Graf, |
| date of birth: | January 5th 1984, |
| place of birth: | 86609 Donauwörth, Germany, |
| citizenship: | german. |

**Course of education**

| | |
|---|---|
| 10.2009 – 11.2012 | Ph.D. Student at the Universität Augsburg, supervisor: *Prof. Dr. Malte Peter*, |
| 08.2011 – 04.2012 | research stay at the University of British Columbia, Vancouver, Canada, |
| 02.2009 – 07.2009 | teacher internship for mathematics and physics, Allgäu Gymnasium Kempten, |
| 10.2003 – 01.2009 | student in mathematics at the Universität Augsburg, |
| 05.2003 | Abitur, Bonaventuragymnasium Dillingen. |